大学数学建模
创新课程系列教材

A Series of Teaching Materials for Innovative Course
of Mathematical Modeling in University

数学建模
优秀赛文评析

Evaluation and Analysis of
Excellent Articles
in Mathematical Modeling

主　　编◎朱家明
副主编◎李　勇　汪　凯　李　强
　　　　王拥兵　刘玉琳
参编人员（按姓氏笔画排序）
　　　　王拥兵　朱家明　刘玉琳
　　　　庄科俊　孙礼娜　李　勇
　　　　李　强　汪　凯

北京师范大学出版集团
安徽大学出版社

内容简介

本书精选数学建模竞赛优秀论文 15 篇,包括中国大学生数学建模竞赛的获奖论文 6 篇,美国大学生数学建模竞赛的获奖论文 5 篇,全国研究生数学建模竞赛的获奖论文 4 篇,每篇均包括竞赛原题再现、获奖论文精选、建模特色点评 3 部分。每一篇获奖论文的特色点评分别从论文特色及不足之处两个方面评价论文的建模方法及写作技巧。

本书可作为高等院校在校研究生、本科生及专科生"数学建模"课程的参考书,也可以作为研究生和本科生参加全国大学生数学建模竞赛、美国大学生数学建模竞赛及全国研究生数学建模竞赛的培训教材,还可以作为从事复杂问题建模工作的工程技术人员的建模参考书。

图书在版编目(CIP)数据

数学建模优秀赛文评析/朱家明主编. —合肥:安徽大学出版社,2021.3
大学数学建模创新课程系列教材
ISBN 978-7-5664-2156-2

Ⅰ.①数… Ⅱ.①朱… Ⅲ.①数学模型－高等学校－教材 Ⅳ.①O141.4

中国版本图书馆 CIP 数据核字(2020)第 237073 号

数学建模优秀赛文评析

朱家明 主编

出版发行	:	北京师范大学出版集团
		安 徽 大 学 出 版 社
		(安徽省合肥市肥西路 3 号 邮编 230039)
		www.bnupg.com.cn
		www.ahupress.com.cn
印 刷	:	合肥现代印务有限公司
经 销	:	全国新华书店
开 本	:	184 mm×260 mm
印 张	:	24.75
字 数	:	602 千字
版 次	:	2021 年 3 月第 1 版
印 次	:	2021 年 3 月第 1 次印刷
定 价	:	69.00 元

ISBN 978-7-5664-2156-2

策划编辑:刘中飞 张明举	装帧设计:孟献辉
责任编辑:张明举 李 雪	美术编辑:李 军
责任校对:宋 夏	责任印制:赵明炎

版权所有 侵权必究

反盗版、侵权举报电话:0551－65106311
外埠邮购电话:0551－65107716
本书如有印装质量问题,请与印制管理部联系调换。
印制管理部电话:0551－65106311

大学数学建模创新课程系列教材
编写委员会

（按姓氏笔画排序）

万诗敏	天津城建大学
王　娟	安徽三联学院
王大星	滁州学院
王拥兵	安庆师范大学
王春利	桂林电子科技大学
石爱菊	南京邮电大学
朱家明	安徽财经大学
刘玉琳	安徽工业大学
刘家保	安徽建筑大学
孙娓娓	阜阳师范大学
李　勇	安徽财经大学
李　强	安徽理工大学
杨洪礼	山东科技大学
杨鹏辉	安徽财经大学
肖华勇	西北工业大学
吴正飞	淮南师范学院
何　新	沈阳师范大学
何道江	安徽师范大学
汪晓银	天津工业大学
宋国强	安徽医科大学
张　永	池州学院
张　海	安庆师范大学
张成堂	安徽农业大学
陈　昊	淮北师范大学
陈华友	安徽大学
陈宜治	浙江工商大学
周本达	皖西学院
周礼刚	安徽大学
庞培林	河北工程大学
胡建伟	黄山学院
盛兴平	阜阳师范大学

总 序

"创新是一个民族进步的灵魂,是一个国家兴旺发达的不竭动力"。高等教育肩负着培养创新人才的重任,对民族和国家的未来发展至关重要。为了有效培养创新人才,高等学校的优秀教育工作者们穷尽其能,可谓"八仙过海,各显神通"。作为高等院校最早的学科竞赛,数学建模竞赛集数学方法、软件编程、科技论文写作于一身,全面培养当代大学生综合科研创新能力,通过在有限时间内以"头脑风暴"方式团队协作,解决来源于真实世界的各类难题,其已经成为培养大学生科研创新的重要途径,在推动高等教育改革中披荆斩棘,在科研创新培养中引领风骚,在科学化时代的风潮浪尖上不断引吭高歌,成为新时代创新人才培养的典范。

为更好地通过数学建模培养创新人才,结合作者二十余年成功的实践经验,特编写"大学数学建模创新课程系列教材"。本系列教材共四种,具体介绍如下。

一、数学建模优秀赛文评析。本书讲解中国大学生数学建模竞赛优秀赛文6篇,美国大学生数学建模竞赛优秀赛文5篇,全国研究生数学建模竞赛优秀赛文4篇,每篇赛文包括竞赛原题再现、获奖论文精选以及建模特色点评3部分。

二、数学建模方法。本书介绍数学建模方法概要、初等数学建模方法、数学规划建模方法、模糊数学建模方法、微分方程建模方法、统计分析建模方法、图与网络分析建模方法等,具体包括方法介绍、案例分析和软件实现3部分。

三、数学建模竞赛论文写作。本书介绍科技论文写作的基本技能、数学建模竞赛论文的规范写作、竞赛后研究论文的写作3部分。

四、数学建模软件编程。本书基于MATLAB软件,分基础知识、专题编程、优秀案例分析3部分。

大学数学建模创新课程系列教材全面系统地介绍了数学建模竞赛的知识、技能和方法,为学生赛前知识储备、赛中查阅参考、赛后总结及深入研究提供方法和技能支持。

"悠悠四年时光短,匆匆千日转瞬远。千琢万磨欲成才,合格容易优秀难。望能点醒梦中人,崎岖山路勇登攀。出类拔萃做翘楚,屡克艰难创新天。"希望本书的读者能够成为具备创新能力的优秀人才,为民族未来的科技发展添砖加瓦。

安徽财经大学大学数学教学研究中心
2021 年 1 月 3 日

前言

中国大学生数学建模竞赛创办于1992年，迄今已走过28个年头。1992年，这项竞赛初次举办时国内只有74所高校的314个队伍942人参加，到2019年共有来自全国及美国和新加坡的1490所院校、42992个队伍（本科39293队、专科3699队）超过120000人报名参赛。参赛校数和队数平均每年分别以15.34%及23.45%的速度增长，该竞赛不仅成为我国高校规模最大的学科性竞赛活动，而且成功地推进以数学建模为核心的数学教学改革，使许多同学"一次参赛，终生受益"。

通过数学建模竞赛可以提高高等学校教育质量及学生创新实践能力并激励学生学习数学，增强学生建立数学模型并运用计算机技术解决实际问题的能力，推动大学数学教学体系、教学内容和教学方法的改革。数学建模竞赛是全国高校规模最大的课外科技活动之一，每年吸引几万名大学生参加，相关的数学建模书籍也大量涌现。但是，对于初次参加数学建模的大学生来说，除了分门别类地训练各种方法以外，更需要了解针对特定问题，可以用什么方法建模和求解，如何对求解结果进行分析和评价等；还需要掌握数学建模竞赛论文写作的基本要领，包括论文基本结构与写作方法以及论文的格式等。基于以上考虑，我们编写了《数学建模优秀赛文评析》。

自1998年组织本科生参加中国大学生数学建模竞赛，2009年组织研究生参加全国研究生数学建模竞赛，2010年组织本科生参加美国大学生数学建模竞赛以来，安徽财经大学数学建模竞赛呈现规模化（国赛达85队）、多样化（大学生国赛、美赛、研究生赛等）、高质量化（获奖比例高）和继续研究化（发表学术论文及申报项目多）的四化方向发展，不仅带动学生参与了创新，而且在研究生推免、出国深造、就业等活动中起到推动作用，为学校"新经管"发展起到良好的支撑作用。

《数学建模优秀赛文评析》是从安徽财经大学22年来参加中国大学生数学建模竞赛、美国大学生数学建模竞赛和全国研究生数学建模竞赛的论文中精选出15篇获奖论文经加工整理而成的。所选择的论文都具有代表性,每篇论文都按照竞赛论文的写作要求,包含论文的摘要、问题的重述、问题的分析、模型的假设与符号说明、模型的建立与求解、模型的分析与检验、模型的评价与改进等内容。本书几乎完整地保留了参赛论文的原貌,在每篇论文后编者都给出了简要的点评。值得指出的是:每一篇论文都是参赛学生在教师的指导下用3天或4天时间完成的参赛论文,尽管论文的写作有些稚嫩,对问题的分析和解决还不太成熟,甚至表述还有不完善的地方,但这是他们集体智慧的结晶,体现了学生的创新能力,能为读者提供参考和借鉴。

本书可作为本科生、专科生"数学建模"课程的教学参考书,也可作为大学生、研究生参加国际数学建模竞赛、中国大学生数学建模竞赛和研究生数学建模竞赛的培训教材,还可作为从事复杂问题建模的工程技术人员的参考用书。

由衷感谢安徽财经大学校领导、教务处、科研处和统计与应用数学学院各级领导对本书出版所给予的大力支持;感谢安徽财经大学数学建模教练组的各位老师,尤其要感谢书中论文的指导老师及参赛学生:冯守平、庄科俊、闫云侠、李勇等,本书的出版应该归功于数学建模教练组各位指导老师的辛勤工作和无私奉献。在此,一并致以诚挚的感谢。

由于编者才疏学浅,书中难免有错误及不妥之处,尤其是点评不一定恰当,敬请各位专家、同行和广大读者不吝指正!

编　者

2020 年 11 月 8 日

目 录

第1部分　中国大学生数学建模竞赛优秀论文评析

第1篇　车道被占用对城市道路通行能力的影响 …………………… 3
◆竞赛原题再现 …………………………… 3
◆获奖论文精选 …………………………… 4
◆建模特色点评 …………………………… 27

第2篇　嫦娥三号软着陆轨道设计与控制策略 …………………… 28
◆竞赛原题再现 …………………………… 28
◆获奖论文精选 …………………………… 29
◆建模特色点评 …………………………… 49

第3篇　"互联网＋"时代的出租车资源配置 …………………… 50
◆竞赛原题再现 …………………………… 50
◆获奖论文精选 …………………………… 50
◆建模特色点评 …………………………… 72

第4篇　小区开放对道路通行的影响 …………………… 73
◆竞赛原题再现 …………………………… 73
◆获奖论文精选 …………………………… 74
◆建模特色点评 …………………………… 97

第5篇　"拍照赚钱"的任务定价 …………………… 98
◆竞赛原题再现 …………………………… 98
◆获奖论文精选 …………………………… 99
◆建模特色点评 …………………………… 119

第6篇　机场的出租车问题 ··· 120
- ◆ 竞赛原题再现 ··· 120
- ◆ 获奖论文精选 ··· 121
- ◆ 建模特色点评 ··· 144

第2部分　美国大学生数学建模竞赛优秀论文评析

第7篇　电动汽车对环境和经济的影响如何？其广泛使用是否可行？······
·· 147
- ◆ 竞赛原题再现 ··· 147
- ◆ 获奖论文精选 ··· 149
- ◆ 建模特色点评 ··· 166

第8篇　它是可持续的吗？ ··· 167
- ◆ 竞赛原题再现 ··· 167
- ◆ 获奖论文精选 ··· 169
- ◆ 建模特色点评 ··· 189

第9篇　我们的星球正在走向干涸吗？ ······························· 190
- ◆ 竞赛原题再现 ··· 190
- ◆ 获奖论文精选 ··· 192
- ◆ 建模特色点评 ··· 218

第10篇　收费后的车流合并问题 ··································· 219
- ◆ 竞赛原题再现 ··· 219
- ◆ 获奖论文精选 ··· 220
- ◆ 建模特色点评 ··· 247

第11篇　能源生产 ··· 248
- ◆ 竞赛原题再现 ··· 248
- ◆ 获奖论文精选 ··· 249
- ◆ 建模特色点评 ··· 270

第3部分　全国研究生数学建模竞赛优秀论文评析

第12篇　我国就业人数或城镇登记失业率的数学建模 …………… 275
- ◆竞赛原题再现 …………… 275
- ◆获奖论文精选 …………… 276
- ◆建模特色点评 …………… 311

第13篇　确定肿瘤的重要基因信息 …………… 312
- ◆竞赛原题再现 …………… 312
- ◆获奖论文精选 …………… 314
- ◆建模特色点评 …………… 333

第14篇　基因识别问题及其算法实现 …………… 334
- ◆竞赛原题再现 …………… 334
- ◆获奖论文精选 …………… 336
- ◆建模特色点评 …………… 359

第15篇　中等收入定位与人口度量模型研究 …………… 360
- ◆竞赛原题再现 …………… 360
- ◆获奖论文精选 …………… 362
- ◆建模特色点评 …………… 381

第 1 部分

中国大学生数学建模竞赛优秀论文评析

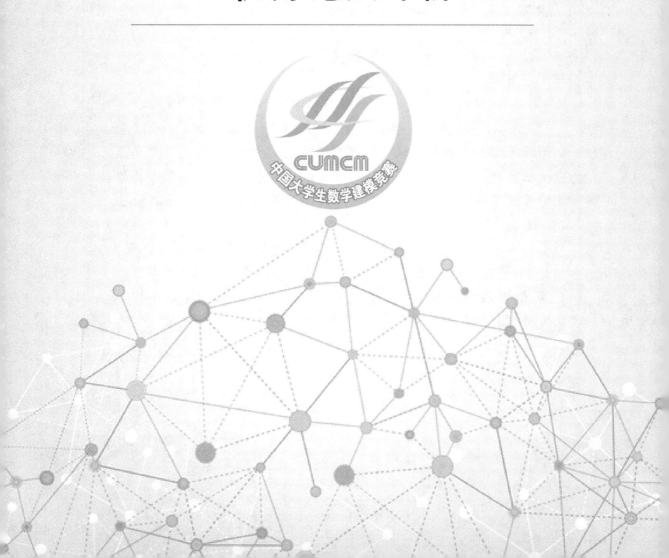

内 容 简 介

一、全国大学生数学建模竞赛简介

中国大学生数学建模竞赛创办于1992年,每年一届,目前已成为全国高校规模最大的基础性学科竞赛,也是世界上规模最大的数学建模竞赛。2018年,来自全国及美国和新加坡的1449所院校/校区的42128个队伍(本科38573队、专科3555队)、超过12万名大学生报名参加本项竞赛。

二、安徽财经大学参赛获奖情况介绍

安徽财经大学从1998年起组织学生参加全国大学生数学建模竞赛,至今已获国家奖80项,其中16项全国一等奖、64项全国二等奖。杨桂元和朱家明总结安徽财经大学1998—2012年这十五年获全国奖的优秀竞赛论文15篇,编辑成第一辑《数学建模竞赛优秀论文评析》,于2013年在中国科学技术大学出版社出版发行。

本书此部分共选2013—2019年全国大学生数学建模竞赛6篇获全国奖的论文,分别为2013年全国一等奖,2014年全国二等奖(当年缺少全国一等奖),2015年全国一等奖,2016年全国一等奖,2017年全国一等奖,2019年推荐全国一等奖的文章。

第 1 篇
车道被占用对城市道路通行能力的影响

◆ 竞赛原题再现

2013 年 A 题 车道被占用对城市道路通行能力的影响

车道被占用是指因交通事故、路边停车、占道施工等因素,导致车道或道路横断面通行能力在单位时间内降低的现象。由于城市道路具有交通流密度大、连续性强等特点,一条车道被占用,也可能降低路段所有车道的通行能力,即使时间短,也可能引起车辆排队,出现交通阻塞。如处理不当,甚至会出现区域性拥堵。

车道被占用的情况种类繁多、复杂,正确估算车道被占用对城市道路通行能力的影响程度,将为交通管理部门正确引导车辆行驶、审批占道施工、设计道路渠化方案、设置路边停车位和设置非港湾式公交车站等提供理论依据。

视频 1(附件 1)和视频 2(附件 2)中的两个交通事故处于同一路段的同一横断面,且完全占用两条车道。请研究以下问题:

问题一、根据视频 1(附件 1),描述视频中交通事故发生至撤离期间,事故所处横断面实际通行能力的变化过程。

问题二、根据问题一所得结论,结合视频 2(附件 2),分析说明同一横断面交通事故所占车道不同对该横断面实际通行能力影响的差异。

问题三、构建数学模型,分析视频 1(附件 1)中交通事故所影响的路段车辆排队长度与事故横断面实际通行能力、事故持续时间、路段上游车流量间的关系。

问题四、假如视频 1(附件 1)中的交通事故所处横断面距离上游路口变为 140 m,路段下游方向需求不变,路段上游车流量为 1 500 pcu/h,事故发生时车辆初始排队长度为零,且事故持续不撤离。请估算,从事故发生开始,经过多长时间,车辆排队长度将到达上游路口。

附件 1:视频 1;
附件 2:视频 2;

附件 3：视频 1 中交通事故位置示意图；
附件 4：上游路口交通组织方案图；
附件 5：上游路口信号配时方案图。
注：只考虑四轮及以上机动车、电瓶车的交通流量，且换算成标准车当量数。
原题详见全国大学生数学建模竞赛网（http://www.mcm.edu.cn）。

获奖论文精选

车道被占用对城市道路通行能力影响的综合分析[①]

摘要：本文针对因交通事故导致城市车道被占用，由此引起的道路通行能力的变化和道路阻塞引起车辆排队通行等问题进行了研究。根据交通波理论、流量守恒定律、排队论等理论，构建了道路通行能力评价和车辆排队等模型，综合运用 SPSS 软件和 Excel 软件进行分析及求解，描述了事故导致实际通行能力的变化过程，不同车道对实际通行能力影响存在差异，车辆排队长度与事故横断面实际通行能力、事故持续时间、路段上游车流量间的关系，车辆排队长度将到达上游路口的时间等。

问题一要求根据视频 1，描述视频中交通事故发生至撤离期间，事故所处横断面实际通行能力的变化过程。将事故发生到撤离期间分为碰撞瞬间、碰撞瞬间至开始拥堵、开始拥堵到撤离三段，建立道路实际通行能力的计算模型，利用 Excel 软件对数据进行分析，分别解出各段实际通行能力，各段的实际通行能力分别为 1 820 pcu/h、1 829 pcu/h、1 794 pcu/h。

问题二要求根据问题一所得结论，结合视频 2，分析说明同一横断面交通事故所占车道不同对该横断面实际通行能力影响的差异。选取各时间段车流量及平均车流量、阻塞时的车辆密度、平均行程速度、实际通行能力为指标，通过以上四项指标证明了同一横断面交通事故所占车道不同对该横断面实际通行能力影响存在差异，并指出存在差异的原因主要是各车道车流量不同和靠右行驶的交通规则的影响。

问题三要求构建数学模型，分析视频 1 中交通事故所影响的路段车辆排队长度与事故横断面实际通行能力、事故持续时间、路段上游车流量间的关系。根据交通波理论和流量守恒定律，可以推导出车辆排队长度公式，以此公式来说明车辆排队长度与事故横断面实际通行能力、事故持续时间、路段上游车流量间的关系，关系式为 $x = |v_f(1-(v_{s2}s_1 + v_{s1}s_2)/k_j)|(t-t_0)$。

问题四要求假如视频 1 中的交通事故所处横断面距离上游路口变为 140 m，路段上

[①] 本文获 2013 年全国一等奖。队员：王兴，刘晶晶，周庆豪；指导教师：冯守平。

游车流量为 1 500 pcu/h,估算当车辆初始排队长度为零且事故持续不撤离时,从事故发生要经过多长时间,车辆排队长度将到达上游路口。问题四可以看作对问题三的简化,完全可以利用问题三所建立的的车辆排队模型求解,相关的数据也可以通过观测视频 1 获取,求得经过 4.19 min,车辆排队长度将到达上游路口。

本文对模型的误差进行了定性分析,此外还利用独立两样本 t 检验验证模型结果的正确性。考虑到小区路口出入车辆对主干道通行能力的影响,给出了模型的改进意见。最后,对模型进行了横向和纵向的推广。

关键词:车道被占用;道路通行能力;交通波理论;车辆排队长度;SPSS;灵敏度分析

1.1 问题的重述

1.1.1 背景知识

1. 车道被占用

车道被占用是指因交通事故、路边停车、占道施工等因素,导致车道或道路横断面通行能力在单位时间内降低的现象。由于城市道路具有交通流密度大、连续性强等特点,一条车道被占用,也可能降低路段所有车道的通行能力,即使时间短,也可能引起车辆排队,出现交通阻塞。如处理不当,甚至出现区域性拥堵。

本文涉及的车道被占用情况是因交通事故产生,视频 1(附件 1)和视频 2(附件 2)中的两个交通事故处于同一路段的同一横断面,且完全占用两条车道。

2. 道路通行能力

道路通行能力是表示道路所能承担车辆通过的能力,具体指道路上某一点某一车道或某一断面处,单位时间内可能通过的最大交通实体(车辆或行人)数,亦称道路通行能量,用"辆/h"或"辆/昼夜"或"辆/秒"表示,车辆多指小汽车,当有其他车辆混入时,均采用等效通行能力的当量标准车辆(小汽车)为单位(pcu)。

本文指同一路段的同一横断面,完全占用两条且只剩下一条道路后的通行能力。

3. 研究意义

车道被占用的情况种类繁多、复杂,正确估算车道被占用对城市道路通行能力的影响程度,将为交通管理部门正确引导车辆行驶、审批占道施工、设计道路渠化方案、设置路边停车位和非港湾式公交车站等提供理论依据。

1.1.2 相关附件

1. 附件 1

视频 1(http://special.univs.cn/service/jianmo/index.shtml)。

2. 附件 2

视频 2(http://special.univs.cn/service/jianmo/sxjmtmhb/2013/0525/969401.shtml)。

3. 附件3

视频1中交通事故位置示意图(如图1.1)。

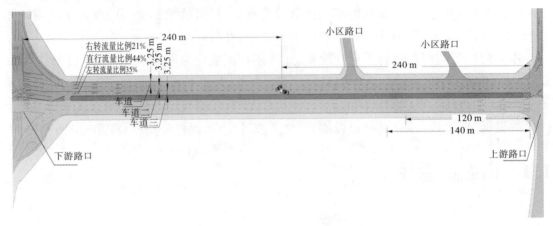

图1.1 视频1中交通事故位置示意图

4. 附件4

上游路口交通组织方案图(如图1.2)。

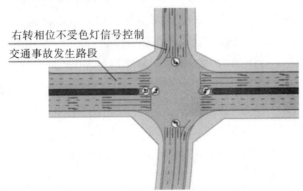

图1.2 上游路口交通组织方案图

5. 附件5

上游路口信号配时方案图(如图1.3)。

相位时间均为30 s，黄灯时间为3 s，信号周期为60 s
相位时间=绿灯时间+绿闪时间(3 s)+黄灯时间

图1.3 上游路口信号配时方案图

注：只考虑四轮及以上机动车、电瓶车的交通流量,且换算成标准车当量数。

1.1.3 具体问题

1. 问题一

根据视频1,描述视频中交通事故发生至撤离期间,事故所处横断面实际通行能力的变化过程。

2. 问题二

根据问题一所得结论,结合视频2,分析说明同一横断面交通事故所占车道不同对该横断面实际通行能力影响的差异。

3. 问题三

构建数学模型,分析视频1中交通事故所影响的路段车辆排队长度与事故横断面实际通行能力、事故持续时间、路段上游车流量间的关系。

4. 问题四

假如视频1中的交通事故所处横断面距离上游路口变为140 m,路段下游方向需求不变,路段上游车流量为1 500 pcu/h,事故发生时车辆初始排队长度为零,且事故持续不撤离。请估算,从事故发生开始,经过多长时间,车辆排队长度将到达上游路口。

1.2 问题的分析

1.2.1 对问题的总体分析

本文主要研究车道被占用对城市道路通行能力的影响,问题涉及两个主要因素:车道被占用(本文因交通事故导致车道或道路横断面通行能力在单位时间内降低的现象)和道路通行能力(本文指原来主干道为三条车道正常通行,但因交通事故三道变一道后的状况),这在问题重述的背景知识里已详细交代,其中车道被占用是条件,道路通行能力是主体,前者对后者的影响是问题的核心。本问题是涉及道路性质、车流量变化、车道占用、道路通行能力变化等多因素的交通管理问题,解决本问题的关键是相关交通研究的机理分析和两个视频实际数据的实验分析法相结合的规划管理问题。

1.2.2 对具体问题的分析

1. 对问题一的分析

问题一要求根据视频1,描述视频中交通事故发生至撤离期间,事故所处横断面实际通行能力的变化过程。观看视频1后了解到,当两辆车发生交通事故后(见图1.1),车道二、车道三受阻,仅有车道一可通行,这种情况下,车道一的车辆则正常向前行驶,而受阻的车道二、车道三上原行驶车辆必须换道转移到车道一绕行。当车流量较少时,交通流是畅通的,但因为出现换道转移使车速放缓,导致道路通行能力有所下降。问题一要求描述视频1中交通事故发生至撤离期间,实际通行能力的变化过程。

2. 对问题二的分析

问题二要求根据问题一所得结论,结合视频2,分析说明同一横断面交通事故所占车道不同对该横断面实际通行能力影响的差异。该问题和问题一不同的是事故发生位置不同,问题一中事故所占车道为车道二和车道三,仅有车道一可通行,而问题二中事故所占车道为车道一和车道二,仅有车道三可通行。为说明交通事故所占车道不同对该横断面实际通行能力影响的差异,本文参照问题一的求解过程,计算出此次横断面的实际通行能力,再与问题一的结果进行比较。

3. 对问题三的分析

问题三要求构建数学模型,分析视频1(附件1)中交通事故所影响的路段车辆排队长度与事故横断面实际通行能力、事故持续时间、路段上游车流量间的关系。根据交通波理论和流量守恒定律,可以推导出车辆排队长度公式,以此公式来说明车辆排队长度与事故横断面实际通行能力、事故持续时间、路段上游车流量间的关系。

4. 对问题四的分析

针对问题四要求假如视频1(附件1)中的交通事故所处横断面距离上游路口变为140 m,路段下游方向需求不变,路段上游车流量为1 500 pcu/h,事故发生时车辆初始排队长度为零,且事故持续不撤离。请估算,从事故发生开始,经过多长时间,车辆排队长度将到达上游路口。问题四可以看作对问题三的简化,完全可以利用问题三所建立的排队长度模型求解,相关的数据也可以通过观测视频1获取。

1.3 模型的假设

1. 时间

不考虑不同时间段车流量的差异,忽略下班高峰对车流量的影响。

2. 车辆

只考虑四轮及以上机动车、电瓶车的交通流量,且换算成标准车当量数。

3. 道路

道路平坦,不考虑坡度对车速的影响,且视距满足理想要求。

4. 驾驶

本文研究时忽略不同驾驶人的驾驶能力差异。

5. 路口

忽略小区路口车辆对干线车流量的冲击。

6. 天气

忽略光线、阴雨、雾霾等天气因素对车速的影响。

1.4 名词解释和符号说明

1.4.1 名词解释

1. 实际通行能力

以理论通行能力为基础,考虑到实际的地形、道路和交通状况,确定其修正系数,再以此修正系数乘以前述的理论通行能力,即得实际道路、交通在一定环境条件下的可能通行能力。

2. 交通波理论

运用流体力学的基本原理,模仿流体的连续性方程,建立车流的连续性方程,把车流密度的变化比拟成水波的起伏,抽象为车流波。

3. 畅行速度

车流密度为零时,车辆行驶的最大速度。

4. 阻塞密度

车流密集到所有车辆都无法移动时的密度。

1.4.2 符号说明

本模型所用的主要变量符号及意义如表1.1所示。

表1.1 主要变量符号及意义

序号	符号	意义
1	T	拥挤持续时间
2	T_1	事故延迟时间
3	T_2	拥挤消散时间
4	T_3	事故检测时间
5	C_B	单条车道的基本通行能力(pcu/h)
6	H_m	最小安全车头时距(s)
7	M_i	第 i 级服务水平的最大服务交通量(pcu/h/ln)
8	C_b	基本路段的通行能力,即理想条件下一个车道所能通行的最大交通量(pcu/h/ln)
9	$(C/V)_i$	第 i 级服务水平最大服务交通量与基本通行能力的比值
10	C_D	单向车行道设计通行能力,采用 i 级服务水平时所能通过的最大交通量(veh/h)
11	n	单向车行道的车道数
12	f_w	车道宽度和侧向净宽对通行能力的修正系数
13	f_p	驾驶人条件对通行能力的修正系数
14	f_{HV}	非标准型车对通行能力的修正系数
15	E_{HV}	非标准型车换算成小客车的车辆换算系数
16	P_{HV}	各类型车交通量占总交通量的百分比
17	n_t	t 时刻到达的交通流量(veh)
18	c_t	t 时刻道路的通行能力(veh)
19	Q	事故路段实际通行能力

1.5 模型的建立与求解

1.5.1 问题一的分析与求解

1. 对问题一的分析

要求描述事故发生到撤离期间,事故所取的横截面通行能力的变换情况。我们对公路交通事故造成的拥堵时间进行分析,将事故开始到撤离这段时间分为碰撞瞬间、碰撞到开始拥挤、开始拥挤到撤离这三段,以 30 s 为时间间隔分别研究各段内的车流量和实际通行能力,建立道路通行能力评价模型,求解出各段的通行能力,分别比较三段内的车流量和实际通行能力,再对三段的车流量和实际通行能力进行横向比较。

2. 对问题一的求解

(1)模型的准备。

①公路交通事故造成的拥堵时间分析。当公路上发生交通事故时,道路通行能力降低,当此处通行能力小于上游到达的交通流量时,车辆在事件点上游排队,并不断向上游延伸,甚至延伸到上游交叉口,当事故清理完毕、道路通行能力恢复,拥挤车流开始消散。公路交通事故影响下的交通拥挤从产生到消散完毕的整个过程如图 1.4 所示,事故延迟时间又可以分成几个部分,如图 1.5 所示。

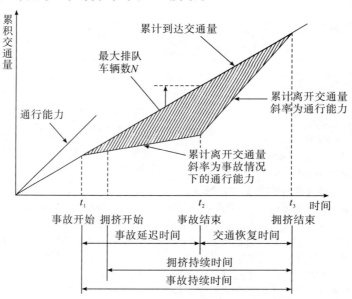

注:t_1——事故发生时刻,t_2——事故结束时刻,t_3——拥堵结束时刻,图中阴影部分是交通拥堵引起的延误。

图 1.4 交通拥挤形成过程

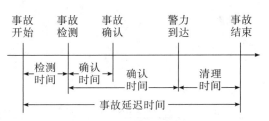

图 1.5　事故延误时间

事故检测是以检测交通异常为机理,从事故发生到事故被检测到之间的时间段拥挤并没有产生或者拥挤比较轻微,因此事故拥挤持续时间并不包括事故检测时间(T_3)。

$$T = T_1 + T_2 - T_3 \qquad (1\text{-}1)$$

一般情况下,当事故发生后,事故检测时间相对于拥挤持续时间和事故延迟时间都比较小,因此可以不考虑,式(1-1)转化为

$$T \approx T_1 + T_2 \qquad (1\text{-}2)$$

T_1由事故类型、事故严重程度等多方面因素决定,在交通事故影响下的交通拥挤扩散范围估计方法中,事故延迟时间的确定对于拥挤扩散的预测非常重要。

②道路通行能力的影响因素。

ⅰ)道路条件:包括车道数、道路设施类型、车道宽度等因素,这些因素都会影响通行能力。

ⅱ)交通条件:影响通行能力的交通条件包括车辆类型以及车道使用和方向分布。

ⅲ)环境条件:指横向干扰、非交通占道和停车位置等因素。

ⅳ)气候条件:风、雨、雪、雾、沙尘暴等恶劣天气会影响道路通行能力。

ⅴ)规定运行条件:主要是指计算通行能力的限制条件,这些限制条件通常根据速度和行程时间、驾驶自由度、交通间断、舒适和方便性以及安全等因素来规定。其运行标准是针对不同的交通设施用服务水平来定义的。

另外,道路周围的地形、地物、景观、驾驶人技术等也对道路通行能力有一定的影响。

(2)模型的建立——模型Ⅰ:道路通行能力评价模型。

通行能力是针对单向车流而言的。通行能力分析是基于具有标准的道路及交通条件的公路路段,如果这些主要条件之一发生了显著变化,则路段通行能力和运行条件也会随之发生变化。道路的通行能力分为基本通行能力、设计通行能力、可能(实际)通行能力三种。

①基本通行能力。基本通行能力又称理论通行能力,是指在规定的道路交通条件下,一条车道的一个断面一小时能够通过的最大车辆数,是理论上能通行的最大交通量,基本通行能力建立在理想条件下:

ⅰ)车道宽≥3.5 m;

ⅱ)从车道边缘至路旁障碍物的距离≥1.75 m;

ⅲ)无限速,无横向车辆及行人的干扰;

ⅳ)车种单一;

ⅴ)平、纵线性满足要求,视距满足要求。

按车头时距计算,其计算公式为:

$$C_B = 3\,600/H_m \tag{1-3}$$

如果没有 H_m 观测值,C_B 可通过查表1.2得到。

表1.2 不同速度的基本通行能力

速度 km/h	50	40	30	20
通行能力 pcu/h	1 690	1 640	1 550	1 380

通过资料查询,我们将道路的通行能力限制在速度为 50 km/h 的时候,则道路的基本通行能力为 1 690 pcu/h。

对于公路的正常路段,其最大服务交通量可以采用计算公式:

$$M_i = C_b \cdot (V/C)_i \tag{1-4}$$

式(1-4)中$(V/C)_i$为第i级服务水平最大服务交通量与基本通行能力的比值,这里考虑设计通行能力达到最大的情况,所以在此采用四级服务水平下的通行能力,取V/C为1。

②可能(实际)通行能力。由于道路的实际情况比理想状态更为复杂,道路中的车辆有各种类型,各路段的车速随着道路条件及车辆行人的干扰情况在变化,需要根据不同情况对基本通行能力进行折减。影响实际通行能力的因素主要有下面4个。

ⅰ)封闭行车道数:出于事故路段安全需要,有时必须关闭1个或多个行车道,这将会大大影响公路事故路段通行能力,导致公路事故道路交通拥挤、堵塞,甚至中断交通。

ⅱ)封闭车道长度:封闭车道长度越长,驾驶人越需要更加谨慎地驾驶通过事故路段,导致车流速度降低,极大降低了事故路段通行能力。

ⅲ)行车道宽度:与公路基本路段通行能力修正一样,公路事故路段开放车道宽度对其通行能力有一定的影响。一般认为,当车道宽度达到某一数值时其通过量能达到理论上的最大值,当车道宽度小于该值时,通行能力降低。

ⅳ)交通事故延迟时间:事故延迟时间是事故下公路通行能力的一个重要影响因素,事故延迟时间是从事故开始时刻到事故清理完毕时刻之间的时间段。

③单向车行道的设计通行能力。

$$C_D = M_i \cdot n \cdot f_w \cdot f_{HV} \cdot f_p = C_b \cdot (V/C)_i \cdot n \cdot f_w \cdot f_{HV} \cdot f_p \tag{1-5}$$

由于道路运行中存在四轮及以上机动车、电瓶车这些不同的车型,在计算道路通行能力[1]时要折换成标准车型计算,非标准车型对通行能力的修正系数为:

$$f_{HV} = \frac{1}{1 + P_{HV}(E_{HV} - 1)} \tag{1-6}$$

四轮及以上机动车、电瓶车等车型的折算系数见表1.3。从视频1的观察中可看出城市该道路中运行的车辆种类主要为公交车,电瓶车,小轿车和出租车等,将车型大小小于公交车的车都归类在小型车中。

表1.3 车辆类型及折算系数

车辆类型	小型车	公交车	电瓶车
折算系数	1	1.5	0.5

事故路段的通行能力是本文中所建立的车流波模型中的一个重要参数,通过标定事故路段的通行能力来确定排队长度及消散时间等量,并且以此来确定相应的交通组织方案,所以对事故路段通行能力的标定是非常重要的。

根据流量守恒方程及图 1.1 中的拥挤扩散过程,可以得到:

$$\sum_{t=t_1}^{t_3}(n_t - c_t) = 0 \qquad (1-7)$$

在式(1-7)中,当 $t_2 < t < t_3$ 时,c_t 等于道路的实际通行能力,与道路类型、道路几何特性等静态因素有关,可以根据道路通行能力手册[2]里的计算公式获得;$t_1 < t < t_2$ 时,发生交通事故,c_t 降低,此时 c_t 为事故路段的道路通行能力。

(3)模型的求解。

①碰撞瞬间。由图 1.4 分析知在事故发生的瞬间车辆仍旧保持原有的同性状态,所以这一瞬间道路的实际通行能力未发生变化,所以此时的实际道路通行能力和道路发生事故前的通行能力是一样的。从 16:38:19 至 16:42:09 这段时间内,以 30 s 为间隔统计每个时间间隔内各类型车的数量,计算碰撞瞬间的车流量和道路容纳能力,见表 1.4。

表 1.4　统计的碰撞前各时段的车辆数和占比

时间	公交车	小型车	电瓶车	标准车	公交车比类	小型车比类	电瓶车比类
16:38:39—16:39:09	1	6	4	9.5	9.09%	54.55%	36.36%
16:39:09—16:39:39	1	7	2	9.5	10.00%	70.00%	20.00%
16:39:39—16:40:09	0	5	0	5	0.00%	100.00%	0.00%
16:40:09—16:40:39	0	9	7	12.5	0.00%	56.25%	43.75%
16:40:39—16:41:09	1	6	6	10.5	7.69%	46.15%	46.15%
16:41:09—16:41:39	2	7	0	10	22.22%	77.78%	0.00%
16:41:39—16:42:09	2	8	0	11	20.00%	80.00%	0.00%

从表 1.4 中可看出每 30 s 折合的标准车辆都很稳定,除了第三段时间的标准车辆数较小,其他的基本都较为稳定,对这段时间标准车辆数进行分析。

表 1.5　标准车统计计算表

车道数	平均值	中值	标准差	极差
3 条	10.5	10.25	1.14	3

由表 1.5 可以看出事故发生前的车流量较为稳定,每分钟经过约 20 辆标准车,碰撞瞬间的车流量约为每分钟 20 辆。每小时车流量为 1200 辆。

我们假设道路的侧向净宽和视距都满足基本道路要求且道路坡度为 0,则 $f_w = 1$,驾驶人条件对道路的修正系数 $f_p = 1$,由于我们不知道该条道路的速度要求是多少,所以我们根据表 1.2 不同速度下的基本通行能力来计算不同速度下的实际通行能力,相关数据代入公式(1-5)计算出此时道路的实际通行能力为 1 820 pcu/h。

②从碰撞到开始拥挤。由视频 1 观察从碰撞到此后的一段时间,这时只有一条车道可供通行,道路开始出现拥堵,但是此时的拥堵可以很快的疏通,在一个红灯的间隔时

间所有的车都能够从车道一离开,不会形成持续的拥堵。详细车辆数和占比见表 1.6。

表 1.6 统计的碰撞到拥挤各时段的车辆数和占比

时间	公交车	小型车	电瓶车	标准车	公交车比类	小型车比类	电瓶车比类
16:42:32—16:43:02	2	6	2	10	20.00%	60.00%	20.00%
16:43:02—16:43:32	1	8	2	10.5	9.09%	72.73%	18.18%
16:43:32—16:44:02	1	10	0	11.5	9.09%	90.91%	0.00%
16:44:02—16:44:32	1	8	5	12	7.14%	57.14%	35.71%
16:44:32—16:45:02	0	9	2	10	0.00%	81.82%	18.18%
16:45:02—16:45:32	0	7	0	7	0.00%	100.00%	0.00%
16:45:32—16:46:02	0	7	3	8.5	0.00%	70.00%	30.00%
16:46:02—16:46:32	1	8	1	10	10.00%	80.00%	10.00%
16:46:32—16:47:02	0	8	4	10	0.00%	66.67%	33.33%
16:47:02—16:47:32	0	6	3	7.5	0.00%	66.67%	33.33%
16:47:32—16:48:02	1	8	2	10.5	9.09%	72.73%	18.18%

对每分钟的标准车数量进行分析,详见表 1.7。

表 1.7 标准车统计计算表

车道数	平均值	中值	标准差	极差
3 条	9.7	10	1.5	5

此时每小时的平均车流量为 1 164 辆,道路的实际通行能力为 1 829 pcu/h。

③ 从拥挤到车辆撤离。这一阶段车辆开始形成堵塞,一次绿灯从上游来的车辆不能全部经过事故横断面时下一次绿灯放行的车辆又行驶过来,而通道的通行能力有限。拥堵的车辆在事故发生至上游路口段排起长龙。

表 1.8 统计的拥挤到车辆撤离各时段的车辆数和占比

时间	公交车	小型车	电瓶车	标准车	公交车比类	小型车比类	电瓶车比类
16:48:02—16:48:32	0	11	3	12.5	0.00%	78.57%	21.43%
16:48:32—16:49:02	0	10	1	10.5	0.00%	90.91%	9.09%
16:49:02—16:49:32	0	10	2	11	0.00%	83.33%	16.67%
16:50:04—16:50:34	1	7	2	9.5	10.00%	70.00%	20.00%
16:50:34—16:51:04	0	11	2	12	0.00%	84.62%	15.38%
16:51:04—16:51:34	0	8	1	8.5	0.00%	88.89%	11.11%
16:52:34—16:53:04	0	9	4	11	0.00%	69.23%	30.77%
16:53:04—16:53:34	1	7	0	8.5	12.50%	87.50%	0.00%
16:53:34—16:54:04	2	9	0	12	18.18%	81.82%	0.00%
16:54:04—16:54:34	0	8	1	9.5	0.00%	88.89%	11.11%
16:54:21—16:54:51	1	9	1	11	9.09%	81.82%	9.09%
16:54:51—16:55:21	0	8	2	9	0.00%	80.00%	20.00%
16:55:21—16:55:51	0	9	3	10.5	0.00%	75.00%	25.00%

由于视频1中有些部分的时间是不连续的,所以把不连续的时间段舍弃,取后面连续的时间段测量通行的各类车辆。根据表1.8的分析可得出,此时道路每小时的平均车流量为1 248辆,道路的实际通行量为1 794 pcu/h。

表1.9 标准车统计计算表

车道数	平均值	中值	标准差	极差
3条	10.4	10.5	1.3	4

结论分析:这段时间的事故横断面的实际通行能力和车流量都未发生太大变化,由表1.5、表1.7、表1.9可看出车辆的通行一直保持在较为平均的水平。

1.5.2 问题二的分析与求解

1. 对问题二的分析

在现实生活中,交通事故的发生是随机的,事故所占车道往往也是不一致的。那么在同一道路的同一横断面处发生事故,所占车道不同对该横断面实际通行能力的影响是否存在差异? 这是我们接下来所需研究的。

问题二与问题一的不同之处有三点:一是交通事故所占车道不同。在问题一中事故所占车道为车道二和车道三,仅有车道一可通行,而问题二中事故所占车道为车道一和车道二,仅有车道三可通行。二是交通事故发生的时间不同。问题一中事故发生时间为16:38,而问题二中事故发生时间为17:35。按照我国的上下班时间,问题二中事故发生时正值下班高峰,车流量较大,拥堵情况更为严重,对该横断面实际通行能力的影响是显而易见的,因此本文不考虑事故发生时间的差异对实际通行能力的影响。三是交通事故的持续时间不同。前者持续时间为15 min,后者持续时间为29 min,但是在衡量横断面通行能力的各项指标中,时间均出现在分母中,消除了事故持续时间的影响。

由于问题一和问题二都是求解横断面实际通行能力,故完全可参照问题一的求解方法计算出问题二中横断面的实际通行能力,在此基础上与问题一的结果相比较。若存在差异,则说明交通事故所占车道不同对该横断面实际通行能力有影响,反之则没有影响。

2. 对问题二的求解

(1)各时间段车流量及平均车流量。

由于两起事故发生在同一横截面,事故发生前和撤离后的横截面实际通行能力相同,因此只需比较交通事故发生至撤离期间横截面的实际通行能力即可。根据视频2的内容,每隔1 min统计一次各类型车辆的数量,并对各时间段进行编号(例如,将17:34:17—17:35:17编号为1,…,将18:02:17—18:03:17编号为29)。

根据公式

$$K = \frac{N_x + 1.5N_y + 0.5N_z}{\Delta t} \times 60$$

上式中：N_x，N_y，N_z 分别表示小型客车、大型客车、电动车的数量，Δt 表示时间间隔，在此取 $\Delta t=1$ min，可以得到各时间段的车流量，结果如图 1.6 所示。

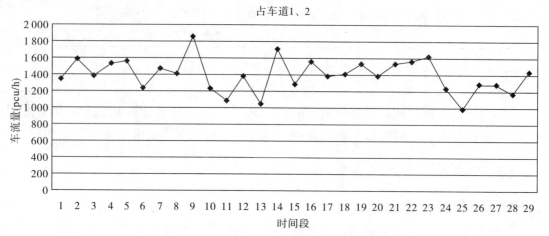

图 1.6 占车道 1、2 时各时间段车流量的变化情况

可以计算出问题二中事故发生至撤离期间（17:34:17—18:03:17）的平均车流量 $K_2=1\ 396$。

同理可得问题一中各时间段车流量及平均车流量，结果如图 1.7 所示。

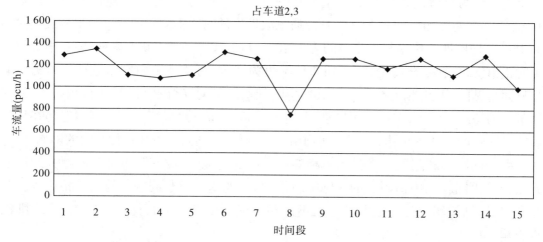

图 1.7 占车道 2、3 时各时间段车流量变化情况

可以计算出问题一中事故发生至撤离期间（16:42:32—17:00:17）的平均车流量 $K_1=1\ 174$。

结果说明： 当事故所占车道为 1、2 时平均车流量为 $K_2=1\ 396$，与所占车道为 2、3 时平均车流量 $K_1=1\ 174$ 比较变大了，说明事故所占车道的不同影响了平均车流量。

(2) 阻塞时的车辆密度。

在视频 1 和视频 2 中都明显出现车辆排队的情况，因此我们可以比较阻塞时道路指定断面单位长度内的车辆数，即阻塞时的车辆密度，计算公式为

$$\rho = \frac{N_x + 1.5 N_y + 0.5 N_z}{L} \times 1\ 000$$

不妨以 120 m 为观测长度，统计道路阻塞时事故发生处向后 120 m 内排队的车辆数，结果如表 1.10 和表 1.11 所示。

表 1.10 事故一发生后阻塞时的车辆密度

时刻	小型客车	电瓶车	大型客车	标准车当量数	密度
16:42:32	8	1	2	11.5	95830
16:47:50	10	1	1	12	100000
16:50:42	22	0	0	22	183330
16:51:44	24	3	1	27	225000
16:52:46	21	1	2	24.5	204170
16:54:03	22	1	2	25.5	212500

表 1.11 事故二发生后阻塞时的车辆密度

时刻	小型客车	电瓶车	大型客车	标准车当量数	密度
17:41:45	13	1	1	15	125000
17:50:04	17	1	0	17.5	145830
17:50:48	21	3	1	24	200000
17:51:50	25	7	1	30	250000
17:54:51	21	5	2	26.5	220830
17:55:53	14	9	4	24.5	204170
17:58:51	22	4	4	30	250000
18:00:04	30	4	2	35	291670
18:02:06	31	1	3	35.5	295830

利用表 1.10 和表 1.11 的数据，计算得出事故一发生后阻塞时的平均车辆密度 $\rho_1 = 170\,140$ pcu/km，事故二发生后阻塞时的平均车辆密度 $\rho_2 = 220\,370$ pcu/km。

结果说明：$\rho_2 > \rho_1$ 表明车道 1、2 被占后车辆的阻塞程度大于车道 2、3 被占时的阻塞程度。其背后原因在于交通规则的影响，在我国，电动车被要求行驶在道路的右侧，即行驶在车道 1，那么当车道 1 被挡住时必然给电动车行驶造成不便，导致车道 1、2 被占后平均车辆密度较大。

(3) 平均行程速度。

在交通流中，观测到的速度分布通常比较分散，参考《公路通行能力手册》的做法，以平均行程速度作为速度标准。计算平均行程速度先选取道路的一段长度为 L，除以车辆通过该路段的平均行程时间，如果有 n 辆车，通过道路的长度为 L，测得车辆的行程时间为 $t_1, t_2, \cdots, t_n$，则平均行程速度可按下式计算。

$$V_S = \frac{L}{\sum_{i=1}^{n} t_i / n} = \frac{nL}{\sum_{i=1}^{n} t_i} \tag{1-8}$$

式中，V_S——平均行程速度；L——公路路段长度；t_i——第 i 辆车通过该路程的时间；n——观测行程时间的次数。

针对视频2,我们取120 m为公路路段长度,随机选取各车道上10辆小型车观测其通过这段路程的时间,观测结果见表1.12。

将数据代入公式(1-8)可得车道1、2、3的平均行程速度分别为8.24 km/h、5.81 km/h、8.52 km/h。

针对视频1,仍然取120 m为公路路段长度,随机选取各车道上10辆小型车观测其通过这段路程的时间,观测结果见表1.13。

表1.12 事故二中各车道通行时间

序号	1	2	3	4	5	6	7	8	9	10
车道2的通行时间	71	36	56	56	59	37	58	73	43	35
车道1的通行时间	62	62	79	68	80	40	103	69	95	85
车道3的通行时间	18	32	68	73	39	38	33	74	18	114

表1.13 事故一中各车道通行时间

序号	1	2	3	4	5	6	7	8	9	10
车道2的通行时间	41	115	87	56	33	94	53	44	108	81
车道1的通行时间	16	18	31	19	42	66	47	43	111	50
车道3的通行时间	43	48	133	88	84	75	130	41	63	84

同理可得车道1、2、3的平均行程速度分别为6.07 km/h、9.75 km/h、5.48 km/h。

结果说明:上述结果说明,不仅事故发生后各车道的平均行程速度存在差异,而且所占车道不同导致两起事故中同一车道的平均行程速度差异明显。

(4)实际通行能力。

视频1和视频2中两起交通事故的共同之处:一是事故发生所处道路条件相同,即车道宽度均为3.25 m,右转、直行和左转流量比例保持不变,道路基本建设状况相同。二是驾驶者的驾驶能力相同,即刹车的反应时间、绕道而行的能力都相同。

考虑到上述共同点,问题二中对实际通行能力的计算方法与问题一相同。以一分钟为间隔分别采集视频一、视频二每分钟内各类车的通过量,经式(1-5)计算出实际的通行能力。进行数据采集,经处理可得结果如表1.14与表1.15。

表1.14 视频1事故后各时间段横断面处实际通行能力   单位:pcu/h

时间段	16:42~16:43	16:43~16:44	16:44~16:45	16:45~16:46
实际通行能力	1807	1802	1827	1877
时间段	16:46~16:47	16:47~16:48	16:48~16:49	16:49~16:50
实际通行能力	2009	1843	1850	2028
时间段	16:50~16:51	16:51~16:52	16:52~16:53	16:53~16:54
实际通行能力	1850	1850	1733	1609

表 1.15 视频 2 事故后各时间段横断面处实际通行能力　　　　　　　单位：pcu/h

时间段	17:34～17:35	17:35～17:36	17:36～17:37	17:37～17:38	17:38～17:39
实际通行能力	1727	1785	1910	1723	1885
时间段	17:39～17:40	17:40～17:41	17:41～17:42	17:42～17:43	17:43～17:44
实际通行能力	1731	1931	1725	1908	1896
时间段	17:44～17:45	17:45～17:46	17:46～17:47	17:47～17:48	17:48～17:49
实际通行能力	1783	1983	1834	1778	1886
时间段	17:49～17:50	17:50～17:51	17:51～17:52	17:52～17:53	17:53～17:54
实际通行能力	1820	1836	1941	2120	1836
时间段	17:54～17:55	17:55～17:56	17:56～17:57	17:57～17:58	17:58～17:59
实际通行能力	1855	1950	1940	1648	1741
时间段	17:59～17:00	17:00～18:01	18:01～18:02	18:02～18:03	
实际通行能力	1729	1886	1646	1690	

由图 1.8 可直观地看出从事故发生到撤离事故横断面的实际通行能力均未发生太大的变化。

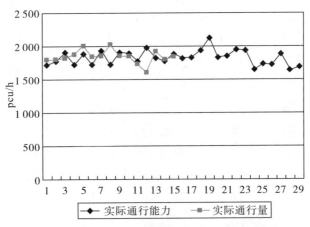

图 1.8 视频 1、2 中事故横断面的实际通行能力

3. 问题的结论

用上述 4 项指标去衡量同一横断面交通事故所占车道不同对该横断面实际通行能力的影响，经过计算，4 项指标在事故一和事故二中的数值各不相同，充分说明所占车道不同影响该横断面实际通行能力。存在差异的原因有两点：一是右转、直行及左转流量比例不同。右转流量比例为 21%，直行流量比例为 44%，左转流量比例为 35%，当车道 2,3 被占时，有 79% 的车辆需要绕道；当车道 1,2 被占时，有 65% 的车辆需要绕道。二是靠右行驶的交通规则的影响。我国的交通规则是靠右行驶，这一规则对大型客车、小型客车的影响较小，因为它们在道路中间行驶，但对电动车的影响是巨大的，电动车被要求行驶在道路的右侧，即行驶在车道 1，那么当车道 1 被挡住时必然给电动车行驶造成不便。

1.5.3 问题三的分析与求解

1. 对问题三的分析

问题三要求构建数学模型,分析视频1中交通事故所影响的路段车辆排队长度与事故横断面实际通行能力、事故持续时间、路段上游车流量间的关系。

对排队问题的另一种常见解决方法是运用交通波理论来求解。交通波理论是运用流体力学的基本原理,模仿流体的连续性方程,建立车流的连续性方程,把车流密度的变化比作水波的起伏,抽象为车流波。

当道路中发生交通事故时,由于道路实际通行能力下降形成交通瓶颈,造成车辆向上游排队的现象,道路的排队可能发生在拥堵道路段,也可能影响上游交叉口对整个交通网络产生影响。我们首先根据交通波理论建立数学模型,将相关变量之间的关系用公式表示出来,再结合视频一,求出具体的参数值。

2. 对问题三的求解——模型Ⅱ 车辆排队模型

(1)模型的准备。

①交通波理论[3]。设有一个交通波以速度 u_w 沿行车道稳定向右传播,波阵面 s 前车流密度为 k_1,速度为 u_1,波传播过后的车流密度为 k_2,速度变为 u_2。以波阵面 s 为界面,将看到原车流以 u_w-u_1 的速度向左流过波阵面,而以 u_w-u_2 的速度从波阵面流出,如图1.9所示。

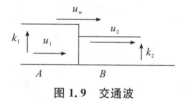

图1.9 交通波

假设车道为单车道,根据质量守恒定律,在波稳定的传播条件下,时间 t 内波阵面右侧流入车辆数应等于从左侧流出的车辆数,由此可得 $k_1(u_w-u_1)t=k_2(u_w-u_2)t$

即
$$u_w = (u_1 k_1 - u_2 k_2)/(k_1 - k_2) \tag{1-9}$$

由于流量 $q=uk$,令波阵面前后的车流量分别为 q_1 和 q_2,则有:
$$u_w = (q_2 - q_1)/(k_2 - k_1) \tag{1-10}$$

此即为交通波的基本方程式。

②流量、车速、密度[4]。根据物理学的基本常识,流量 q、车速 v 和密度 k 的关系为 $q=vk$。1935年,Greenshields 通过对观测数据的统计分析,提出车速与密度的线性模型为:
$$v = v_f(1 - k/k_j) \tag{1-11}$$

其中 v_f 是密度 $k=0$ 时的车速,即理论上的最高车速,称为畅行速度,k_j 是速度 $v=0$ 时的密度,称为阻塞密度。

(2)模型的求解。

根据交通波理论和流量、车速、密度三者间的关系,即联立方程(1-9)、(1-10)、(1-11)可得

$$u_w = v_f \left(1 - \frac{k_1 + k_2}{k_j}\right)$$

一般来说事故发生后,部分车道被占,该横截面的实际通行能力下降为 s_1,车辆密度

上升为 k_{s1}，车速下降为 v_{w1}；事故持续时间为 t_0；事故撤离后到车队消散前该横截面的实际通行能力上升为 s_2，车辆密度上升为 k_{s2}，车速上升为 v_{w2}；排队长度为 x，上游车流量为 q_1，行驶速度为 v_1。

根据守恒定律：流入量－流出量＝数量变化，假设消散波和集结波相遇的时间为 t，则有

$$u_{集} t = u_{消} (t - t_0)$$

又 $u_{集} = \left| \dfrac{q_1 - s_1}{k_1 - k_{s1}} \right| = \left| v_f \left(1 - \dfrac{k_1 + k_{s1}}{k_j}\right) \right|$，$u_{消} = \left| \dfrac{s_1 - s_2}{k_{s1} - k_{s2}} \right| = \left| v_f \left(1 - \dfrac{k_{s1} + k_{s2}}{k_j}\right) \right|$，故

$$t = \dfrac{k_1 - k_{s1} - k_{s2}}{k_1 - k_{s2}} t_0$$

则

$$x = u_{消}(t - t_0) = \left| v_f \left(1 - \dfrac{k_{s1} + k_{s2}}{k_j}\right) \right| (t - t_0)$$

而 $k_{s1} = s_1 / v_{s1}$，$k_{s2} = s_2 / v_{s2}$，$k_1 = q_1 / v_1$，从而

$$x = \left| v_f \left(1 - \dfrac{v_{s2} s_1 + v_{s1} s_2}{k_j}\right) \right| (t - t_0)$$

结合视频 1，不难发现从事故发生到撤离期间出现了好几次排队，并且排队长度各不相同，而 $s_1, k_{s1}, v_{w1}, t_0, s_2, k_{s2}, v_{w2}, v_1$ 这些量变化不大，甚至可以视为定量，那么只能是上游车流量 q_1 的变化在影响排队长度 x。

根据上游路口交通组织方案，可知上游车流量 q_1 的变化主要受右转车辆、直行车辆和左转车辆的影响。其中，右转相位不受色灯信号控制，这说明右转车辆连续不断的随机到达，而直行、左转车辆受色灯信号控制，呈周期性变化。右转车辆的到达情况如图 1.10 所示。

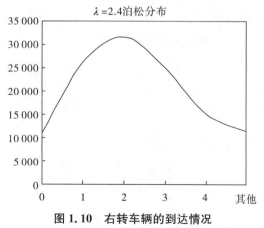

图 1.10　右转车辆的到达情况

从图 1.10 可以看出，右转车辆的到达大约服从 $\lambda = 2.4$ 的泊松分布，故上游右转车流量 $q_a = \lambda T = 2.4 T$（T 为从事故发生开始计时的时间）。

直行和左转车辆受到色灯信号影响的具体表现为，信号灯控制的存在使得交通流在交叉口产生周期性的间断，红灯期间会形成排队，每30 s红灯时间内聚集的车辆数是随机的，但在绿灯时间内通过停车线的车辆数与时间成正相关，直行和左转车辆的到达情况如图1.11所示。

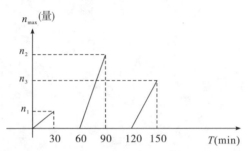

图1.11　直行和左转车辆的到达情况

1.5.4　问题四的分析与求解

1. 对问题四的分析

问题四要求：假如视频1（附件1）中的交通事故所处横断面距离上游路口变为140 m，路段下游方向需求不变，路段上游车流量为1 500 pcu/h，事故发生时车辆初始排队长度为零，且事故持续不撤离。请估算，从事故发生开始，经过多长时间，车辆排队长度将到达上游路口。

问题四实质上是对问题三所建立数学模型的实际应用，同时也是对问题三的简化，因此可以利用问题三中的模型求解问题四，具体的数据可以通过对视频1的观察和统计得出，再将这些数据代入问题三所建模型（交通波模型）即可。

2. 对问题四的求解

在求解问题三时，排除了两个小区路口对路段上游车流量的影响，而问题四将交通事故所处横断面距离上游路口变为140 m，排除了其中一个小区路口对路段上游车流量的影响，我们假设这一变化对排队长度没有影响。

联立下列公式

$$t = \frac{k_j - k_{s1} - k_{s2}}{k_1 - k_{s2}} t_0 \quad \text{与} \quad x = u_{消}(t - t_0) = \left| v_f \left(1 - \frac{k_{s1} + k_{s2}}{k_j}\right) \right| (t - t_0)$$

得

$$x = \left| v_f \left(1 - \frac{k_{s1} + k_{s2}}{k_j}\right) \right| \left(\frac{k_j - k_{s1} - k_{s2}}{k_1 - k_{s2}} - 1\right) t_0$$

变化此公式，得

$$t_0 = \frac{1}{v_f} \left| \frac{k_j}{k_j - k_{s1} + k_{s2}} \right| \left(\frac{k_1 - k_{s2}}{k_j - k_1 - k_{s1}}\right) x$$

由于问题四中假设事故持续不撤离，所以按照上式计算得到的 t_0 表示经过 t_0 时间车辆排队长度达到上游路口。为避免混淆，我们用 t' 代替 t_0。

对视频1的观察和我国道路交通的实际情况,能够得到公式中各参数的具体数值:

(1)事故发生至撤离前:实际通行能力 $s_1=1\,174$ pcu/h,车速 $v_{w1}=7.1$ km/h,车流密度 $k_{s1}=s_1/v_{w1}\approx 166$ pcu/h;

(2)事故撤离至车队消散前:实际通行能力 $s_2=1\,652$ pcu/h,车流行驶速度 $v_{w2}=16.5$ km/h,在此期间的车流密度 $k_{s2}=s_2/v_{w2}=100.12$ pcu/h≈ 101 pcu/h;

(3)事故发生时:上游车流量 $q_1=1\,500$ pcu/h,车速 $v_1=43.2$ km/h。结合我国道路的实际限速情况,事故发生路段的畅行速度一般取 $v_f=80$ km/h,阻塞密度 $k_j=334$ pcu/km,$x=0.14$ km。

将上述数据代入公式 $t'=\left|\dfrac{k_j}{k_j-k_{s1}+k_{s2}}\right|\left(\dfrac{k_1-k_{s2}}{k_j-k_1-k_{s1}}\right)x$,可得 $t'=0.069$ h,即经过 4.15 min,车辆排队长度将到达上游路口。

1.6 模型的进一步讨论

在一开始的模型假设中,我们忽略小区路口(匝道)车辆对干线车流量的冲击,但在实际生活中这种冲击确实存在,它会影响到交通干线的通行能力和阻塞时的排队长度。因此,当排队长度波及匝道和上游主线,并造成道路阻塞时,事故上游主线和匝道上的通行能力计算模型[1]为:

$$Q'_{\text{主}}=\min\left(C_D,\dfrac{Q_{\text{主}}}{Q_{\text{匝}}+Q_{\text{主}}}\min(Q_f,C_{\text{合}})\right),\ Q'_{\text{匝}}=\min\left(C_{rD},\dfrac{Q_{\text{匝}}}{Q_{\text{匝}}+Q_{\text{主}}}\min(Q_f,C_{\text{合}})\right)$$

式中,Q_f 表示合流区下游事故段的通行能力;C_D 表示主线的可能通行能力;C_{rD} 表示匝线的可能通行能力;$Q_{\text{匝}}$ 表示匝线上车辆的到达率;$Q_{\text{主}}$ 表示主线上车辆的到达率。

利用这个模型就可以有效地排除匝道的影响,准确地计算主线的通行能力,但是由于缺乏相关数据且时间紧迫,我们仅指出利用此计算公式改进本文所建模型的可能性。

1.7 模型的检验、误差分析和灵敏度分析

1.7.1 模型的检验

在求解问题二时,我们以两起事故发生至撤离期间各时间段车流量和平均车流量为指标,通过指标数值的不同证明了同一横断面交通事故所占车道不同对该横断面实际通行能力有影响。其实,除了这种比较方法外,还可以用 t 检验来说明数据间的差异,因此我们利用 SPSS 软件对两起事故发生至撤离期间各时间段车流量进行 t 检验,并检验求解问题二所用方法的正确性。

(1)正态性检验——Q-Q 图法。

正态 Q-Q 概率图:它是指以样本的分位数为横坐标,以按照正态分布计算的相应理论分位数为纵坐标,把样本表现作为直角坐标系的散点所描绘的图形。如果数据服从正态分布,则样本点应为一条围绕第一象限对角线的直线。

正态去势 Q-Q 图:它是指以样本的实际分位数为横坐标,以样本的实际分位数与按照正态分布计算的相应理论分布分位数的差为纵坐标,把样本表现作为直角坐标系的散点所描绘的图形。如果数据服从正态分布,残差散点基本在 $Y=0$ 上下均匀分布。

以事故二中的车流量为例,利用 SPSS 软件绘制出相应的正态 Q-Q 概率图和正态去势 Q-Q 图,如图 1.12 和图 1.13 所示。

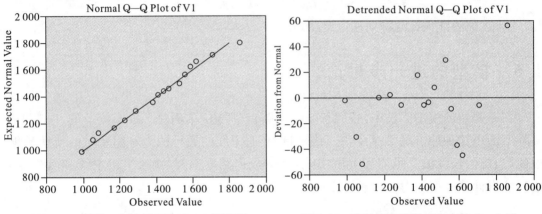

图 1.12　事故二车流量正态 Q-Q 概率图　　图 1.13　事故二车流量正态去势 Q-Q 图

由图 1.12 和图 1.13 可知,事故二的车流量服从正态分布。同理,事故一的车流量也服从正态分布。

(2)独立两样本 t 检验。

独立两样本 t 检验用于两个独立样本间的比较。所谓独立样本是指抽取其中一个样本对抽取另一个样本没有任何影响,两个样本各自接受相同的测量,样本数可以相等也可以不等。

条件:①两样本必须是独立的;②样本来自的总体要服从正态分布。

首先,经过前面的正态性检验发现,两起交通事故的车流量都服从正态分布,首先满足独立两样本 t 检验的条件②;其次,两起交通事故的发生时间不一样,对车流量进行的时间也不一样,所以这两个样本是独立的,符合条件①。既然样本符合独立两样本 t 检验的条件,那么就可以利用 SPSS 软件对车流量进行独立两样本 t 检验,检验结果见表 1.16 和表 1.17。

表 1.16　两起事故车流量分组统计结果 Group Statistics

因素	事故	N	Mean	Std. Deviation	Std. Error Mean
车流量	1	15	1174.00	156.652	40.447
	2	22	1381.36	215.634	45.973

表 1.17 两起事故车流量独立两样本 t 检验结果 Independent Samples Test

		Levene's Test for Equality of Variances		t-test for Equality of Means						
		F	Sig.	t	df	Sig. (2-tailed)	Mean Difference	Std. Error Difference	95% Confidence Interval of the Difference	
									Lower	Upper
车流量	Equal variances assumed	1.273	0.267	−3.189	35	0.003	−207.364	65.028	−339.377	−75.350
	Equal variances not assumed			−3.386	34.809	0.002	−207.364	61.234	−331.699	−83.028

观察表 1.16 可知,两起事故车流量的方差不齐,再看表 1.17 中方差不齐时的情况知 t 值为 −3.386,双尾 t 检验的显著性概率是 0.002,远远小于 0.05,说明两起事故中车流量存在显著性差异,即同一横断面交通事故所占车道不同对通过该横断面的车流量有影响。

综上所述,无论是以两起事故发生至撤离期间各时间段车流量和平均车流量为指标,还是独立两样本 t 检验,结果是一致的:同一横断面交通事故所占车道不同对通过该横断面的车流量有影响。

1.7.2 误差分析和灵敏度分析

(1)本文的误差主要来源于对车辆个数的统计,这是由于在车流量大且车速快的情况下,仅凭人工计数很容易导致错误。

(2)对畅行速度、阻塞密度的估计可能与实际不符,这也是产生误差的原因之一。

为说明车辆个数统计误差对模型结果的影响,我们研究了道路实际通行能力对于电瓶车的统计误差的灵敏度[6]。

由于道路运行中存在四轮及以上机动车、电瓶车这些不同的车型,在计算道路通行能力时要折换成标准车型计算,非标准型车对通行能力的修正系数为:

$$f_{HV} = \frac{1}{1 + P_{HV}(E_{HV} - 1)}$$

对所给视频进行车辆数的统计时,因为视频模糊等因素,使得统计的各类汽车所占比出现误差,又小型客车的 $E_{HV}=1$,故仅考虑大型客车和电瓶车,结合视频,我们看出,大型客车很难统计错,电瓶车由于体积和视频画质较容易出错,多统计或少统计,导致误差,以多统计了电瓶车的情况为例,利用 Matlab[5] 进行分析,得到数据如图 1.14 所示。

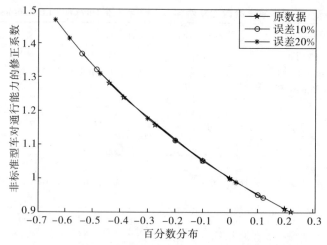

图 1.14 道路实际通行能力对电瓶车百分数的灵敏度

可看出,电瓶车多统计导致 f_{HV} 增大,所以模型对电瓶车的统计误差较灵敏,即道路实际通行能力对于电瓶车的统计误差较敏感。

1.8 模型的评价与推广

1.8.1 模型的评价

1. 优点

(1)本文所建模型贴合实际,充分考虑了现实生活中影响实际通行能力的各项因素。
(2)本文从多个角度对同一问题进行分析,结论合理可信。
(3)本文思路清晰,对问题的分析较为详细。
(4)本文多处使用图表说明问题,图文并茂,简单易懂。
(5)本文所建模型具有广泛的实用性,易于向其他领域推广。

2. 缺点

(1)本文所使用的数据大部分依靠人工统计,虽然确保了数据的真实性,但也难免会有误差。
(2)由于本文考虑的实际因素太多,因此所建模型较为复杂。

1.8.2 模型的推广

1. 横向推广

针对问题一道路通行能力的应用:①确定新建道路的等级、性质、主要技术指标和线形几何要素;②确定现有道路负荷情况,针对问题提出改进方案或措施;③作为交通枢纽的规划,设计及交通设施配置的依据(如交叉口类型选择,信号周期设计);④城市路网规划,分路网规划设计的依据。

针对问题三车辆排队模型的应用推广:①用于计算信号交叉口的车辆排队长度;②用于确定各种服务平台处的排队长度;③用于评价道路的通行状况或服务平台的服务能力。

2. 纵向推广

问题三中车辆排队模型不仅可以用于计算因交通事故而造成的车辆排队长度,也可以计算上游路口信号灯处的排队长度,甚至可以解决匝道对主道车辆通行的影响。所以,问题三所建模型在交通通行问题上有着广泛的应用。

参考文献

[1] 谢陈峰. 高速公路事故条件下通行能力及改善措施研究[D]. 西安:长安大学,2010.

[2] 周荣贵,钟连德. 公路通行能力手册[M]. 北京:人民交通出版社,2017.

[3] 孔惠惠,秦超,李新波,李引珍. 交通事故引起的排队长度及消散时间的估算[J]. 铁道运输与经济,2005(05):65—67.

[4] 姜启源,谢金星,叶俊. 数学建模(第四版)[M]. 北京:高等教育出版社,2011.

[5] 吴礼斌,闫云侠. 经济数学实验与建模[M]. 天津:天津大学出版社,2009.

[6] 运筹学教材编写组,运筹学[M]. 北京:清华大学出版社,2005.

建模特色点评

❋论文特色❋

◆**标题定位**:"车道被占用对城市道路通行能力影响的综合分析"将原赛题"车道被占用对城市道路通行能力影响"增加了综合分析,融问题与方法特色为一体,贴切、恰当。

◆**方法鉴赏**:①建模贴合实际,充分考虑现实生活中影响实际通行能力的各项因素;②能从多个角度对同一问题进行分析,结论合理可信;③技术面较好,各种软件使用较熟练,模型具有广泛的实用性,易于向其他领域推广。

◆**写作评析**:本文思路清晰,对问题的分析较为详细,多处使用图表说明问题,图文并茂,简单易懂。

◆**其他解读**:论文还给出了灵敏度分析和误差分析,并对模型给出了合理适当的评价。

❋不足之处❋

阅读文献较少,对问题的分析缺少特色,没有对其进行改进。

第2篇
嫦娥三号软着陆轨道设计与控制策略

◆ 竞赛原题再现

2014年A题　嫦娥三号软着陆轨道设计与控制策略

嫦娥三号于2013年12月2日1时30分成功发射,12月6日抵达月球轨道。嫦娥三号在着陆准备轨道上的运行质量为2.4 t,其安装在下部的主减速发动机能够产生1 500 N到7 500 N的可调节推力,其比冲(即单位质量的推进剂产生的推力)为2 940 m/s,可以满足调整速度的控制要求。在四周安装有姿态调整发动机,在给定主减速发动机的推力方向后,能够自动通过多个发动机的脉冲组合实现各种姿态的调整控制。嫦娥三号的预定着陆点为19.51 W,44.12 N,海拔为−2 641 m(见附件1)。

嫦娥三号在高速飞行的情况下,要保证准确地在月球预定区域内实现软着陆,关键问题是着陆轨道与控制策略的设计。其着陆轨道设计的基本要求:着陆准备轨道为近月点15 km,远月点100 km的椭圆形轨道;着陆轨道为从近月点至着陆点,其软着陆过程共分为6个阶段(见附件2),要求满足每个阶段在关键点所处的状态;尽量减少软着陆过程的燃料消耗。

根据上述的基本要求,请你们建立数学模型解决下面的问题:

(1)确定着陆准备轨道近月点和远月点的位置,以及嫦娥三号相应速度的大小与方向。

(2)确定嫦娥三号的着陆轨道和在6个阶段的最优控制策略。

(3)对于你们设计的着陆轨道和控制策略做相应的误差分析和敏感性分析。

附件1:问题的背景与参考资料;
附件2:嫦娥三号着陆过程的六个阶段及其状态要求;
附件3:距月面2 400 m处的数字高程图;
附件4:距月面100 m处的数字高程图。

原题详见2014年全国大学生数学建模竞赛A题。

获奖论文精选

嫦娥三号着陆轨道设计与最优控制模型[①]

摘要：本文研究嫦娥三号软着陆过程的轨道设计与优化控制问题，借助物理运动学方程，活力积分公式，最小中值差法，非线性规划，分别构建拟抛物线动力学模型、月心坐标定位模型、障碍地形识别选取模型、阶段轨道设计模型，使用 Matlab、Excel 软件，求解了模型中所涉及的各问题并给出了结论。

为了进行合理的假设和模型的简化，先在忽略月球自转情况下对抛物线过程建立二维坐标系，再对模型进行了这方面的改进。下面对具体问题进行概括。

针对问题一，根据拟抛物线动力学方程从主减速段逆推求解出近月点速度为 1 689.9 m/s 且沿切线方向，同时得到二、三阶段的轨道方程和着陆点与近月点的距离。利用活力积分公式求得远月点的速度为 1 613 m/s 且沿切线方向。建立着陆准备轨道的椭圆方程，与距离公式联立可得近月点和远月点的月心坐标，转换成经纬坐标后得到近月点和远月点的位置分别为 19.06°W，29.77°W，海拔 14.77 km 和 160.93°E，41.25°S，海拔 99.86 km。

针对问题二，利用最小中值差法在保证一定精度的前提下识别题目所给地形高程图的障碍区域与安全区域，并根据就近原则从安全区域中选择最合适的降落点，满足燃料使用的控制。粗避障区域：降落点坐标为 (1043,1104)，且相对粗糙度系数为 70.95835。精避障区域：降落点坐标为 (406,878)，且相对粗糙度系数为 37.91182。后就选取的降落点和相应的状态要求，运用运动学方程设置降落轨道，由于问题一中已求出前三个阶段的轨道方程，因此此处仅针对后三个阶段进行轨道设计，且得到各阶段消耗燃料量分别为 40.56 kg，33.9 kg，5 kg。

针对问题三，考虑到实际测量的误差与不稳定性，针对问题二中安全区域的选取，使用正态分布模拟可能存在的选择偏差。然后通过调整探测器的识别精度落实其对敏感性大小的分析，即微调备选降落区域的面积大小来观测安全区域选取结果的变动。再对轨道方程中所设置的反推力进行灵敏度分析，针对数据的上下浮动，深入探究轨迹的变动。

本文后续考虑月球自转对主减速段运动过程的三维位移与所用时间的影响，对模型进行了改进，并把三维坐标系运用在星体定位和深海探测等领域。同时通过对模型的推广，分析了在其他领域的应用，并综合评价了模型的优缺点。

关键词：软着陆；轨道设计；拟抛物线；活力积分；障碍地形识别；Matlab

[①] 本文获 2014 年全国二等奖。队员：汪思铭，纪元昕，王天琛；指导教师：苏涵。

2.1 问题的重述

2.1.1 基本背景知识

1. 探月工程概况

探月工程是我国航天事业第三座里程碑,是我国科技不断进步的重要标志。工程规划为三个阶段,简称为"绕""落""回",旨在绕月进行全球探测、降落到月球表面局部详细侦测并将采集样本带回地球。探月工程也将对社会产生间接效益,比如特殊的矿产和能源将对生物制品和新材料的研发提供有力支持。

2. 嫦娥三号的参数与特征

嫦娥三号是我国探月工程第二阶段的登月探测器,并首次实现了我国地外天体软着陆任务。相对于美国和前苏联的月球探测器,嫦娥三号有其自身特点:它由着陆器和月球车组成,其中着陆器总质量约 4 t(携带推进剂约 2 t),月球车质量约 120 kg(负载重量 20 kg);安装目前最大推力的发动机,可产生从 1500 N 到 7500 N 的可调节推力;顶部设有一台极紫外相机,将对地球空间等离子体层实施大视域、一次性的极紫外成像。嫦娥三号现已成功在虹湾区域着陆,突破了在动力下降、月地间遥控操作、测控通信等多方面的重大关键技术,实现了中国航天领域的科技创新。

2.1.2 着陆过程的阶段与要求

航天器在降落过程中通过逐渐降低速度和调整姿态方向,使得在接触星球表面的瞬时速度达到较小值,可以不受损坏地降落到星体表面,实现安全着陆。减速方式有多种,比如通过变轨利用大气层逐步减速,安装推力发动机进行反向推进。嫦娥三号的软着陆过程分为六个阶段,相应要求分别是:

(1)在近月点 15 km、远月点 100 km 的椭圆形着陆准备轨道,找到近月点和远月点的位置及嫦娥三号对应的速度和方向。

(2)距离月面 3~15 km 的主减速阶段,需要将速度降到 57 m/s。

(3)从 3 km 处进入快速调整阶段,主要是调整姿态和衔接上下段,使距离月面 2.4 km 处水平速度为 0 m/s 且主减速发动机推力竖直向下。

(4)距离月面 0.1~2.4 km 是粗避障阶段的区间,要求避开大的陨石坑,在着陆点上方 100 m 处悬停并初步确定落月点。

(5)距离月面 30~100 m 是精避障阶段的区间,同样要求避开较大陨石坑,再次确认着陆点,实现在着陆点上方 30 m 处水平速度为 0 m/s。

(6)从 30 m 处进入缓速下降阶段,需要控制嫦娥三号在距离月面 4 m 处相对月面静止,之后关闭发动机使其自由落体到着陆点。

2.1.3 相关资料与数据

(1) 问题的背景与参考资料(详见题目附件1)。
(2) 嫦娥三号着陆过程的六个阶段及其状态要求(详见题目附件2)。
(3) 距月面2 400 m处的数字高程图(详见题目附件3)。
(4) 距月面100 m处的数字高程图(详见题目附件4)。

2.1.4 要解决的问题

1. 问题一

确定着陆准备轨道近月点和远月点的具体位置,以及嫦娥三号经过这两点时相应的速度大小和方向。

2. 问题二

设计嫦娥三号的着陆轨道并制定在软着陆过程的六个阶段的最优控制策略。

3. 问题三

对设计的着陆轨道和控制策略进行误差分析和敏感性分析。

2.2 问题的分析

2.2.1 对问题的总体分析与建模思路

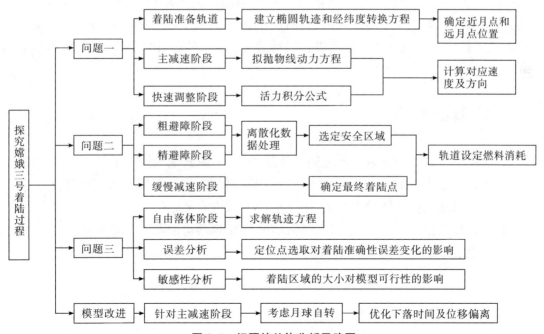

图2.1 问题的总体分析思路图

2.2.2 对具体问题的初步分析与处理方法

1. 对问题一的分析

本题先解决嫦娥三号在近月点的速度问题,由于已知距离月面 3 km 时的速度为 57 m/s,及主减速段的运动轨迹可近似为抛物线,通过运动学原理逆推近月点的情况,利用活力积分公式求得远月点的速度,加之后续阶段的简化计算可求得着陆点与近月点的空间距离;建立月心直角坐标系和经纬度转换方程,联立着陆准备轨道的椭圆方程与空间距离公式计算出近月点与远月点的月心坐标,最终转换为经纬度,实现位置的确定。

2. 对问题二的分析

问题要求设计六个阶段的着陆轨道并给予每个阶段相应的最优控制策略,在这里控制策略主要针对在保证各阶段要求的状态前提下,所消耗的燃料是最小的。对于粗避障与精避障阶段,可通过 Matlab 将其地形工程图转化为易于识别的灰度图像,利用最小中值差法识别安全区域与故障区域。为了有效控制燃料的使用,我们选择离嫦娥三号中心点最近的安全区域降落,并按要求的运动状态设计相应的轨道方程。对于最后两个阶段,由于路程较短,运动状态较简易,我们直接利用运动学原理求解运动轨道方程。

3. 对问题三的分析

在对模型误差的分析中,考虑到现实情况存在着许多的不稳定性,因此针对问题二中的安全区域选取,本文使用正态分布的形式来模拟现实中的不稳定性,从而对于降落点的误差有了定量的分析。由于嫦娥三号在月球附近人为难以干涉,因而在模型的敏感度方面,从可以事先设定的因素考虑,对降落区域的大小进行分析,更具有实际意义。再对轨道方程中的所设置的反推力进行灵敏度分析,针对数据的上下浮动,深入探究轨迹的变动。

2.3 模型的假设

(1)假设只有月球和嫦娥三号之间的作用力,不考虑其他天体的干扰。
(2)假设月地表面的小型岩石、特殊地质和坑洞对探测器降落不产生影响。
(3)假设月地表面整体是水平的,即平坦的地域均为平原。
(4)假设探测器机载计算机运算速度极快,即探测器定位降落区耗时可忽略。
(5)假设月球是一个均匀的球体。

2.4 名词解释与符号说明

2.4.1 名词解释

1. 拟抛物线

指近似于抛物线形态的下落运动,但在速度、加速度等描述值上不一定满足标准抛物线运动的特征,如匀速水平运动与匀加速竖直运动。

2. 定位点

嫦娥三号从安全区域中定位的精确降落点。

3. 预安全区域

指从地形高程图转化为灰度图像中识别的无障碍可选区域。

2.4.2 符号说明

表 2.1 主要变量符号及意义

序号	符号	符号说明
1	$F_i^{(1)}$	第 i 个阶段的竖直方向反推力
2	$F_i^{(2)}$	第 i 个阶段的水平方向反推力
3	t_i	第 i 个阶段的下落时间
4	h_i	第 i 个阶段的竖直下落高度
5	l_i	粗、精避障区域选定的安全区域与嫦娥三号投影位置的水平距离
6	$v_i^{(1)}$	第 i 个阶段的竖直下落速度
7	$v_i^{(2)}$	第 i 个阶段的水平位移速度
8	$\tan\theta$	第 i 个阶段竖直方向速度与水平方向速度的比值
9	m_i	第 i 个阶段消耗的燃料质量值

2.5 模型的建立与求解

2.5.1 问题一的分析与求解

1. 对问题一的分析

我们将问题一再拆分为两个具体问题,分为确定近月点、远月点的位置和嫦娥三号相应的速度大小与方向,并将运用的方法和思路形成图 2.2。

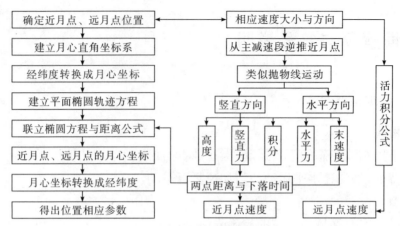

图 2.2 问题一的分析思路图

2. 对问题一的求解

模型Ⅰ 拟抛物线动力学模型

(1) 模型的准备。

①建模思路。鉴于嫦娥三号在着陆过程中距离月面 3 km 时的速度(57 m/s),以及已知竖直的海拔高度,我们采用运动学原理,借助逆推法求得近月点开始降落的速度与方向。根据题意,嫦娥三号的着陆过程近似一个抛物线运动,但该抛物线依据实际情况可分为两个阶段:第一阶段为从 15 km 至 3 km 的变减速下落运动,第二阶段为 3 km 至 2.4 km 的水平匀减速运动。而由于当嫦娥三号到达高度 2.4 km 时,已经基本在着陆点正上方,因此该点的经纬坐标便与着陆点一致,即可根据前两段运动逆推出抛物线经过的水平距离。

(2) 模型的建立与求解。

①第一阶段(主减速阶段)竖直方向运动。根据题意,嫦娥三号以着陆准备轨道的近月点处水平的切线方向速度开始降落,此时推力发动机沿嫦娥三号运动方向的反向延长线方向给予反推力。由于反推力随速度方向的变化而变化,大小亦在变化,求解较为复杂。因此为了简化问题,本文直接将反推力分解为竖直方向的推力 $F_1^{(1)}$ 与水平方向的推力 $F_1^{(2)}$。我们设立初始的总反推力 $F_1 = 7\,235$ N,则 $F_1^{(1)} = 3\,730$ N, $F_1^{(2)} = 6\,200$ N。

对于第一阶段的下落过程,为了逐步控制好速度,$F_1^{(1)}$ 与 $F_1^{(2)}$ 均为变力,且力由大逐渐变小,实现速度的逐步控制。本文设置变化规则为 $F_1^{(1)} = 3730 - 2t$,即时间变化呈线性,则可建立运动方程如下:

$$\begin{cases} a_1^{(1)}(t) = g_0 - \dfrac{F_1^{(1)} - 2t}{M_0 - ((\sqrt{F_1^{(1)2} + F_1^{(2)2} - 4F_1^{(1)}t - 4F_1^{(2)}t + 8t^2}/v_e) \times t)}, \\ v_1^{(1)}(t) = \int_0^t a_1^{(1)}(t) \mathrm{d}t, h_1 = 0 + \int_0^t v_1^{(1)}(t) \mathrm{d}t = 12000. \end{cases}$$

由于嫦娥三号下落过程涉及均匀变力,匀变速抛物线运动公式已不再适用,则通过积分法求解。h_1 为第一阶段的竖直海拔高度,竖直方向的初始速度为 0 m/s。考虑到燃料的耗用,嫦娥三号自身质量在减小,根据题目所给出的比冲公式:$F_1=v_e m$ 知,(式中 m 为单位时间燃料消耗的公斤数)$m \times t$ 即为下落过程消耗的燃料质量,$F_1^{(1)}-2t$ 为变化的反推力,则反推力产生的与月球引力加速度 g_0 反向的加速度即为:

$$\frac{F_1^{(1)}-2t}{M_0-((\sqrt{F_1^{(1)\,2}+F_1^{(2)\,2}-4F_1^{(1)}t-4F_1^{(2)}t+8t^2}/v_e)\times t)}。$$

初始 $F_0=3\,730\text{ N},M_0=2.4\times 10^3\text{ kg},v_e=2\,940\text{ m/s}$,代入方程,求得降落到 $3\,000\text{ m}$ 海拔处所需时间 $t_0=549\text{ s}$,竖直方向末速度 $v(t_0)=52.7\text{ m/s}$。

②第一阶段(主减速阶段)水平方向运动。对于下落运动的水平方向,加速度即为水平方向上的反推力产生的加速度,由于近月点的初始速度即水平速度,因此在水平方向上的减速需要以较大的反推力产生较大的加速度使其尽快减速,才能达到理想的效果。建立运动方程如下:

$$\begin{cases} a_1^{(2)}(t)=-\dfrac{F_1^{(2)}-2t}{M_0-((\sqrt{F_1^{(1)\,2}+F_1^{(2)\,2}-4F_1^{(1)}t-4F_1^{(2)}t+8t^2}/v_e)\times t)}, \\ v_1^{(2)}(t)=v_0+\int_0^{t_0}a_1^{(2)}(t)\mathrm{d}t, X_1=\int_0^{t_0}v_1^{(2)}(t)\mathrm{d}t。 \end{cases}$$

该方程组中的 $v_2(t_0)$ 是嫦娥三号达到 $3\,000\text{ m}$ 时的水平速度,由于该速度已由题目要求确定为 57 m/s,且竖直方向速度已求出为 52.7 m/s,则根据矢量公式:

$$(v_1(t_1))^2=(v_1^{(1)}(t_1))^2+(v_1^{(2)}(t_1))^2。$$

求得 $v_1^{(2)}(t_1)=21.7\text{ m/s},F_1^{(2)}=6\,200\text{ N}$,其他值 M_0,v_e 不变,则由该方程组求出近月点初始速度为 $v_0=1\,689.9\text{ m/s}$。

此时第一段抛物线过程的各变量已求得,该过程所经过的水平距离 $X_1=538\,990\text{ m}$。

该主减速阶段的轨迹曲线图如图 2.3 所示。

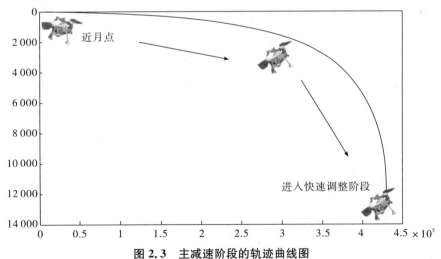

图 2.3 主减速阶段的轨迹曲线图

由图 2.3 可看出,该阶段的轨道运动过程较为光滑平整,符合实际。

表 2.2 更直观地呈现了该阶段的各项描述值。

表 2.2 主减速阶段各描述值表

描述值	$F_1^{(1)}$(N)	$F_1^{(2)}$(N)	t_1(s)	m(kg)	$\tan\theta_1$
数值(初始)	3730	6200	549	1050	2.43

③第二阶段运动情况及着陆点与近月点的空间距离。第二段的抛物线过程是从 3 km 到 2.4 km,在该过程中需要快速调整姿态,此时嫦娥三号距离着陆点正上方仅有小段距离,本文认为主推力发动机仅对竖直方向给予恒定的反推力,使其与月球引力相持衡,即加速度为 0 的匀速下落;水平方向上的反推力使得水平方向速度逐渐减小至 0。为了简化问题,由于这段运行时间较短,距离较小,则假设该反推力是恒定的,即加速度恒定,则建立运动学方程如下:

$$\begin{cases} t_2 = \dfrac{h_2}{v_1^{(1)}(t_1)}, \\ a_2^{(2)}(t) = \dfrac{0 - v_1^{(2)}(t_1)}{t_1}, \\ X_2 = \dfrac{0 - v_1^{(2)}(t_1)^2}{2 \times a_2^{(2)}(t_1)} \end{cases}$$

第二阶段嫦娥三号下降了 600 m,即 $h_2=600$ m,求得 $t_2=11.3$ s。对于水平方向的匀减速运动,该阶段的初始水平速度即为上一阶段末速度的水平方向速度,求得该水平方向上的加速度 $a_2^{(2)}(t)=1.92$ m/s²,再借助匀减速运动的运动学公式,求得该阶段抛物线运动所横跨的水平距离 $X_2=122.6$ m。

综合上述两个阶段,可求得两次抛物线运动横跨的总水平距离 $X=X_1+X_2=539\ 112.6$ m,也可近似代表着陆点到近月点的水平距离(暂不考虑后阶段水平距离的微弱变化),根据竖直距离 15 000 m,由勾股定理可得着陆点到近月点的空间距离为 539 321.2 m。

两段下落总时间 t 为:$t=t_1+t_2=561.3$ s 且近月点的速度 v_0 为 1 689.9 m/s,方向沿水平切线。

④活力积分公式及远月点速度。为求得椭圆轨迹上远月点的瞬时速度,使用活力积分公式:

$$v^2 = G(M+m)\left(\dfrac{2}{r} - \dfrac{1}{a}\right)$$

将公式运用于本题中,G 是万有引力常量,值为 6.67×10^{-11} N·m²/kg²;M 是月球质量,值为 7.3477×10^{22} kg;m 是嫦娥三号的质量,由于其与月球质量相比实在太小可以忽略不计;r 是月心到嫦娥三号(将嫦娥三号看作一个质点)的距离;a 是椭圆轨道方程的长半轴,v 是椭圆轨道上某点的速度。

已知月球平均半径 $R=1\ 737\ 013$ m,近月点距离 $r_1=15\ 000$ m,远月点距离 $r_2=100\ 000$ m,计算可得:

$$a = 2R + r_1 + r_2/2 = 1\ 794\ 513\text{ m}$$

同时将嫦娥三号在远月点时的 r 值(分别为 1 752 013 m 和 1 837 013 m)代入公式，得到速度 v 为 1 613 m/s，方向沿轨道切线方向。

模型Ⅱ　月心坐标定位模型

(1)模型的准备。

假设只有月球和嫦娥三号之间的作用力，不考虑其他天体的干扰，建立月心直角坐标系：以月球的球心 O 为原点，Oz 轴垂直月球赤道面指向月球北极，Oy 轴在月球赤道面内，Ox 轴与其他两轴垂直并构成右手坐标系。A 点表示嫦娥三号在月球面上的着陆点，如图 2.4 所示。

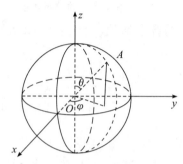

图 2.4　月心直角坐标系 (O,x,y,z)

以月球总对着地球的那面中央子午线为零经线，以月球赤道为零纬线，可以建立经纬坐标与月心直角坐标的相互转换方程：

$$\begin{cases} v = \dfrac{a}{(1-e^2\sin^2\varphi)^{0.5}}, \\ X = (v+h)\cos\varphi\cos\lambda, \\ Y = (v+h)\cos\varphi\sin\lambda, \\ Z = ((1-e^2)v+h)\sin\varphi。 \end{cases} \quad (2\text{-}1) \qquad \begin{cases} \varphi = tg^{-1}\dfrac{Z+e^2 v\sin\varphi}{(X^2+Y^2)^{0.5}}, \\ h = X\sec\lambda\sec\varphi - v, \\ \lambda = tg^{-1}\dfrac{Y}{X}。 \end{cases} \quad (2\text{-}2)$$

式(2-1)、式(2-2)中：X,Y,Z 为月心直角坐标系中某点的坐标；φ,λ 为某点的纬度和经度；h 为某点的海拔即相对于椭球面的高度；v 为纬度 φ 处的卯酉圈曲率半径；a 为椭球的长半轴即月球的平均半径；e 为椭球即月球的离心率，可用扁率 f 计算 $e^2=2f-f^2$。

为简化计算，该方程的参数以千米为单位。将嫦娥三号在月球表面的着陆点 A 的纬度 φ、经度 λ、海拔 h(19.51°W, 44.12°N, −2.641 km)代入方程(2-1)，计算可得 A 点在月心直角坐标系中的坐标为(1 356.7 km, 1 053.9 km, 237.58 km)。

(2)模型的建立与求解。

在着陆准备轨道上，嫦娥三号绕月球作椭圆运动。取月球与嫦娥三号共面的正投影，研究在该平面上以月心为右焦点的椭圆轨迹方程。为了更直观简要地说明其运动情况，作出其轨迹的平面示意图如图 2.5 所示。

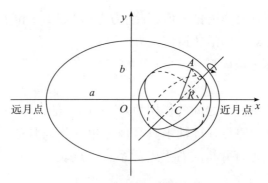

图 2.5 椭圆运动轨迹示意图

由图 2.5 可计算椭圆方程的各个参数,即长半轴 a、短半轴 b 和半焦距 c。

$$\begin{cases} c = a - R - r_1, \\ a = \dfrac{2R + r_1 + r_2}{2}, \\ b = a^2 - c^2, \end{cases} \rightarrow \begin{cases} a = 1\,794.513 \text{ km}, \\ b = 1\,794.01 \text{ km}, \\ c = 42.486 \text{ km}。 \end{cases}$$

故椭圆轨迹方程为 $x^2/1794.513^2 + y^2/1794.01^2 = 1$。

由模型I得到的着陆点 A 与近月点距离为 539.32 km,且已知在月心坐标系中点 A 坐标为 (1 356.7 km, 1 053.9 km, 237.58 km),当选取平面为正投影且月心、近月点、远月点与椭圆圆心在一条直线上,可联立距离公式与椭圆轨迹方程得到近月点的月心直角坐标:

$$\begin{cases} \dfrac{x^2}{1\,794.513^2} + \dfrac{y^2}{1\,794.01^2} = 1, \quad x, y \geqslant 0, \\ (x - 1\,356.7)^2 + (y - 1\,053.9)^2 = 539.32^2, \end{cases} \rightarrow \begin{cases} x = 624.35 \text{ km}, \\ y = 1\,475.5 \text{ km}。 \end{cases}$$

且 $z = Z' + 15 = 252.58$ km,故近月点坐标为 (624.35 km, 1 475.5 km, 252.58 km),远月点坐标易得为 (−524.35 km, −1 475.5 km, −252.58 km)。代入方程 (2-2) 即可求得近月点的位置为 (19.06°W, 29.77°N, 14.78 km),即西经 19.06°,北纬 29.77°,海拔 14.78 km;远月点的位置为 (160.93°E, 41.25°S, 99.86 km),即东经 160.93°,南纬 41.25°,海拔 99.86 km。

2.5.2 问题二的分析与求解

1. 对问题二的分析

问题要求设计六个阶段的着陆轨道并给予每个阶段相应的最优控制策略,在这里控制策略主要针对在保证各阶段要求的状态前提下,所消耗的燃料是最少的。在问题一中我们基本已给出前两个阶段的运动方程,由于各种速度、推力、时间的限制,燃料的消耗值较难进一步调整,因此在这里不再做考虑。对于之后的粗避障与精避障阶段,根据附件中的数字高程图,通过 Matlab 调整其影像显现,根据易处理识别的灰度图像,利用最小中值差法辨别故障区域,并调整其遍历识别精度,确认较为精确的区域。为了有限控制燃料的使用,我们选择离嫦娥三号中心点最近的安全区域降落,并按要求的运动状态设计相应的轨道方程。对于最后两个阶段,由于路程较短,运动状态较简易,我们直接利用运动学原理求解运动轨道方程。

2. 对问题二的求解

模型Ⅲ 障碍地形识别选取模型

(1)模型的准备。

①月地表避障情况分析。月地表情况较为复杂,其海拔高度起伏波动较大,但是在预先选择的着陆区域内,大部分地表较为平坦,因此可以得到一个平均海拔,在此海拔以下的地形称为陨石坑,在此海拔以上的地形称为巨型岩石,在嫦娥三号选择着陆区域时,需要保证两点:一是着陆区域不应有大型陨石坑;二是在误差范围内着陆点附近要保证没有巨型岩石和陨石坑。

②粗避障区段。嫦娥三号进入粗避障区段,其相对水平速度为 0 m/s,距离月地表面 2 400 m,此时将根据月面成像所得的高程图来初步定位着陆区域。嫦娥三号进入粗避障区所获得的高程图的范围大小为 2 300 m×2 300 m,而后需要在距离月地表面 100 m 时悬停于一个 100 m×100 m 的着陆区域正上空;根据高程图,建立地形粗糙度模型,根据该模型选择出一个"较为光滑"且距离嫦娥三号投影点较近的 100 m×100 m 着陆区域,此着陆区域应不含大型陨石坑及大型岩石;根据此着陆区域来设计飞行路线,计算嫦娥三号飞行的各项数据。

③精避障区段。在距离月地表面 100 m 时,嫦娥三号悬停在空中,其相对速度为 0 m/s,同样,嫦娥三号将根据月面成像绘制一幅高程图来精确定位着陆点。此时的高程图大小为 100 m×100 m;在此高程图中,在先前建立的地形粗糙模型基础上,把模型的精度提高,可以定位出一个平坦且距离嫦娥三号投影点较近的着陆点;在着陆点确定后,便可以设计飞行路线,计算嫦娥三号飞行的各项数据。

④最小中值差法。该算法将地表高程图的所有数据点作为一个样本集,其中的样本包括两部分:在地形平面上的点,称为局内点;不在地形平面上的点,称为局外点,超过着陆器障碍容忍极限的局外点即为障碍点。

(2)模型的建立。

将高程图中的每一个地形按照像素块进行微元化,设所有的地形微元海拔分别为 $a_{11},a_{12},\cdots,a_{1n},a_{21},a_{22},\cdots,a_{2n},\cdots,a_{nn}$,则可以将高程图转化为矩阵 A:

$$A=\begin{pmatrix} a_{11} & a_{12} & \cdots & a_{1n} \\ a_{21} & a_{22} & \cdots & a_{2n} \\ \vdots & \vdots & & \vdots \\ a_{n1} & a_{n2} & \cdots & a_{nn} \end{pmatrix}$$

本文为了简化计算,根据公式:

$$b_{ij}=\frac{\max a-a_{ij}}{\max a-\min a} \quad i,j=1,2,\cdots,n$$

计算,对矩阵 A 进行无量纲均值化,得到均值化矩阵 B:

$$B = \begin{pmatrix} b_{11} & b_{12} & \cdots & b_{1n} \\ b_{21} & b_{22} & \cdots & b_{2n} \\ \vdots & \vdots & & \vdots \\ b_{n1} & b_{n2} & \cdots & b_{nn} \end{pmatrix}$$

通过最小中值差法找出矩阵 B 中相对稳定的数值 $\bar{b}$，即相对平坦地形的数据；为了计算出平坦的着陆区域，需要将矩阵 B 分成大小为 $r \times r$ 矩阵 $(T_{11}, T_{12}, \cdots, T_{n-r+1\ n-r+1})$ 来进行检验。而这些小矩阵的左上角边界点分别为 $b_{11}, b_{12}, \cdots, b_{1\ n-r+1}, b_{21}, b_{22}, \cdots, b_{2\ n-r+1}, \cdots, b_{n-r+1\ 1}, b_{n-r+1\ 2}, \cdots, b_{n-r+1\ n-r+1}$；根据矩阵 B 中所找的稳定数值 $\bar{b}$ 计算出每一个矩阵 $T_{ij}, i,j=1,2,\cdots,n-r+1$ 所对应的相对粗糙系数 m_{ij}：

$$m_{ij} = \sum_{k=1}^{r} \sum_{l=1}^{r} (t_{lk} - \bar{b}), t \in T_{ij}, i,j=1,2,\cdots,n-r+1$$

最后，对所求得的每一个矩阵 T 的相对粗糙系数进行筛选，找出相对稳定的数值点，即找出相对平坦的着陆区域。

(3) 模型的求解。

① 粗避障筛选。首先在 Matlab 中将高程图读入并转换成矩阵，对矩阵进行均值化处理，根据最小中值差法求出相对稳定数值 $\bar{b}=0.5586$。

编写 Matlab 程序，得到地形相对粗糙度直观图如图 2.6 所示。

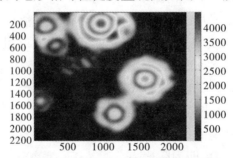

图 2.6　2400 米高度地形粗糙度直观图

由于地形粗糙度模型的区域精确度达到 1 m，则相邻的平坦区域会影响筛选，因此需要相对分散地选取其中部分区域进行筛选，利用 Matlab 分块筛选出其中较为平坦的几个区域，得到备选着陆点相对粗糙系数、坐标及投影距离的数值见表 2.3，示意图如图 2.7（黑色方块代表备选着陆区）。

表 2.3　粗避障下备选着陆点相关参数表　　　　　　　　　　（单位：m）

系数	位置坐标		对嫦娥三号投影距离
106.7728	635	592	759.3346
68.33028	1367	627	566.2314
58.4433	1635	515	799.0307
68.84239	462	840	754.6151

续表

系数	位置坐标		对嫦娥三号投影距离
70.95835	1043	1104	116.4689
55.43101	2099	1472	1002.14
76.65046	773	1606	591.6629
65.90844	1020	1428	306.8941
55.04587	1411	2174	1056.739

考虑到燃料消耗的控制问题,在选取着陆区域相对平坦的情况下,采用就近原则,定位着陆区域尽量离嫦娥三号的投影距离最小,因为距离越远,行进时间相对越长,耗用的燃料数相对较大。最终,由区域块坐标定位出一个最优着陆点,其坐标为(1043,1104),相对粗糙度系数为70.95835,如图2.8。

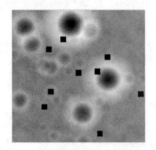

图2.7　2 400 m高度筛选区域图

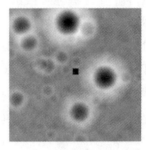

图2.8　最优着陆区定位图

②精避障筛选。在该阶段同样运用地形粗糙度模型,但对模型的精度有所提高,将着陆区域定为5 m×5 m,以确保嫦娥三号可以更加安全地着陆在月球表面。

同样,开始先用最小中值差法计算出相对稳定值$\bar{b}=0.5818$。使用Matlab对模型进行求解(见附录程序)并对平坦区域进行筛选,可以初筛选出9个备选着陆点,如图2.9(黑色方块代表备选着陆区),详细数据见表2.4。

表2.4　精避障下备选着陆点相关参数表　　　　　　　　　　　　(单位:m)

系数	位置坐标		对嫦娥三号投影距离
36.24636	162	209	573.8566
35.98545	390	163	476.8747
35.58545	795	146	550.1871
63.93	1	350	662.69
36.34636	372	562	480.9976
35.81091	856	559	584.9851
39.76273	183	611	558.8447
37.91182	406	878	471.5522
36.03636	735	678	520.0839

最后,根据就近原则,并剔除相对粗糙度系数相对较大的着陆点,选取着陆点,其坐

标为$(406, 878)$,相对粗糙度系数为37.91182,如图2.10。

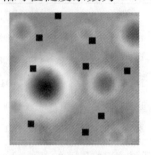
图2.9　100 m高度筛选区域图

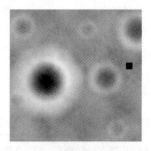
图2.10　最优着陆点定位图

模型Ⅳ　阶段轨道设计模型

(1)建模的思路。

选取了粗避阶段与精避阶段的安全区域后,即确定了这两个阶段嫦娥三号所要经过的位置,且此时位置的选取已在一定程度上满足了最小燃料的耗用。之后对于设计具体的运动轨道,我们建立非线性规划,将下落时间与所需推力作为目标函数,求其最小值,约束条件根据题目要求的状态来建立。

(2)模型的建立与求解。

①粗避障阶段模型。嫦娥三号进入该阶段时,其质量为$M_1 = M_0 - m = 1\,350$ kg,水平速度已在快速调整阶段降为0,竖直方向下降的初速度为$v_3^{(1)} = 52.7$ m/s,该阶段下落的竖直距离$h_3 = 2\,400$ m。由于此时嫦娥三号已经进入下落的稳定阶段,因此设计竖直方向上的反推力是恒定的,且一直减速直至悬停于精避障阶段的初始时段,即此时速度为$v_3 = 0$ m/s。对于水平方向上的反推力,由模型Ⅲ的结果可知,我们通过事先的地形高程图已选定最合适的安全区域,因此规避较为直接简易,则设定不变的水平反推力,给嫦娥三号适当的初速度,使其做类似抛物线运动降落到选定的安全区域。

由模型Ⅲ的结果知,该粗避障区域的安全区域与嫦娥三号的投影位置的水平距离$l_1 = 116.47$ m,竖直下落距离与飞行的初始竖直速度均已知,且需满足末速度为零的要求,因此为了求解该段作用的水平反推力,竖直反推力和下落时间,可直接联立方程组求解,不需进行规划,具体方程组如下。

$$\begin{cases} \int_0^{t_3} \dfrac{F_3^{(2)}}{M_1 - (\sqrt{F_3^{(1)\,2} + F_3^{(2)\,2}}/v_e) \times t}\,\mathrm{d}t - v_3^{(2)} = 0, \\ \int_0^{t_3} g_0 - \dfrac{F_3^{(1)}}{M_1 - (\sqrt{F_3^{(1)\,2} + F_3^{(2)\,2}}/v_e) \times t}\,\mathrm{d}t - v_3^{(1)} = 0, \\ \int_0^{t_3}\int_0^{t} \dfrac{F_3^{(2)}}{M_1 - (\sqrt{F_3^{(1)\,2} + F_3^{(2)\,2}}/v_e) \times t}\,\mathrm{d}t\,\mathrm{d}t = l_1, \\ \int_0^{t_3}\int_0^{t} \left(g_0 - \dfrac{F_3^{(1)}}{M_1 - (\sqrt{F_3^{(1)\,2} + F_3^{(2)\,2}}/v_e) \times t} \right)\mathrm{d}t\,\mathrm{d}t = h_3. \end{cases}$$

上述方程组中的方程,反映了在恒定的竖直反推力下的变减速下落运动(质量在变化),与水平方向上先加速后减速运动,且水平反推力大小恒定,方向相反,式中的 $v_3^{(2)}$ 是水平加速到的最大速度。$F_3^{(1)}$,$F_3^{(2)}$ 分别为该阶段的竖直与水平方向上的反推力,v_e 依然是恒定不变的比冲。由 Matlab 编程求解得:

$$F_3^{(1)} = 1\,316.24\,\text{N}, F_3^{(2)} = 415.55\,\text{N}, t_3 = 86.5\,\text{s}$$

该阶段的运动轨迹如图 2.11。

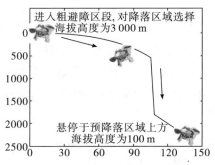

图 2.11 粗避障阶段轨迹曲线图

由图 2.11 可看出,该阶段的运动过程中间靠后某一段时间竖直方向速度较大,水平方向速度很小,位移不明显,即在该时间嫦娥三号基本呈竖直下落运动,可见此段较为通畅,无明显障碍物。整个轨迹曲线较为光滑平整,符合实际。

同时计算得出,该段消耗的燃料质量 $m_1 = 40.56\,\text{kg}$,此时嫦娥三号探测器的总质量为 $M_2 = 1\,309.4\,\text{kg}$。

为更直观地呈现数据,我们列表得到该阶段各项描述值,见表 2.5。

表 2.5 粗避障阶段各描述值表

描述变量	$F_3^{(1)}$(N)	$F_3^{(1)}$(N)	t_3(s)	m_1(kg)	$\tan\theta_3$
数值	1 316.24	415.55	86.5	40.56	0

②精避障阶段模型。与粗壁障阶段不同的是,该阶段仅要求水平末速度减为 0,对竖直末速度没有具体要求,因此该阶段轨道设计具有一定灵活性,可建立适当的目标函数与约束条件进行非线性规划。目标函数即是为了使飞行的时间与所需的反推力尽可能减小(总消耗的燃料数为总时间与单位时间消耗燃料公斤数的乘积),约束条件即根据该阶段的各状态要求。

该阶段的初始速度为 0 m/s,为了避开障碍落向安全区域,依旧给嫦娥三号探测器一个推力,分解为竖直方向与水平方向的反推力,与粗避障阶段相同,仍旧设计为恒力,简化问题。针对燃料消耗的优化,我们建立目标函数:

$$z = w_1 t_4 + w_2 F_4,\text{且}\ F_4 = \sqrt{F_4^{(1)^2} + F_4^{(2)^2}}$$

根据题意,水平末速度再次减为 0,嫦娥三号在水平方向上经历先加速再减速过程,竖直方向上为继续实现下落设置为变加速运动。由模型Ⅲ,精避障阶段选定的安全区域与嫦娥三号投影位置的水平距离 $l_2 = 4.72\,\text{m}$,该段竖直下落高度 $h_4 = 70\,\text{m}$,则具体约束条件如下。

$$\begin{cases} t_3 - 2 \times t_3^1 = 0, \\ 1\,500 \leqslant \sqrt{F_4^{(1)^2} + F_4^{(2)^2}} \leqslant 7\,500, \\ \int_0^{t_3^1} \int_0^t \dfrac{F_4^{(2)}}{M_2 - (\sqrt{F_4^{(1)^2}+F_4^{(2)^2}}/v_e) \times t} \mathrm{d}t\mathrm{d}t = \dfrac{l_2}{2}, \\ \int_0^{t_4} \int_0^t \left(g_0 - \dfrac{F_4^{(1)}}{M_2 - (\sqrt{F_4^{(1)^2}+F_4^{(2)^2}}/v_e) \times t}\right)\mathrm{d}t\mathrm{d}t = h_4 \, . \end{cases}$$

上述约束条件中,由于水平方向上的反推力大小是恒定的,但方向在 t_3^1 时刻反向(从而实现由加速变为减速)。由于加速与减速运动是对称的,则在 t_3^1 时刻实现了一半的水平距离,即 $l_2/2$。同样,根据题意也对合力的大小给予边界约束。最后,由 Matlab 编程求得结果:

$$F_4^{(1)} = 1\,995.5\,\mathrm{N}, F_4^{(2)} = 4.6\,\mathrm{N}, t_3 = 50\,\mathrm{s}$$

该阶段的运动轨迹见图 2.12。

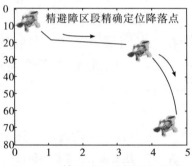

图 2.12 精避障区域轨迹曲线图

由图 2.12 可看出,嫦娥三号在中间的一段时间竖直方向上速度较小,位移运动不明显,而在水平方向上速度较大,且该段是较为不平整的一段,即可能为避开障碍物的运动时段,曲线总体上是较为光滑的,符合实际。

同时,计算得出该段消耗的燃料量 $m_2 = 33.9\,\mathrm{kg}$,此时嫦娥三号探测器的质量 $M_3 = 1\,275.5\,\mathrm{kg}$。

为更直观地呈现数据,我们选取部分时刻的速度制表,见表 2.6。

表 2.6 精避障阶段各描述值表

指标变量	$F_4^{(1)}$(N)	$F_4^{(2)}$(N)	t_4(s)	m_2(kg)	$\tan\theta_4$
数值	1 995.5	4.6	50	33.9	1

③缓速下降阶段。当嫦娥三号进入该区段时,其质量为 M_3,水平速度为 $0\,\mathrm{m/s}$,竖直方向下降的初速度为 $v_4^{(1)} = 2.42\,\mathrm{m/s}$(即为上一阶段的竖直方向的末速度),该段下落距离为 26 m;在此阶段嫦娥三号将进行竖直方向的减速,当距离月地表面 h_1 时需要将速度降至 $v_5^{(1)} = 0\,\mathrm{m/s}$,此时因为无障碍规避,所以不考虑水平方向运动。

在此阶段中,嫦娥三号的质量 M_4 满足:$M_4 = M_4^0 - t \times \Delta m$,$\Delta m$ 为单位时间消耗的燃料质量;

则 $v_6^{(1)}(t)$ 的函数满足：

$$v_5^{(1)} = v_4^{(1)} - a_5 t = v_4^{(1)} - \frac{F_5 - M_4 g_0}{M_4} t,$$

a_5 为加速度，F_5 为发动机推力，假定恒不变；

设此阶段所用时间为 t_5 对 $v_5^{(1)}$ 积分，可以得到：

$$\begin{cases} h_5 = \int_0^{t_1} v_5^{(1)} \, \mathrm{d}t = \int_0^{t_1} v_4^{(1)} - \frac{F_5 - (M_4^0 - t \times \Delta m) g_0}{M_4^0 - t \times \Delta m} t \, \mathrm{d}t, \\ v_4^{(1)} = \int_0^t a_5 \mathrm{d}t = \int_0^t \frac{F_5 - (M_4^0 - t \times \Delta m) g_0}{M_4^0 - t \times \Delta m} \mathrm{d}t. \end{cases}$$

接着由 Matlab 对公式进行计算，解得此时 $t_5 = 5.83 \, \mathrm{s}$，此时飞行器质量 $M_4 = 1\,270.5 \, \mathrm{kg}$，方向竖直向下。由于该阶段无水平方向上的运动，且时间很短，则不再呈现具体的轨迹曲线图。

具体的各项描述值见表 2.7。

表 2.7 缓速下降阶段各描述值表

描述变量	$F_5^{(1)}$(N)	$F_5^{(2)}$(N)	t_5(s)	m_3(kg)	$\tan\theta_5$
数值	2 022	0	5.83	5	1

之后嫦娥三号在此阶段关闭所有发动机，无燃料消耗，其下落加速度为 g_0，为自由落体运动，其速度 v_6 满足公式：$v_6 = g_0 t_6$。

运动方向竖直向下，耗时约为 $t_6 = 2.236 \, \mathrm{s}$。

由于该阶段运动较为简单，不再赘述。

2.5.3 问题三的分析与求解

1. 对问题三的分析

考虑到实际测量的误差与不稳定性，则针对问题二中的安全区域选取，使用正态分布的形式模拟可能存在的选择偏差。后通过调整探测器的识别精度落实其对敏感性大小的分析，即微调备选降落区域的面积大小来观测安全区域选取结果的变动。再对轨道方程中所设置的反推力进行灵敏度分析，针对数据的上下浮动，深入探究轨迹的变动。

2. 对问题三的求解

本文在问题二中，为了定位相对平坦的降落地点，建立地形粗糙度模型。在此过程中，运用最小中值差法得到一个相对稳定的地形指标，从而建立相对地形粗糙度系数来对地形进行数据化处理，最后对平坦区域进行选择。其中的地形指标为近似值，其误差会使得地形数据不准确，会影响平坦区域选择的准确性。

(1) 误差分析。

在该模型中，探测器对于降落区域的定位精确到了米(m)，属于高精度定位，然而在区域定位时，为了计算的方便，所定位的降落区域(定位点)用一个面积为 $1 \times 1 \, \mathrm{m}^2$ 的小方块代替(该模型选取边界最右上角的小方块)，会使得探测器在降落时与最优的降落

地点出现偏差,如果该偏差超过了误差范围,将会严重影响到探测器的安全,现对其误差做一个定量的分析:

假设探测器降落所允许的最大误差为 θ(探测器中心点离最近的障碍物距离,单位:m);探测器预降落区域为 S_1;探测器因各种不可控因素导致降落偏离原定降落点的距离满足正态分布 $L \sim N(0,1)$,偏离超过 L_m 预降落点的概率几乎为 0;在预降落区域 S_1 一定是安全的。

根据假定可以作出安全区域的俯视图,见图 2.13。

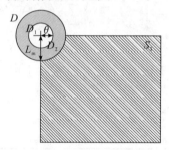

图 2.13 安全区域俯视图

根据图 2.13 我们可以发现,当探测器降落在降落可能区域与安全区域交集区域和误差范围 θ 内区域时是安全的,根据概率分布:

设降落可能区域为 D,误差允许区域(不包含与安全区域交集)为 D_1,安全区域与降落可能区域的交集为 D_2,距离预定降落点的水平距离为 x,竖直距离为 y;

则探测器落在 D 中的概率为:

$$P_D = \iint_D f(x,y)\, \mathrm{d}x\mathrm{d}y$$

由极坐标公式变换($x = L\cos\delta, y = L\sin\delta$)可以求得:

$$P_D = \int_0^{2\pi} \mathrm{d}\delta \int_0^{L_m} \frac{1}{2\pi} \exp\left(-\frac{L^2}{2}\right) L\, \mathrm{d}L = 1 - \exp\left(-\frac{L_m^2}{2}\right)$$

同理,探测器落在 D_1 中的概率为:

$$P_{D_1} = \iint_{D_1} f(x,y)\, \mathrm{d}x\mathrm{d}y = \int_0^{\frac{3\pi}{2}} \mathrm{d}\delta \int_0^{\theta} \frac{1}{2\pi} \exp\left(-\frac{L^2}{2}\right) L\, \mathrm{d}L = \frac{3}{4}\left[1 - \exp\left(-\frac{\theta^2}{2}\right)\right]$$

探测器落在 D_2 中的概率为:

$$P_{D_2} = \iint_{D_2} f(x,y)\, \mathrm{d}x\mathrm{d}y = \int_0^{\frac{\pi}{2}} \mathrm{d}\delta \int_0^{L_m} \frac{1}{2\pi} \exp\left(-\frac{L^2}{2}\right) L\, \mathrm{d}L = \frac{1}{4}\left[1 - \exp\left(-\frac{L_m^2}{2}\right)\right]$$

则可以计算出探测器安全着陆的概率为:

$$P_c = \frac{P_{D_1} + P_{D_2}}{P_D} = \frac{\int_0^{\frac{3\pi}{2}} d\delta \int_0^{\theta} \frac{1}{2\pi} \exp\left(-\frac{L^2}{2}\right) L dL + \int_0^{\frac{\pi}{2}} d\delta \int_0^{L_m} \frac{1}{2\pi} \exp\left(-\frac{L^2}{2}\right) L dL}{\int_0^{2\pi} d\delta \int_0^{L_m} \frac{1}{2\pi} \exp\left(-\frac{L^2}{2}\right) L dL} =$$

$$\frac{\frac{3}{4}\left[1 - \exp\left(-\frac{\theta^2}{2}\right)\right] + \frac{1}{4}\left[1 - \exp\left(-\frac{L_m^2}{2}\right)\right]}{1 - \exp\left(-\frac{L_m^2}{2}\right)}$$

令 $1 - \exp\left(-\frac{\theta^2}{2}\right) = \alpha$，$1 - \exp\left(-\frac{L_m^2}{2}\right) = \beta$，$P_c = \frac{k_1 \alpha + k_2 \beta}{\beta} = k_1 \frac{\alpha}{\beta} + k_2$

上式中：α，β 与 θ 和 L_m 有关，然而 θ 和 L_m 通常由探测器的科技水平决定。如果有条件改进探测器，则可以通过改变 α，β 来提高安全水平；通常地，若想提高安全性，可以改进算法，比如本文模型便可以通过改变定位点来改变 k_1 和 k_2 值，比如将定位点向边界转移，或者向安全区域内部转移，增加了算法复杂度，但可以提高安全性能。

(2) 敏感性分析。

本文针对精避障区段的安全区域大小 S，在一定范围 θ 内变化时，对于降落点选取的影响作出敏感性分析，其结果如图 2.14。

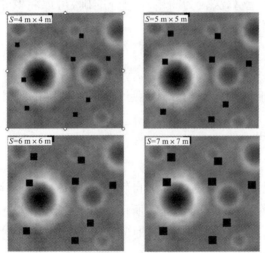

图 2.14　不同精度预安全区域比较图

根据图 2.14 所示的结果，可以发现，S 的取值变化影响了降落区域的选择精度：当 S 的值较大时，探测器降落允许的误差可以增加，其选择精度会增加；然而，在有限的范围内进行选择时，当 S 的值很大时，将很难选择出一个合适的降落区域；相反地，当 S 的值很小时，探测器降落时允许的误差很小，需要十分精确的定位，因此其选择精度就降低了。所以，在选择安全区域大小时，应当参照高程图的复杂度以及其实际范围的大小。

2.6 模型的评价与推广

2.6.1 模型的优点

(1)本文巧妙运用流程图,将建模思路和具体算法完整清晰地展现出来;
(2)在对天体运动的处理上,合理地简化模型,方便其求解;
(3)将难以直接使用的图片数据转化为数值数据,处理得非常巧妙;
(4)对误差分析进行了多方位、定量的分析,为减小误差的产生提供方向。

2.6.2 模型的缺点

(1)模型中为使计算简便,使所得结果更理想化,忽略了一些次要的影响,比如空间中其他天体对嫦娥三号的影响;
(2)在最后模型改进的时候,提出了思路和解法,但是由于时间有限,并没有将最后的具体结果计算出来。

2.6.3 模型的推广

(1)月心坐标定位模型算法较为简单却十分准确,可以将其推广到其他星体的定位或是海底潜艇定位等领域;
(2)地形粗糙度模型所需技术支持较高,在有足够数据的情况下,可精确地对一个区域进行数字化处理。可以运用在地形勘测、海底地形勘探等较为先进的领域。此外,这个模型可以进行简化,推广于大规模农业信息调查,大数据的统计领域。

2.7 模型的改进

在问题一、二中,我们都没有考虑月球自转将造成嫦娥三号运动的时间变化和位移偏离。由于主减速阶段在整个着陆过程中所用时间最长、位移变动最大,因此我们将针对主减速阶段在受到月球自转影响下进行时间和位移参数的改进。

在月球自转引起的作用力下,嫦娥三号的运动轨迹不再是一个平面内的抛物线,而变为复杂的空间曲线,此时前文中二维的拟抛物线动力学模型将不再适用,我们选择经纬度为切入点,建立在月心直角坐标系下的三维位移—时间模型。

三维位移—时间模型

$$\begin{cases} x = vt\cos\alpha + v\omega t^2 \sin\varphi\cos\beta, \\ y = vt\cos\beta - v\omega t^2(\cos\beta\cos\lambda + \sin\varphi\cos\alpha), \\ z = vt\cos\lambda + v\omega t^2 \cos\varphi\cos\beta - \dfrac{gt^2}{2}, \\ t = \dfrac{2v\cos\lambda}{g - 2v\omega\cos\varphi\cos\beta}. \end{cases}$$

上式中：v——运动轨道上某点的瞬时速度，m/s；t——主减速过程中的降落时间，s；λ——运动轨道上某点的经度；φ——运动轨道上某点的纬度；ω——月球自转的角速度，为常值；g——月球重力加速度，为常值。

当忽略自转因素时，$t = (2v\cos\beta + v\cos\alpha)/g$，利用模型Ⅰ的数据反解出 $\cos\alpha, \cos\beta$，代入三维位移—时间模型，即可得到定位点的月心坐标和时间。

参考文献

[1] 王鹏基，曲广吉. 月球软着陆飞行动力学和制导控制建模与仿真[J]. 中国科学，2009,39(3):521—527.

[2] 梁东平，柴洪友，陈天智. 月球着陆器软着陆动力学建模与分析综述[J]. 航天器工程，2011,20(6):104—105.

[3] 张洪华，关轶峰，黄翔宇等. 嫦娥三号着陆器动力下降的制导导航与控制[J]. 中国科学：技术科学，2014,44(4):377—384.

[4] 吴礼斌. 经济数学实验与建模[M]. 天津：天津大学出版社，1999.

◆ 建模特色点评

❋**论文特色**❋

◆标题定位："嫦娥三号着陆轨道设计与最优控制模型"与原赛题"嫦娥三号软着陆轨道设计与控制策略"相比较，既紧扣赛题，又包含最优模型，标题定位准确、简洁、有特色。

◆方法鉴赏：流程框图构架思路清晰，将建模思路和具体算法完整清晰地展现出来；在天体运动的处理上，合理地简化模型，方便其求解过程；将难以直接使用的图片数据转化为数值数据，数据处理得非常巧妙；对误差进行了多方位、定量地分析，为减小误差的产生提供了方向。

◆写作评析：总体写作思路清晰、语言简洁流畅，技术面好，写作格式规范。

◆其他解读：论文写作较为全面，有误差分析、模型的评价、模型的改进与模型的推广。

❋**不足之处**❋

没有对模型改进作深入的研究，缺少灵敏度分析。

第 3 篇
"互联网+"时代的出租车资源配置

◆ 竞赛原题再现

2015 年 B 题 "互联网+"时代的出租车资源配置

出租车是市民出行的重要交通工具之一，"打车难"是人们关注的一个社会热点问题。随着"互联网+"时代的到来，有多家公司依托移动互联网建立了打车软件服务平台，实现了乘客与出租车司机之间的信息互通，同时推出了多种出租车的补贴方案。

请你们搜集相关数据，建立数学模型研究如下问题：

(1) 试建立合理的指标，并分析不同时空出租车资源的"供求匹配"程度。
(2) 分析各公司的出租车补贴方案是否对"缓解打车难"有帮助？
(3) 如果要创建一个新的打车软件服务平台，你们将设计什么样的补贴方案，并论证其合理性。

原题详见全国大学生数学建模竞赛 2015 年 B 题。

◆ 获奖论文精选

"互联网+"时代出租车补贴是否有效[①]？

摘要：本文针对基于"互联网+"下的出租车补贴方案是否有效问题，以北京市为例，运用主成分分析、聚类分析等方法，借助 MATLAB、SPSS、EVIEWS 等软件构建了出租车资源供求匹配评价模型、资源调度优化等模型，综合分析了打车难现状，并研究打车

---
① 本文获 2015 年全国一等奖。队员：黄丽华、贾思钰、王锐杰；指导教师：杨鹏辉。

软件服务平台的补贴方案对打车难易程度的影响,并对模型进行了合理性分析。

针对问题一,首先研究常用指标体系,从出租车供给——需求体系出发,利用主成分分析法,建立了供求匹配评价模型,然后从时间、空间两个角度进行分析,构建时空分布模型,得到了时间角度北京市日常一天中 6～8 点、12～13 点及 16～18 点三个时间段出租车资源匹配程度较差,属于等级三;空间角度三环内以及学院路附近匹配程度较差,其余地方平均匹配程度一般,打车困难程度属于一般水平。

针对问题二,首先分析影响"打车难"的主要因素,建立司乘推荐模型;然后对传统和基于打车软件服务平台推出的两种补贴方案进行对比分析;最后求出北京市未实施补贴方案时打车成功率为 45%,传统补贴方案下为 55%,打车软件服务平台推出的补贴方案使打车成功率提高到 86%,提高了 41%,因此打车软件服务平台推出的补贴方案对"缓解打车难"更有帮助。

针对问题三,首先分析现有补贴方案的优缺点,结合实际情况,建立激励机制约束模型,创建出具有自动搜寻最优补贴方案功能的打车软件服务平台,据此设计出新的补贴方案;最后选取北京市五个典型地区进行研究,通过对比分析补贴方案的合理性,可知在原打车成功率 75% 的水平上,实施新的补贴方案后,各地区打车率均有提高,特别是学院路,打车成功率提高了 9.7%,说明新设计的补贴方案具有合理性,可以较好地解决乘客"打车难"问题。

本文还对模型进行了改进,建立时间序列模型预测未来出租车需求量,此外还对出租车资源"供求匹配"程度评价模型进行了灵敏度检验,实验结果证明了该模型稳定性高,具有较高的可靠性。最后,通过分析模型的优缺点,对模型进行客观评价,并将激励机制约束模型应用于公司激励员工方面,将模型进一步推广。

关键词:出租车补贴方案;主成分分析法;激励机制约束模型;灵敏度分析;MATLAB

3.1 问题的重述

3.1.1 背景知识

1. 背景介绍

随着我国经济的飞速发展和人民生活水平的极大提高,出租车以其能够根据乘客需要提供灵活、方便、快速、直达的运输服务和可以很好地满足人们的专有乘车空间和私密要求等特点,迅速占领了城市客运交通的一方市场,其需求量在不断激增,发展规模不断扩大。但与此同时,合理控制出租车的发展规模也成为一个热点问题,出租车规模没有得到合理控制则会导致经营者利润低下、行业不稳定、道路交通资源浪费、城市交通拥挤等诸多问题。

当下我国出租车市场涌现出多款热门打车软件,其中拥有用户最多的是滴滴打车和快的打车,打车软件的流行打破了传统的路边扬招模式,给出租车行业带来了不可忽视的影响,打车软件的出现为解决行业矛盾提供了新思路,但因为体制不成熟,打车软件也存在管理和盈利模式上的缺陷。因此,合适的城市出租车拥有量是保证出租车行业资源优化配置、健康、有序发展以及城市公共交通体系协调发展的前提。

2. 问题的产生

快的打车等打车软件的流行虽为客户与司机都带来了方便,实现了乘客与出租车司机之间的信息互通,但由于最近兴起,在体制等方面不成熟,为出租车行业带来规模扩大太快、交通资源浪费等问题,因而研究不同时空出租车资源的"供求匹配"程度、分析各公司的出租车补贴方案对"缓解打车难"有无帮助对合理分配出租车资源,提高资源利用率等有重要意义。

3.1.2 要解决的问题

1. 问题一

分析影响出租车资源配资指标,研究不同时空出租车资源的"供求匹配"程度。

2. 问题二

明确北京各公司的出租车补贴方案,并就补贴方案对"缓解打车难"有无帮助进行分析。

3. 问题三

根据问题一、二建立的模型,为一个新的打车软件服务平台设计补贴方案,并论证补贴方案的合理性。

3.2 问题的分析

3.2.1 问题的总分析

一、分析题目所给信息,找出相关指标,从时间、空间角度,结合补贴方案,对出租车资源进行分析;

二、根据上述所建立的模型,建立补贴方案,验证合理性;

三、画出整体框图,如图 3.1 所示。

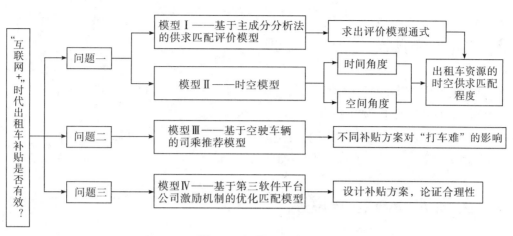

图 3.1　问题总思路图

3.2.2　对具体问题的分析

1. 对问题一的分析

问题一要求我们自己查找数据,根据题目要求,寻找合理的指标建立模型分析在不同时间、空间条件下,出租车的供给与需求的匹配程度,即供求是否平衡。针对问题一,我们首先查阅相关数据,分析哪些数据对出租车的载客时间有影响。分析选取出合适的指标。根据所选取的指标的属性将其归属到需求或者供应指标体系,然后利用主成分分析法,筛选出所有指标中最重要的指标。其次利用筛选出的数据,计算需求和供给函数主成分,我们利用两者函数的主成分求得两函数的综合值,然后利用两者函数综合值的差值反映供给——需求是否平衡,体现两者匹配程度相差多少。

2. 对问题二的分析

问题二要求我们查找各个打的软件公司的补贴方案,分析这些补贴方案是否有利于缓解现在"打车难"的问题。针对问题二,我们首先查找各个打的软件公司的补贴方案以及传统补贴方案。提取这些方案中的有效信息并将其数字化。然后利用这些数据建立一个衡量"打车难"的模型求解打车成功率,并利用该模型求解在传统补贴方案、各个打的软件公司的补贴方案下的打车成功率,利用打车成功率的高低分析各个打车软件公司的补贴方案与传统补贴方案是否可以缓解打车难问题。

3. 对问题三的分析

问题三要求我们在现有的打车平台的基础上创建一个新的打车软件服务平台,设计一个新的补贴方案,使得乘客出行打车更加方便,并且对其正确性进行论证。针对问题三,我们利用问题一与问题二的数据结果,根据委托—代理中的激励机制的运行规律,创建一个具有自动寻优补贴方案的智能打车软件服务平台,利用奖励的方式鼓励出租车司机接"差单"(交通拥挤区的打车订单),鼓励乘客将等车时间延长,从而使得打车成功率增加。最后通过对比分析实施补贴方案前后打车成功率的变化,验证其打车服务平台是否具有合理性。

3.3 模型的假设

(1) 假设补贴方案的变化是随机无周期的；
(2) 假设天气的变化对补贴方案是没有影响的；
(3) 假设出租车运费上调只用来消化运营成本，不是增加利润；
(4) 假设北京市油价在研究期间不变。

3.4 名词解释与符号说明

3.4.1 名词解释

1. 供给偏离值

它是指在一经济主体中，需求量与供给量差值的绝对值，用于衡量供求关系。

2. 帕累托最优

帕累托最优亦称帕累托效率，是资源分配的一种理想状态。假定有固定的一群人和固定可分配的资源，从一种分配状态到另一种状态的变化过程中，在没有使任何人境况变坏的前提下，不可能使得至少一个人的境况变得更好，即帕累托最优状态时不可能再有帕累托改进的余地。帕累托改进是达到帕累托最优的路径和方法，不存在帕累托改进时即为帕累托最优状态。

3. 激励相容约束

由于代理人和委托人之间存在目标函数不一致、不确定性强和信息不对称等问题，因此会出现代理人损害委托人利益的现象，从而造成逆向选择和道德风险两种后果。为避免出现这种结果，委托人受代理人的激励相容约束，在任何激励合同下，代理人总会选择使自己期望效用最大化的行动。

3.4.2 主要符号说明

表 3.1 主要符号说明

序号	符号	符号说明
1	D	乘客总的出租车需求数量
2	W_i	单位时间内的出租车空驶数量
3	A_n	所研究的 n 个交通区
4	G	时间空驶率
5	R	空间空驶率
6	L_n	研究范围内出租车总量
7	T_{zi}	乘客在第 n 个交通区的平均等候时间
8	P_l	出租车起步平均价

续表

序号	符号	符号说明
9	Q	单位时间内通过某路段的车流量
10	C	出租车司机载客周转
11	p	打车成功率
12	δ	补贴费
13	r	基础补贴
14	H	综合匹配度
15	α	公司累计补贴能力系数
16	S_b	司机的保留收入

3.5 数据的收集与处理

3.5.1 数据的来源

1. 区域的选取

我们研究的是北京市出租车运行情况,由于北京市区域大、车流量大、不易于统计等特点,不便于全方位研究,因而我们对北京市按环路随机选取具有代表性的不同区域进行研究。为了便于模型的建立求解,从上述区域中选取具有代表性的地点:朝阳路、陶然亭、白石桥、学院路、建国门。

2. 数据的收集

本文在数据收集过程中,采用了智能打车软件服务平台"苍穹"智能出行平台,该平台依赖于智能服务,基于其用户的大数据进行实时动态监测,采用路网数据采集方法[1],极大提高了其目标点的精度,基于苍穹智能出行平台服务软件,我们收集到了上述地区出租车的基本运行情况和所需的基础数据,如各个地区出租车供给量,并通过网站后台浏览器控制平台对所需数据进行了抓取,将其导入 EXCEL 软件,极大简化了我们在建模时收集所需数据的问题。

3.5.2 数据的预处理

由于在数据采集过程中所能采集到的数据是一些基础数据,这些数据并不能完全解决本问题,因此需要通过数据预处理,挖掘出有用的数据。

本文在解决出租车资源的"供求匹配"程度时,所需重要指标有时间空驶率、空间空驶率、乘客等候时间等,而在收集这些指标时存在一定难度,没有较好的观测方法可以统计出所需数据,因此我们通过已收集到的数据,挖掘数据之间的关系,找出其他所需指标数据。在解决这些指标数据时我们采用以下方法。

1. 空驶率

空驶率是指在未载客出租车与其总行驶的出租车的比值,而在收集该项指标数据时存在很大难度,因此我们通过利用打车的难易程度与乘客的等车时间以及出租车的行驶量计算空驶率。

首先我们以一天为时间周期,按一小时为间隔时间,将一天分为 $t_1,t_2,t_3,\cdots,t_{24}$ 共 24 个时间点,并假设这一天的乘客打车难易程度为困难、一般、容易,依次用 3、2、1 表示。然后统计出每小时各个地点的出租车行驶数量 W_i,则可得到不同时段出租车行驶数量。

根据打车的难易程度与乘客的等车时间以及出租车的行驶量我们可以计算得到空驶率,见表 3.2。

表 3.2 各个地区不同时间的空驶率

	1	2	3	4	5	6	7	8
朝阳路	0.215	0.324	0.237	0.179	0.214	0.185	0.156	0.168
陶然亭	0.178	0.198	0.181	0.176	0.226	0.164	0.146	0.237
白石桥	0.224	0.245	0.199	0.223	0.235	0.185	0.175	0.174
学院路	0.444	0.228	0.192	0.226	0.242	0.302	0.357	0.228
建国门	0.361	0.214	0.238	0.467	0.387	0.223	0.389	0.468

2. 乘客等候时间

乘客等候时间是影响人们选择出租车出行的重要因素,其时间的长短在很大程度上决定了人们对出租车的需求量,如果等候时间超出乘客所能接受的范围,他们便不会选择出租车作为出行工具,因此研究该指标在解决本文相关问题时具有重要意义。

同理,按上述方法将一天分为 24 个时间点,则在单位时间点内,如果第 m 辆出租车为空驶状态,$C=1$;反之 $C=0$;其中出租车总的空驶时间为:

$$G_V = \sum_{m=1}^{L} \int_0^1 L_m \cdot C_m \mathrm{d}t = W$$

则在单位时间点内其出租车的时间空驶率与总空驶时间存在以下关系:

$$G = W/L \Rightarrow W = G \cdot L$$

经过统计分析发现,乘客等候时间 T_{zi} 与出租车的空驶数量 W 和出租司机寻找乘客时间 P 存在一定关系,其中:

$$T_{zi} = \alpha \frac{S_i}{W_i P_i} \tag{3-1}$$

计算得到的结果见表 3.3。

表 3.3 各地点不同时间条件下的等车时间表

	1h	2h	3h	4h	5h	6h	7h	8h
朝阳路	0.1304	0.2480	0.2923	0.7316	0.3285	0.4875	0.2245	0.4007
陶然亭	1.2053	0.3845	1.9756	3.0475	1.1351	1.9446	0.4541	0.3297
白石桥	0.2569	0.7964	1.3862	0.8153	0.6547	1.3514	0.9329	0.8675
学院路	0.0961	0.3432	0.8189	1.0435	0.8454	1.1963	0.3997	0.4774
建国门	0.2105	0.4972	0.1927	0.1406	0.0948	0.3059	0.2735	0.1776

3.6 模型的建立与求解

3.6.1 问题一的分析与求解

1. 对问题一的分析

本问题要求我们选取影响出租车供求量的主要指标,并分析不同时空条件下出租车资源的"供求匹配"程度问题。针对这一问题,我们的思路是首先依据国内外目前常用指标体系,最终确定本文研究问题所需要的指标,并利用主成分分析法选取主要指标,建立出租车资源的"供求匹配"程度评价模型;然后以北京市内 5 个典型地区作为研究地点,分别在时间、空间方面对出租车资源的"供求匹配"程度进行了分析。

2. 对问题一的求解

模型Ⅰ——基于主成分分析法的供求匹配评价模型

(1)模型的准备。

①模型的原理。主成分分析法[2]是将多目标体系中相关性较大的指标进行筛选剔除,使最终留下来的指标两两不具有相关性,并进行重新组合为一组相互无关但又能代表实际问题的指标,起到简化问题的作用,是一种降维思想方法。

②模型概述。主成分分析法[1]的基本思想通过一个简单例子进行说明:

假设有 n 门课程 $x_1, x_2, x_3, \cdots, x_n$,其中 $c_1, c_2, c_3, \cdots, c_n$ 为各门课程的权重,则:$s = c_1 x_1 + c_2 x_2 + \cdots + c_n x_n$

一般情况下,在评价一个学生的综合成绩时,由于课程数较多在计算过程中较为复杂,因此我们可以设 $X_1, X_2, X_3, \cdots, X_n$ 为样本观测的随机变量,令 $P = Var(c_1 x_1 + c_2 x_2 + \cdots + c_n x_n)$,通过计算我们要找到能使 P 值达到最大,因为方差在一般情况下反映了数据的差异程度,所以我们可以通过 P 值找到 n 个变量中最大变异变量,一般我们规定其约束条件为:$c_1^2 + c_2^2 + \cdots + c_n^2 = 1$

在上述约束条件下,P 值的最优解是 n 维空间向量的一个单位向量,它代表一个主成分方向。

③建模思路。本文建模的思路是首先选出评价指标建立需求和供应指标体系,并对各指标数据进行标准化处理,即使数据无量纲化。其次,利用经标准化处理后的指标

数据结合 MATLAB 软件进行主成分分析,通过计算相关系数矩阵 R,将指标体系中相关性较大的指标筛选出来,并对其进行剔除重组,建立一个新的指标变量集合,最后利用所求得的主成分指标变量建立出租车资源"供求匹配"的程度评价模型。

(2)模型的建立。

①首先根据国内外目前常用指标体系,最终确定本文研究问题所需要的指标,并根据所选取指标建立出租车"供求匹配"程度评价指标体系,如图3.2所示。

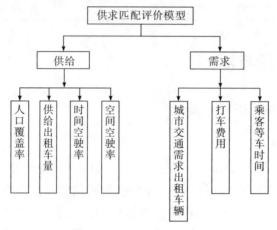

图 3.2 供给—需求指标图

②对原始数据进行标准化处理。假设参与主成分分析的指标变量有 n 个,记为 x_1, $x_2, x_3, \cdots, x_n$,评价对象有 m 个,则第 i 个评价对象的第 j 个指标的取值为 q_{ij},将各指标进行标准化处理,并将标准化处理后的值记为 $\tilde{q}_{ij}$,即:

$$\tilde{q}_{ij} = \frac{q_{ij} - \sigma_j}{s_j}, \quad i=1,2,\cdots,n, j=1,2,\cdots,m \tag{3-2}$$

其中,$\sigma_j = \frac{1}{m}\sum_{i=1}^{m}q_{ij}$,$s_j = \sqrt{\frac{1}{m-1}\sum_{i=1}^{m}(q_{ij}-\sigma_j)^2}$,$j=1,2,\cdots,m$,$\sigma_j$ 为第 j 个指标的平均值,s_j 为第 j 个指标的标准差。

③计算相关系数。在求相关系数时,先求出各指标的相关系数矩阵 R,其 $R=(r_{ij})_{m\times n}$,即:

$$r_{ij} = \frac{\sum_{k=1}^{m}\tilde{q}_{ik}\cdot\tilde{q}_{kj}}{m-1}, i=1,2,\cdots,n; j=1,2,\cdots,m \tag{3-3}$$

需注意的是 $r_{ij}=r_{ji}$,$r_{ii}=1$,其中:r_{ij} 是第 i 个与第 j 个指标的相关系数。

④计算相关系数矩阵的特征值和特征向量。

根据特征方程 $|R-\lambda_i|=0$,计算特征方程根和特征向量,并将其按从大到小排列得 $\lambda_1 \geqslant \lambda_2 \geqslant \lambda_3 \geqslant \cdots \lambda_m \geqslant 0$,则特征向量为 $r_1, r_2, \cdots, r_m$。

其中 $\gamma_j = (\gamma_{1j}, \gamma_{2j}, \cdots, \gamma_{mj})^T$,它是由特征向量组成 m 个新的指标变量得到的。

$$\begin{cases} y_1 = \gamma_{11}x\sim_1 + \gamma_{21}x\sim_2 + \cdots + \gamma_{m1}x\sim_m, \\ y_2 = \gamma_{12}x\sim_1 + \gamma_{22}x\sim_2 + \cdots + \gamma_{m2}x\sim_m, \\ \cdots\cdots\cdots\cdots\cdots \\ y_m = \gamma_{1m}x\sim_1 + \gamma_{2m}x\sim_2 + \cdots + \gamma_{mn}x\sim_m. \end{cases}$$

其中，y_1 为第 1 主成分，$\cdots$，y_m 为第 m 主成分。

⑤分别选取需求、供给函数的主成分。分别计算出特征值 $\lambda_j(j=1,2,\cdots,m)$ 的信息贡献率和累计贡献率，即：

$$a_j = \lambda_j / \sum_{k=1}^{m}\lambda_k, j=1,2,\cdots,m, b_i = \lambda_j / \sum_{k=1}^{m}\lambda_k, j=1,2,\cdots,m, \tag{3-4}$$

其中，a_j 为主成分 y_j 的信息贡献率，同时：

$$\beta = \sum_{k=1}^{p}\lambda_k / \sum_{k=1}^{m}\lambda_k,$$

其中，$p(p \leqslant m)$ 为选取主成分的个数，β 为主成分 $y_1,y_2,y_3,\cdots,y_p$ 的累计贡献率，一般取 $\beta = 0.85, \beta = 0.90, \beta = 0.95$ 时，选取前 p 个指标 $y_1,y_2,y_3,\cdots,y_p$ 为 p 个主成分，然后用其代替原来的 m 个指标。

⑥建立"供求匹配"程度评价模型。根据上述方法所选取的需求和供给函数主成分，我们利用两者函数的主成分求得两函数的综合值，然后利用两者函数构建出供给—需求平衡模型，最终，建立综合评价模型，对出租车资源"供给匹配"程度进行定性评价分析。

综合评价公式为：

$$H = \left| \sum_{j=1}^{p}a_j y_j - \sum_{i=1}^{p}b_i y_i \right|, \tag{3-5}$$

其中，a_j 为供给函数的第 j 个主成分的信息贡献率，b_i 为需求函数的第 i 个主成分的信息贡献率。

(3) 模型的求解。

按照上述模型构建方法，我们逐一对模型进行了求解。

①首先我们根据所建立的出租车资源"供求匹配"程度评价指标体系，可知需求函数下的指标变量有城市交通出租车需求量、打车费用、乘客等待时间；供给函数下的指标变量有人口覆盖率、供给出租车量、时间空驶率、空间空驶率。

②对所选指标数据进行标准化处理，首先对供给函数指标变量进行标准化处理。同时也将供给函数指标变量标准化处理。

③根据处理后的数据我们利用 SPSS 软件对相关系数矩阵进行求解：

同时可以得到 R_1 的特征值为 2.34，R_2 的特征值，3.16；相应的 R_1 中的特征向量为 $(0.31,0.42,0.27)$，R_2 的特征向量为 $(0.21,0.37,0.19,0.23)$。

④根据上述所求得的需求、供给函数的矩阵特征值和特征向量，分别求出特征值的信息贡献率和累计贡献率。其中需求、供给函数中主成分的累计贡献率分别为：

$$\beta_1 = 0.82, \beta_2 = 0.71$$

因此得到影响乘客对出租车需求量的主要指标有乘客等待时间,影响出租车公司供给出租车量的主要指标主要有时间空驶率和空间空驶率,最终可建立"供求匹配"程度评价模型:

$$H = \left| \sum_{j=1}^{p} a_j y_j - \sum_{i=1}^{p} b_i y_i \right| \tag{3-6}$$

模型Ⅱ——时空模型

(1)问题的求解。

首先我们根据研究地区的"供求匹配"程度评价值,利用 SPSS 软件对其值进行了聚类分析,得到了匹配程度等级表,见表3.4。

表 3.4　匹配程度等级表

等级	一	二	三	四
数据范围	0.006~0.056	0.056~0.156	0.156~0.356	0.356~0.556

从时间分析

根据截取的 5 个典型地区的数据,我们利用公式(3-6)可以得到这 5 个地点的主要指标值评价表。观察表 3.4 我们难以得到供求偏离程度随时间的变化规律,因此,为了研究出租车的供求程度随时间的变化规律,我们利用 EXCEL 绘制了 5 个地区的出租车供求程度随时间的变化图(见图 3.3)。

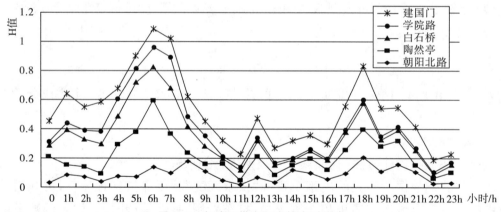

图 3.3　各地区供求程度偏离时间表

观察图 3.3,我们可以知道,在早上 7、8 点,中午 12、13 点以及晚上 18 点的时候,出租车的供给与需求偏离程度较高,属于等级三,在这几个点打车难、空车少,因为这些点是上下班的高峰期,出租车的需求量大增,而出租车的总数即供给量不变,则会导致出租车的供求不平等;晚上 12 点至凌晨 4 点,人流量少,值夜班的出租车也相对较少,导致出租车的需求供给也出现了相对偏差,属于等级二,而且出租车司机根据习惯,认为半夜人少,导致出租车的数量也相对较少,出现打车难的问题;像白天其他时间,出租车司机的行驶路线根据习惯往人流量多的地方走,导致一些人流量少的地方难打车,出租车供求也相对不平衡。

从空间分析

为了研究出租车的供求程度在空间上的分布规律,我们首先要找到各个地区的坐标,确定各个地区的地理位置,然后再计算各个地区上出租车的供求程度的分布。由于在地图上测量各个地区的距离不方便,即使用标尺测量出来也存在一定的误差,而各个监测点的经纬度我们可以准确的测量得到测量值(见表3.5)。

表 3.5　各地区经纬度表

	朝阳路	陶然亭	白石桥	学院路	建国门
纬度	39.922273	39.884625	39.951939	39.994950	39.914524
经度	116.549815	116.380902	116.332443	116.360101	116.442268

易知一地点的经纬度是指该地点在地球表面的坐标,由于北京市面积与地球表面积相比非常小,我们认为北京市近似一个平面,因此,我们可以以经纬度为其横纵坐标,由此我们可以得到每个地区的相对位置,即我们知道了各监测点的空间分布。为了形象的刻画各个地区打车匹配程度,我们利用表3.5求出各个地点的出租车平均供求偏离程度(见表3.6)。

表 3.6　供求程度偏离总表

朝阳路	陶然亭	白石桥	学院路	建国门
0.084713	0.128879	0.131133	0.045863	0.129308

再结合各个地点的坐标,我们利用 MATLAB 软件,得出匹配程度的空间分布图(见图3.4)。

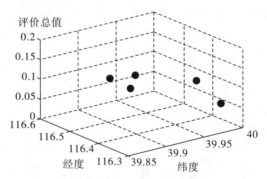

图 3.4　匹配程度的空间分布图

观察图3.4我们可以发现,在学院路的匹配程度值是最低的,属于等级一,说明在学院路是较容易打到车的,因为学院路人流量较大,出租车司机认为在学院路能找到更多的乘客,因此司机更倾向于往学院路跑;而建国门、陶然亭和白石桥处于北京市三环路附近,远离市中心,人流量较少,在空车状态下,出租车司机认为在这些地方不容易找到乘客,同时会增加燃油量,因此这两个地点的匹配偏离程度高,属于等级三;朝阳路人流量较大,出租车多,但出租车的需求量也较高,供需偏离程度一般,打车困难程度为一般水平,属于等级二。

3.6.2 问题二分析与求解

1. 对问题二的分析

本问题要求我们在了解了各公司的出租车补贴方案的基础上,分析各方案是否对"缓解打车难"有帮助。针对这一问题,我们首先通过分析"打车难"的影响因素,建立基于空驶量的衡量"打车难"的司乘推荐模型;其次,依据不同公司推出的出租车补贴方案,分别对打车软件公司提供的补贴方案和传统的补贴方案进行处理,将其补贴方案量化,根据量化的数据建立衡量"打车难"的模型,比较分析打车软件公司提供的补贴与传统的补贴方案可缓解"打车难"的程度,即是否可以减少出行人的打车等待时间,即提高打车成功率。

2. 对问题二的求解

模型Ⅲ——基于空驶辆的司乘推荐模型

(1)模型的准备。

①模型的原理。该模型是由 Phithakkitnukoon 等人率先提出的,该模型主要运用于预测在一定区域内出租车的空驶量[3]。该模型是在利用出租车的历史 GPS 数据的基础上,采用朴素贝叶斯分类器来建立模型;之后 Zheng 等人提出来采用非齐次泊松分布模拟建立出租车行为模型,该模型是经上述一系列改进之后得到的,它主要是通过统计道路上空出租车的行驶时间和驶出这条道路的乘客等候时间的历史数据,模拟出一种可以根据不同位置乘客等候时间向乘客推荐有效打车位置的系统。

②建模思路。根据对题目的分析,我们画出了针对本问题的思路图(见图3.5)。

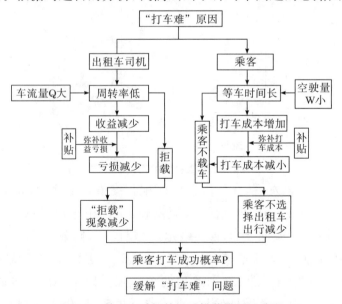

图 3.5 基于空驶辆的司乘推荐模型思路图

(2)模型的建立。

①建立"周转率—车流量"联合关系。周转率 C 代表着司机在一天内拉客的人

数[4]，它在一定程度上影响了载客率，这也是出租车司机经常存在"拒载""挑客"情况的原因，而经过日常观察可以发现城市交通区的车流量很大程度上影响"周转率"的大小，即在车流量大的地方，交通处于拥堵状态，不利于出租车司机接客，因此也会在车流量大的地方存在出租车司机"拒载""挑客"的情况。

基于上述分析，我们先分析车流量情况，假设路段长度为 L，令 t_0 时刻路段下的空驶出租车量为 W_1，从 t_0 到 t_1 时刻进入新行驶路段的空驶出租车辆为 W_2，则时间间隔为：$\Delta t = t_1 - t_0$

将所有路段上出现的出租车数量记为 V，则：$V = W_1 + W_2$。

为了更加形象的分析问题，我们作车流量模型图（见图 3.6）。

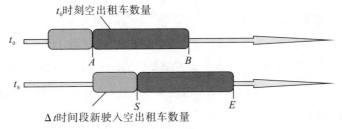

图 3.6　车流量模型图

依据图 3.6，我们将在 Δt 时间段内进入检测路段的出租车平均速度记为 $\tilde{v}$，假定其在 Δt 时间段车流量 Q 是固定不变的，则可以得到：

$$Q = \frac{W \cdot \tilde{v} \cdot \Delta t}{L + \tilde{v} \cdot \Delta t} \tag{3-7}$$

其次假设司机的总收益为 S，则可以得到：

$$S = C \cdot (1 - G_n) \cdot P_l + (l_e \cdot \Delta P) + v \tag{3-8}$$

其中，P_l 为出租车起步平均价；ΔP 为超出规定里程的费用；v 为出租车司机的补贴费用。

依据式(3-7)、式(3-8)我们可以得出车流量和周转率之间的关系，即：

$$C = \frac{v - S + l_e \cdot \Delta P}{(1 - G_n) \cdot Q} \tag{3-9}$$

②建立"乘客等候时间—空驶量"联合关系。乘客等候时间 T 是衡量出租车服务水平的一个重要指标，它直接反映了乘客对出租车出行工具的选择偏好性，它在一定程度上受城市出租车总拥有量、道路交通情况、交通道路布局等因素的影响。

经过我们日常生活观察可知，乘客等候时间的长短直接影响其是否打车，一般情况下等候时间越长，选择出租车的概率越低，而在一定程度上，其与空驶量 W 存在较大关系，一般空驶量越大，乘客等候时间越短；反之越长。因此我们可以得出两者相关关系式，即：$T = \delta Z_i / W_i \cdot t_f \cdot \pi$，其中，$Z_i$ 为第 i 交通区车辆达到率（辆/h）；t_f 为出租车的平均搜索乘客时间（min）；π 为乘客乘车补贴费。

③建立打车概率函数。假设人们日常出行时打车的概率为 P，P 在一定程度上与车流量和空驶量存在关系，因此我们可以得出打车概率表达式为：

$$P = 1 - W/Q_{max} \times 100\% \tag{3-10}$$

而通过上述所建关系我们可以发现，车流量 Q 与出租车司机补贴费和空驶量与乘客乘车补贴费之间存在一定关系，经过推导我们可以得出打车概率与补贴费用的关系：

即 $\begin{cases} C = \dfrac{v - S + l_e \cdot \Delta P}{(1 - G_n) \cdot Q}, \\ T = \delta \dfrac{Z_i}{W_i \cdot t_f \cdot \pi}, \\ P = 1 - \dfrac{W}{Q_{max}} \times 100\%, \end{cases}$ $\Rightarrow P = 1 - \dfrac{\delta \dfrac{Z_i}{T \cdot t_f \cdot \pi}}{\dfrac{v - S + l_e \cdot \Delta P}{(1 - G_n) \cdot C}} \times 100\%$

(3) 模型的求解。

在分析各公司的补贴方案对缓解"打车难"是否有帮助时，我们首先根据北京市的出租车运营状况和乘客日常打车情况，确定了在未实行补贴方案前且属于"打车难"的情况下所对应的打车成功率 P_0，以及在 P_0 基础上的出租车日载客平均周转率 C_0、乘客平均等候时间 T_0、交通区车流量 Q_0 以及通过该交通区的空驶出租车数量 W_0 等，现给出一般情况下的指标数据值（见表 3.7）。

表 3.7 一般情况下的指标数据值

P_0	C_0	T_0	Q_0	W_0
45%	20 次	4.5min	360(辆/h)	198 辆

为了分析各公司的补贴方案对"打车难"的影响，我们首先找出各公司的补贴方案：

① 传统补贴方案分析。在网上搜索相关资料，我们找到了传统出租车公司向出租车司机的补贴方案：

a. 每车每年补贴运营费 17 733.6 元；

b. 每车每天补贴运营费 49.26 元；

c. 每车每年补贴燃油费用为 9 326 元；

d. 每车每天补贴燃油费用 25.5 元。

则依据上述方案可知，传统补贴方案只是针对出租车司机进行补贴，利用上述数据我们可得传统方案下总补贴费 $\delta = \delta_0 + \delta_1 = 49.26 + 25.5 = 74.76$ 元，则将其代入式(3-9)（周转率与车流量）得到：

$$Q_1 = \frac{v - S + l_e \cdot \Delta P}{(1 - G_n) \cdot C} = 434$$

代入打车成功概率模型得：

$$P_1 = \left(1 - \frac{W}{Q_{max}}\right) \times 100\% = 1 - \frac{198}{434} \times 100\% = 55\%$$

则与未实施补贴方案相比，打车成功率提高了 10%，打车等待时间更少了，出租车资源得到了有效利用。

② 基于打车软件平台下的补贴方案分析快的打车和滴滴打车软件补贴方案变更情况。

观察上述两家打车软件公司的补贴方案,可知打车软件补贴不仅针对出租车司机,还对乘客进行补贴,因此我们利用两家补贴方案变更数据,计算统计出了平均补贴水平,并最后利用所建模型求出经补贴之后的打车成功率。

首先分析快的打车补贴方案,快的打车补贴是以出租车每次接单数为基础进行补贴,即每接一单成功后给司机和乘客都进行补贴。通过统计补贴金额,快的打车乘客补贴均值为每单9.8元,给司机补贴均值为8.7元。

则根据$\delta = T \cdot (1-G_n) \cdot m_{司机补贴单价}$;$\upsilon = T \cdot (1-G_n) \cdot m_{乘客补贴单价}$,其中,$m_{司机补贴单价}$为向司机每单平均补贴金额,$m_{乘客补贴单价}$为向乘客每单平均补贴金额。

按上述分析可得,$\delta_1 = T \cdot (1-G_n) \cdot m_{司机补贴单价} = 164$元,$\upsilon_1 = T \cdot (1-G_n) \cdot m_{乘客补贴单价} = 196$元。

则代入打车成功概率模型得:
$$P_1 = (1 - W/Q_{max}) \times 100\% = 84\%。$$

可知与未实施补贴方案相比,打车成功率提高了39%;与传统补贴方案相比,打车成功率提高了29%。

同理,滴滴打车与快的打车补贴方案一样,也是同时向乘客和出租车司机进行补贴,通过统计变化的补贴金额,滴滴打车乘客补贴均值为每单12.5元,给司机补贴均值为9.2元。

按上述分析可得,$\delta_2 = T \cdot (1-G_n) \cdot m_{司机补贴单价} = 184$元,$\upsilon_2 = T \cdot (1-G_n) \cdot m_{乘客补贴单价} = 206$元。

则代入打车成功概率模型得:
$$P_2 = (1 - W/Q_{max}) \times 100\% = 87\%。$$

可知与未实施补贴方案相比,打车成功率提高了42%;与传统补贴方案相比,打车成功率提高了32%;与快的打车补贴方案相比,打车成功率提高了3%。

由上述分析所得数据,我们利用EXCEL软件画出各个补贴方案下的打车成功率变化折线图(见图3.7)。

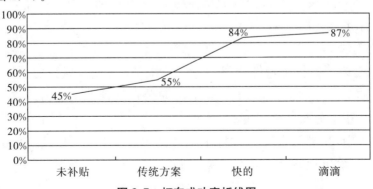

图3.7 打车成功率折线图

观察图3.7,通过分析传统补贴方案和基于打车软件公司的补贴方案对打车成功率的影响,发现基于打车软件公司的补贴方案比传统补贴方案提高的打车率要高出

39%~42%,说明对乘客进行补贴,能提高乘客的耐心,愿意花更多的时间去等车;而且变更了对司机的补贴方案,调动了司机接单的积极性,只要接单就有较高的利益,省去了因交通拥堵而出现的"拒接"现象,提高了出租车资源利用率。从结果我们也可以看出,在各公司推出补贴方案后,基于滴滴打车公司补贴方案,打车成功率达到87%,基于快的打车补贴方案,打车成功率达到84%,但是从一定程度上讲上述两家公司补贴方案并没较好解决高峰路段打车难问题,因此其补贴方案还存在改进空间。而且从基于快的打车与滴滴打车方案的打车成功率不同我们可以看出,相似的补贴方案下的打车成功率也不同,说明在保证打车公司的利益基础上,肯定存在一套较好的补贴方案使打车成功率最高。

3.6.3 问题三的分析与求解

1. 对问题三的分析

本问题要求我们在现有打车软件服务平台的情况下创建一个新的打车服务平台,设计出新的补贴方案,并论证其方案的合理性。针对这一问题,我们的解题思路是首先利用委托—代理中的激励机制模型,创建一个具有自动寻优补贴方案的智能打车软件服务平台。然后利用上述新的打车服务平台,设计出优化处理后的"补贴方案"。最后,我们以北京市为例,选取其五个典型地区进行研究,依据新的打车服务平台得出各地区补贴方案,并通过对比分析实施补贴方案前后打车成功率的变化验证其打车服务平台是否具有合理性。

2. 对问题三的求解

模型Ⅳ——基于第三软件平台公司激励机制的优化匹配模型

(1) 模型的准备。

委托代理关系[5]是指委托人想使代理人按照其利益抉择行动,但委托人对代理人的行动没有直接透视权,能观测到的只是一些由代理人的行动和其他的外生随机因素共同决定的变量,委托人受代理人的激励相容约束[3],在任何激励合同下,代理人总会抉择使自己期望效用最大化的行动。

(2) 模型的建立。

① 假设 w 是第三方打车软件公司的一个一维补贴变量,则我们可以得到出租车司机所拿到的补贴金额 μ,即:$\mu = \alpha w + \varphi$,其中,α 是公司累计补贴能力系数,φ 是方差为 σ^2、均值为零的服从正态分布的随机变量。

在研究过程中,我们假定第三方打车软件公司为委托人,其风险是中性的,出租车司机为代理人,风险可规避。则其线性关系为:

$$S(\mu) = \gamma + \beta(\mu - \mu_0), \tag{3-11}$$

其中,β 为第三方打车软件公司的激励强度,γ 是第三方打车软件公司给予出租车司机基础补贴。

因为第三方打车软件公司风险是中性的,在给定 $S(\mu)=\gamma+\beta(\mu-\mu_0)$,可以得到其期望收入即 $E_v(\mu-S(\mu))=E(\gamma+\beta(\mu-\mu_0))=-\gamma+\beta\mu+\alpha w(1-\beta)$。

②假设司机的效用函数的风险性为中性,$a=-e^{-\varepsilon\kappa}$,其中 ε 是绝对风险规避量,κ 为司机的实际收入。同时,我们假设司机因第三方软件补贴程度增加,其运营成本增加,记为 $d(w)$,其成本公式为

$$d(w)=cw^2/3,$$

其中,c 为成本系数,其值越大,出租车司机因接受第三方软件公司的补贴而使运营成本降低,因此司机的实际收入为

$$q=S(\mu)-d(w)=\gamma+\beta(\mu-\mu_0)-cw^2/3。 \tag{3-12}$$

则其等价收入为

$$E\left(q-\frac{1}{2}\varepsilon\sigma^2\beta^2\right)=\gamma+\beta(\mu-\mu_0)-cw^2/3-\frac{1}{2}\varepsilon\sigma^2\beta^2,$$

其中,E_q 为司机的确定性等价,$\varepsilon\sigma^2\beta^2/2$ 为司机的风险成本。

令 S_b 为司机的保留收入,则如果司机的确定性等价收入小于 θ 时,司机将可以不接受公司给予的补贴,因此,司机的参与约束 IR 为

$$\gamma+\beta(\mu-\mu_0)-cw^2/3-\frac{1}{2}\varepsilon\sigma^2\beta^2\geqslant\theta。$$

③由题可知,第三方打车软件与司机出于对称信息条件,因此激励约束 IC 无作用,任何水平下的 w 一般情况都可以满足参与约束 IR 的限制条件内,因此,我们问题转化为选择 γ,β 和 w 中的最优解,

即 $\max\limits_{\gamma,\beta,w} E v=-\gamma+\beta\mu+\alpha w(1-\beta)$,

$$st.(IR)\gamma+\beta(\mu-\mu_0)-cw^2/3-\frac{1}{2}\varepsilon\sigma^2\beta^2\geqslant\theta。$$

在最优情况下,将参与约束通过固定项 γ 代入目标函数,即可重新表述为

$$\max\limits_{\beta,w}\beta(\mu-\mu_0)-cw^2/3-\frac{1}{2}\varepsilon\sigma^2\beta^2-\theta+\beta\mu+\alpha w(1-\beta),$$

将上述结果代入司机的参与约束 IR 得,

$$\gamma'=\theta+\frac{1}{3}d(w')^2=\theta+\frac{\alpha^2}{2d}, \tag{3-13}$$

其中,γ' 为可观测条件下的帕累托最优解[6],即为该最优解是在代理人不承担任何风险的情况下,第三方打车软件补贴给司机的金额正好等于司机的保留收入与运营成本之和。这时该软件对出租车司机的累计补贴是满足司机和乘客利益最大化的解,它是由第三方软件根据乘客与司机所处的地理信息,系统自动累加加权,最终使司机与乘客满足利益最大化。

(3)模型的求解。

本问题要求我们在之前打车软件平台基础上创建一个新的打车软件平台,并设计出新的补贴方案。我们通过建立第三方打车软件激励机制的优化匹配模型,使其系统

通过分析乘客与出租车司机的所处地理位置,根据乘客与司机的距离及乘客所处交通路段的情况自动模拟出累计加权补贴,最终使得乘客成功打到车,提高其打车成功率。

①设计新的补贴方案。本文根据之前所采集的数据,分析并给出在激励机制模型下不同情况的补贴方案,将其记录在表3.8。

表3.8 激励机制模型下不同情况下的补贴方案表

补贴政策	1 km＜乘客与出租车距离 L≤5 km
车流量 Q≤180(l/h)	不补贴
180≤Q≤300	一单补贴10到15元
300≤Q≤450	补贴按"距离×交通流量=综合值"累计补贴,即综合值每增加 h 时,补贴增加 d 元

②验证新补贴方案的合理性。首先我们选取了北京市五个典型地区作为研究地区,根据问题二的结果,收集统计各个打车软件公司补贴方案下的相关数据,记录在表3.9内。

表3.9 各地区相关补贴下的数据表

	车流量(Q)	供应量(W)	空驶量(N)	打车成功率(P)
朝阳路(A)	490	55	13	0.772
陶然亭(B)	380	196	37	0.812
白石桥(C)	375	55	11	0.803
学院路(D)	500	147	73	0.503
建国门(E)	405	77	15	0.805

根据上述数据,利用本文第三方打车软件的奖励机制优化匹配模型,自动搜寻不同交通情况下的最优补贴方案。将北京市五个地区的数据代入上述模型后得到其最优化补贴方案。

根据代入数据所得,在激励机制中约束条件为 $\gamma+\beta(\mu-\mu_0)-cw^2/3-\frac{1}{2}\varepsilon\sigma^2\beta^2 \geq \theta$,通过系统模拟运算得到,参与约束 IR 为 200 元/天,则将约束条件代入下式:$\max\limits_{\beta,w}\beta(\mu-\mu_0)-cw^2/3-\frac{1}{2}\varepsilon\sigma^2\beta^2-\theta+\beta\mu+\alpha w(1-\beta)$ 得结果如下:

A 地区补贴方案为:$\gamma'_A=m_{司机补贴单价}+m_{乘客补贴单价}=18$,即向出租车司机补贴3元,向乘客补贴15元,累计补贴方案为3元,在高峰路段每次乘客叫单所增加的补贴金额为1元。

B 地区补贴方案为:$\gamma'_A=m_{司机补贴单价}+m_{乘客补贴单价}=15$,即向出租车司机补贴2元,向乘客补贴13元,累计补贴方案为8元,在高峰路段每次乘客叫单所增加的补贴金额为2元。

C 地区补贴方案为:$\gamma'_A=m_{司机补贴单价}+m_{乘客补贴单价}=19$,即向出租车司机补贴3元,向乘客补贴16元,累计补贴方案为6元,在高峰路段每次乘客叫单所增加的补贴金额为2元。

D 地区补贴方案为：$\gamma'_A = m_{司机补贴单价} + m_{乘客补贴单价} = 14$，即向出租车司机补贴 4 元，向乘客补贴 10 元，累计补贴方案为 4 元，在高峰路段每次乘客叫单所增加的补贴金额为 1 元。

E 地区补贴方案为：$\gamma'_A = m_{司机补贴单价} + m_{乘客补贴单价} = 15$，即向出租车司机补贴 2 元，向乘客补贴 13 元，累计补贴方案为 5 元，在高峰路段每次乘客叫单所增加的补贴金额为 1 元。

将上述五个不同地区具体补贴方案代入模型Ⅲ，我们可以得到补贴之后的打车成功率，并与之前补贴方案下的打车概率进行对比，如表 3.10。

表 3.10　北京市 A～E 不同地区补贴前后打车概率对比表

	朝阳路(A)	陶然亭(B)	白石桥(C)	学院路(D)	建国门(E)
改善方案	0.772	0.812	0.803	0.503	0.805
软件公司	0.8	0.846	0.8124	0.6	0.846

根据上表所得数据，我们利用 EXCEL 软件，作出了上述五个地区补贴方案前后打车概率变化折线图，见图 3.8。

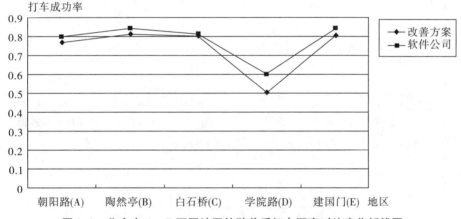

图 3.8　北京市 A～B 不同地区补贴前后打车概率对比变化折线图

观察图 3.8 可知，根据我们拟定的补贴方案效果与软件公司补贴方案效果相对比，打车成功率在一定程度上都有了提高。朝阳路成功率提高了 2.8%，陶然亭提高了 3.4%，白石桥提高了 0.94%，学院路提高了 9.7%，建国门提高了 4.1%。说明补贴方案的拟订十分合理，被乘客与出租车司机所接受，提高了打车成功率，在一定程度上缓解了乘客平常出门"打车难"的问题，也使得出租车司机的收益得到了提高，并给打的软件公司带来更多的收益。增加高峰路段每次乘客叫单所增加的补贴金额，在一定程度上缓解了早晚高峰期打车难的问题，降低了出租车司机"拒接"的概率，使出租车资源得到了更加有效的利用，也说明我们拟订的方案是正确的。

3.7 模型的改进与推广

3.7.1 模型的改进

由于时间关系，上述论文中出租车需求量都是根据“苍穹”网站中获取的数据，为了方便论文的进一步研究，现对出租车的需求量给出预测模型。

模型的建立与求解：时间序列预测模型

对上述论文中出租车的需求量运用 EVIEWS 软件作时序图得图 3.9，由图 3.9 容易看出，该组数据不平稳。为了满足预测模型的条件，对数据进行对数 LOG 处理，设得到平稳序列 $y1$，对 $y1$ 的平稳性进行单位根检验，结果见图 3.10，由检验可知：该序列单位根检验值的绝对值大于给出的显著性水平 1%—10%的 ADF 临界值，落在拒绝域内，拒绝原假设，故该序列为平稳序列。

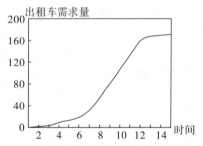

图 3.9　出租车的需求量时序图

Augmented Dickey-Fuller Unit Root Test on Y1

Null Hypothesis: Y1 has a unit root
Exogenous: Constant
Lag Length: 0 (Automatic based on SIC, MAXLAG=2)

		t-Statistic	Prob.*
Augmented Dickey-Fuller test statistic		-6.736696	0.0001
Test critical values:	1% level	-4.057910	
	5% level	-3.119910	
	10% level	-2.701103	

*MacKinnon (1996) one-sided p-values.
Warning: Probabilities and critical values calculated for 20 observations and may not be accurate for a sample size of 13

Augmented Dickey-Fuller Test Equation
Dependent Variable: D(Y1)
Method: Least Squares
Date: 05/30/15 Time: 22:53
Sample (adjusted): 2 15

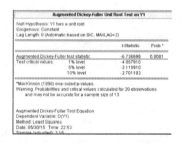

图 3.10　LOG 变换后的平稳性检验图

接着对 $y1$ 作自相关、偏自相关分析得(见图 3.11)。

Date: 05/30/15 Time: 22:47
Sample: 1 15
Included observations: 14

Autocorrelation	Partial Correlation		AC	PAC	Q-Stat	Prob
		1	0.735	0.735	9.3208	0.002
		2	0.509	-0.070	14.152	0.001
		3	0.348	-0.002	16.617	0.001
		4	0.170	-0.147	17.266	0.002
		5	-0.013	-0.154	17.270	0.004
		6	-0.165	-0.122	18.028	0.006
		7	-0.275	-0.095	20.453	0.005
		8	-0.341	-0.061	24.794	0.002
		9	-0.372	-0.062	30.978	0.000
		10	-0.362	-0.035	38.311	0.000
		11	-0.320	-0.029	45.948	0.000
		12	-0.261	-0.034	53.585	0.000

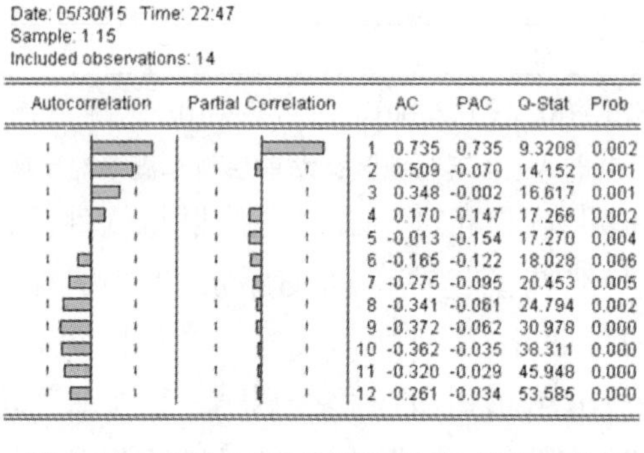

图 3.11　对数 LOG 序列自相关、偏自相关图

由图 3.11 可以看出，自相关图呈阻尼正弦波衰减，偏自相关图滞后一节截尾，分别建立序列 AR(1)、MA(1)、ARMA(1,1)，并比较三种序列的 AIC 和 SC 值，得表 3.11。

表 3.11　AR(1)、MA(1)、ARMA(1,1)模型的 AIC 和 SC 的比较

	AIC	SC
AR(1)	−0.59	−0.51
MA(1)	2.54	2.63
ARMA(1,1)	−0.92	−0.79

由表 3.11 知,由于 AIC:2.54>−0.59>−0.92,SC:2.63>−0.51>−0.79,易知 ARMA(1,1)模型较为合理,能更好地对数据进行拟合。同时可以得到模拟 ARMA(1,1)模型表达式:

$$\text{ARMA}(1,1): x_t - 5.10 = 0.74(x_{t-1} - 5.10) + \varepsilon_t + 1.00\varepsilon_{t-1}$$

为保证模型的完整性,对序列 ARMA(1,1)进行残差检验,由检验可知,P 检验值大于临界值 0.05,落在不拒绝域内,不能拒绝原假设,即该序列为白噪声序列,即可用该序列表达式对出租车需求量进行预测。

3.7.2 模型的推广

(1)主成分分析法在日常生活中被广泛应用,经常被用在动力学模拟、数学建模、人口统计等方面。

(2)第三方打车软件平台是目前基于"互联网+"平台的 O2O 经营模式,因此该软件在市场上被广泛应用。

(3)激励机制约束模型——本文利用该模型来寻找最优补贴方案,而该模型在其他方面还具有较高应用,特别是可用于公司内部寻找激励员工的策略。

3.8 灵敏度分析

问题一中我们得到了综合评价公式[7]为:

$$H = \left| \sum_{j=1}^{p} a_j y_j - \sum_{i=1}^{p} b_i y_i \right|$$

其中:a_j 为主成分 y_j 的信息贡献率,针对 a 的不同取值我们运用 Matlab 软件结合上表作图进行灵敏度分析。

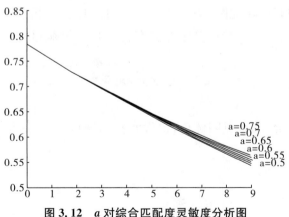

图 3.12 a 对综合匹配度灵敏度分析图

由灵敏度分析可知,当出租车的供应量确定下来时,等量的 a 值变化引起的综合匹配度变化很小。此时贡献率由 $a=0.5$ 到 $a=0.75$ 变化,贡献率变动 0.2,综合匹配度的变化才为 0.005,这是很小的差距。

故此次灵敏度分析效果较好,即贡献率的细微变动对综合匹配度的影响不大。

参考文献

[1] 陈炼红.基于GPS浮动车采集数据的出租车运行特点研究[D].上海:同济大学,2008.

[2] 朱建平.应用多元统计分析[M].北京:科学出版社,2012:93－106.

[3] 刘仰东.一种基于车流量的司乘推荐模型[J].科研信息化技术与应用,2015,03:4－6.

[4] 王宇.对城市"打的难"现象的剖析——基于西安市出租车市场供求失衡问题的分析[J].价格理论与实践,2011,11:15－16.

[5] 刘耀霞.出租车行业利益主体关系研究[D].成都:西南交通大学,2008.

[6] 高鸿业.20世纪西方微观和宏观经济学的发展[J].中国人民大学学报,2000,01:4－11.

[7] 杨桂元,朱家明.数学建模竞赛优秀论文评析[M].合肥:中国科技大学出版社,2013,9.

◆ 建模特色点评

❋ 论文特色 ❋

◆标题定位:从赛题"'互联网＋'时代的出租车资源配置"到"'互联网＋'时代出租车补贴是否有效?",以问题的形式提出赛题,形式活泼,有一定的吸引力。

◆方法鉴赏:模型注重数据的收集与处理,对后期建模量化分析有较好的铺垫作用。文章每一部分均有思路框图,条理性好。通过主成分分析进行降维处理,运用贝叶斯分类器和优化匹配法进行建模,整体方法有针对性。灵敏度分析较好,模型的改进与推广有特色。

◆写作评析:写作规范,技术面好,图文并茂。

❋ 不足之处 ❋

缺少对模型的评价,没有点明优点,也没有指出缺点所在,这是本文的最大不足。

第 4 篇
小区开放对道路通行的影响

◆ 竞赛原题再现

2016 年 B 题　小区开放对道路通行的影响

2016年2月21日,国务院发布《关于进一步加强城市规划建设管理工作的若干意见》,其中第十六条关于推广街区制,原则上不再建设封闭住宅小区,已建成的住宅小区和单位大院要逐步开放等意见,引起了广泛的关注和讨论。

除了开放小区可能引发的安保等问题外,议论的焦点之一是:开放小区能否达到优化路网结构,提高道路通行能力,改善交通状况的目的,以及改善效果如何。一种观点认为封闭式小区破坏了城市路网结构,堵塞了城市"毛细血管",容易造成交通阻塞。小区开放后,路网密度提高,道路面积增加,通行能力自然会有提升。也有人认为这与小区面积、位置、外部及内部道路状况等诸多因素有关,不能一概而论。还有人认为小区开放后,虽然可通行道路增多了,但相应地,小区周边主路上进出小区的交叉路口的车辆也会增多,也可能会影响主路的通行速度。

城市规划和交通管理部门希望你们建立数学模型,就小区开放对周边道路通行的影响进行研究,为科学决策提供定量依据,为此请你们尝试解决以下问题:

1. 请选取合适的评价指标体系,用以评价小区开放对周边道路通行的影响。
2. 请建立关于车辆通行的数学模型,用以研究小区开放对周边道路通行的影响。
3. 小区开放产生的效果,可能会与小区结构及周边道路结构、车流量有关。请选取或构建不同类型的小区,应用你们建立的模型,定量比较各类型小区开放前后对道路通行的影响。
4. 根据你们的研究结果,从交通通行的角度,向城市规划和交通管理部门提出你们关于小区开放的合理化建议。

原题详见全国大学生数学建模竞赛 2016 年 B 题。

获奖论文精选

小区开放对道路通行的影响

摘要：本文针对小区开放对道路通行的影响，使用了层次分析法、权重赋值法、构建函数方程等方法，分别建立了评价指标体系、函数关系模型以及 BPR 路阻函数等模型。运用 EXCEL、MATLAB 等软件，我们建立了小区开放对道路通行的评价指标体系；构建出小区开放对周边道路通行影响的总函数关系方程；然后通过总函数关系定量计算，再比较位于城市不同区域、具有不同结构的小区开放前后对道路通行的影响；最后，从交通通行的角度，向相关部门提出关于小区开放的合理化建议。

本文的突出特色是不仅将小区结构分为 A 型（密网—格网状）、B 型（疏网—格网状）、C 型（密网—环网状）以及 D 型（疏网—环网状）四种类型，而且按照城市道路网密度由高到低依次将城市各区域分为城市核心区（Ⅰ型）、中间城区（Ⅱ型）、城市郊区（Ⅲ型），结合城市的区域类型和小区的结构类型综合研究小区开放前后对道路通行的影响。

另一突出特点是在考虑了非机动车和行人对机动车行驶的干扰及交叉口延误时间因素后，对 BPR 路阻函数模型进行了改进。

针对问题 1，建立合适的评价指标体系，用以评价小区开放对周边道路通行的影响。首先，通过查阅大量文献资料，选取了诸多指标，初步建立繁琐的评价指标模型；进一步，使用层次分析法，利用 MATLAB，对指标进行筛选，得到小区开放对道路通行的影响的指标主要有道路宽度、通行量、出入口数量、车流量、道路等级、延误时间和交叉口饱和度。

针对问题 2，建立关于车辆通行的数学模型，用以研究小区开放对周边道路通行的影响。首先，从小区内部因素、小区外部因素及路阻因素三个方面着手，分别构建三个函数关系方程；然后，按照一定的权重对各因素赋值；最终，建立小区开放对周边道路通行影响的总函数关系模型。$Y = 0.6738 Y_1 + 0.2255 Y_2 - 0.1007 Y_3 = 0.6738 \cdot P_0 \cdot \varphi_i \cdot (0.3120s - 0.0850) + 0.2255 \cdot \theta_i \cdot \dfrac{10 + 5\eta}{0.055} - 0.1007 \cdot \left[1 + 2.5 \left(\dfrac{p}{c}\right)^{4.0} + c'\right] / 0.055$。

针对问题 3，应用建立的总函数关系模型，定量比较位于城市不同区域、具有不同结构的小区开放前后对道路通行的影响（=1,2,3，表示城市的区域类型；=1,2,3,4，表示小区的结构类型）。搜集大量数据，代入方程，则得到结论：当小区处于城市核心区时，选择开放 C 型结构最佳；当小区处于中间城区时，选择开放 A 型结构最佳；当小区处于城市郊区时，可根据实际车流量情况，选择开放 C 型结构小区或封闭小区。

针对问题 4，从交通通行的角度，向城市规划和交通管理部门提出关于小区开放

---

① 本文获 2016 年全国一等奖。队员：吴梦晗，李咏馨，厉培培；指导教师：闫云侠。

的建议。综合以上的研究结果,主要给出针对城市规划和交通管理两方面的建议(共11条)。

最后,我们对结果进行了误差分析,调整部分参数对结果的影响给出了灵敏度分析,同时还对模型的优点和不足之处进行了评价,并在横向和纵向上对模型适用性进行了推广,最后对改进模型中相关量做了一定的构想。

关键词:小区开放;评价指标体系;函数关系模型;BPR 路阻函数模型;灵敏度分析;MATLAB

4.1 问题的重述

4.1.1 引言

1. 背景介绍

当今社会,随着国家政府不断深入推动城市化进程,城市发展速度飞快,建设活动的步伐也迅速迈进,尤其体现为对居民住区建设的改进。为进一步加强对城市规划建设的管理,传统的封闭式住宅小区因其自身的缺陷将不再符合时代发展的需求,因此将封闭式小区开放化将成为当下住区开发建设的主流模式。

2. 提出问题

将传统封闭式小区开放后,可能会引发小区内安保问题,也可能会影响主路通行速度,但同样也会在一定程度上改善城市的交通状况,提高路网密度,增加道路面积,提升通行能力。因此我们应就小区开放对周边道路通行的影响进行研究,从而评价出小区开放后是否改善了原有交通状况及改善效果如何。

3. 研究的意义

开放封闭式小区,推广街区制是对当代居民传统居住环境的一次较大变革,小区规划与我们的日常生活息息相关,尤其是对我们周边道路通行有影响,因此,通过结合实际对其进行研究,有利于制定合理有效的住区开发建设方案,从而实现优化街区建设、促进邻里关系、促进城市内部道路公共化、增大城市路网密度及提高道路通达性和土地利用率等目标。

4.1.2 相关数据

(1)GB 50220.1995,《城市道路交通规划设计规范》;
(2)CJJ 37.2012,《城市道路工程设计规范》;
(3)GB 50180,《城市居住区规划设计规范》。

4.1.3 具体问题

1. 问题一

选取合适的评价指标体系,评价小区开放对周边道路通行的影响。

2. 问题二

建立关于车辆通行的数学模型,研究小区开放对周边道路通行的影响。

3. 问题三

选取不同类型的小区,结合小区结构及周边道路结构、车流量等因素,应用问题二中建立的模型,定量比较这些小区开放前后对道路通行的影响。

4. 问题四

根据上述研究结果,从交通通行的角度向城市规划和交通管理部门提出关于小区开放的合理化建议。

4.2 问题的分析

4.2.1 具体问题的分析和对策

1. 问题一的分析和对策

问题一要求我们选取合适的指标体系来评价小区开放对周边道路通行的影响,考虑到小区开放后会有很多因素对其造成影响,我们考虑把这些因素归类为三个方面,分别是小区内部因素、小区外部因素和其他路阻因素,通过查阅大量资料,建立合适的指标体系。为评价小区开放后对周边道路通行的影响,考虑运用层次分析法,建立合理的数学模型并求解,整理并总结求解结果。

2. 问题二的分析和对策

问题二要求我们建立关于车辆通行的数学模型,用以研究小区开放对周边道路通行的影响。在问题一研究的基础上,我们继续从三个方面着手,构建与小区开放对周边道路通行总影响的函数关系模型。分别建立其对小区开放后道路通行影响效果的函数模型,然后将其按照权重进行结合,最终构建出小区开放对周边道路通行总体影响的总函数关系模型。

3. 问题三的分析和对策

问题三要求我们对具有不同结构、周边道路结构及车流量的小区开放前后对道路通行的影响进行定量分析。经查阅大量资料,我们将城市小区结构分为 A 型、B 型、C 型以及 D 型,按照城市道路网密度由高到低依次将城市各区域分为 Ⅰ 区、Ⅱ 区、Ⅲ 区。应用问题二中建立的模型,代入数据进行计算,定量比较出各类型小区开放前后对道路通行的影响,并得出当小区位于城市的不同区域时,能使小区开放后对道路通行影响最佳的小区结构类型。

4. 问题四的分析和对策

问题四要求我们就有关小区开放问题,向城市规划和交通管理部门提出建议,结合上述研究结果,查阅大量资料,结合生活实际,从交通通行的角度给出合理化的建议。

4.2.2 对问题的总体分析及解题思路

小区开放对道路通行的影响问题,是定性到定量地研究各类型小区开放前后对周边道路通行状况的影响问题。对本问题处理要分四个步骤进行:第一,选取合理指标,建立小区开放对周边道路通行的影响的评价指标体系;第二,针对指标体系的三个方面,分别建立函数模型,然后将其按照权重进行结合,构建出基于车辆通行模型,且能衡量小区开放对周边道路通行总体影响效果的总函数关系方程式;第三,在城市的不同分区,定量计算不同小区开放前后对道路通行的影响,判断得各城市区位哪一种类型的小区最适合进行开放,并给出决策方案;第四,整理结论,从交通通行的角度,向城市规划和交通管理部门提出关于小区开放的合理化建议。

本文的总体解题思路如图4.1所示。

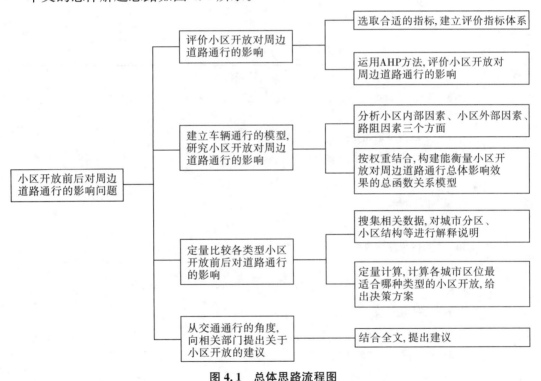

图4.1 总体思路流程图

4.3 模型的假设

(1)假设忽略交通外环境对道路通行能力的影响(外环境、气候等);
(2)假设不考虑Braess现象的发生;

(3) 假设 A、B、C、D 型小区面积相等,外形相同;
(4) 假设小区开放后内部道路均视为城市支路。

4.4 名词解释与符号说明

4.4.1 名词解释

1. 交叉口饱和度

交叉口测得的实际交通量与该交叉口设计时的理论通行能力的比值。

2. 车道宽度折减系数

车道宽度增加或者减少将会影响车道的通行能力,系数表示为该车道宽度下的通行能力与标准车道宽度下通行能力的比值。

3. 路阻

车辆在道路行驶过程中各种阻抗因素的集合。

4.4.2 符号说明

表 4.1 主要变量符号及意义

序号	符号	符号说明
1	Y_{ij}	小区开放前后对周边道路通行影响($i=1,2,3$,表示城市的区域类型;$j=1,2,3,4$,表示小区的结构类型)
2	φ_i	第 i 类小区内部道路密度系数
3	P_0	一条车道上某点的理论车流量
4	α_1	车道折减系数
5	α_2	车道宽度折减系数
6	s	单条车道宽度
7	θ_i	第 i 类街区道路密度系数
8	h	单位时间
9	F	单位长度
10	L	车辆道路占用长度
11	$l_{车距}$	设定的平均安全车距
12	$l_{车身}$	设定的平均车身长度
13	v	车速
14	η	道路等级
15	p	通过某路段的实际交通量
16	c	路段的理论通行能力
17	c'	模型的修正系数

续表

序号	符号	符号说明
18	a,b	模型的待定参数
19	t	路阻存在情况下,通过某一段道路所需的实际时间
20	t_0	存在任何阻碍因素情况下,通过某一段道路所需的时间
21	ω	权重系数
22	Z	交叉口饱和度

4.5 模型的建立与求解

4.5.1 问题一的分析及求解

1. 对问题一的分析

问题一要求我们选取合适的指标体系来评价小区开放对周边道路通行的影响,考虑到小区开放后会有很多因素对其造成影响,我们把这些因素归类为主要的三个方面,分别是小区内部因素、小区外部因素和其他路阻因素。

为了更具体地分析并解决问题,我们将这三方面继续细分,决定选取小区面积、小区道路宽度、小区道路通行量、小区出入口数量、小区地理位置、小区周边道路车流量、小区周边道路等级、延误时间及交叉口饱和度这九个量作为评价指标,如图 4.2。在选取了这些指标后,我们建立了层次分析模型,运用 MATLAB 软件进行求解,通过对运行结果的分析,最终给出小区开放后对周边道路通行影响的评价。

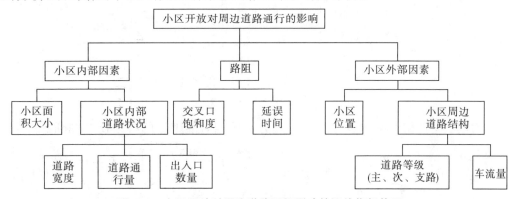

图 4.2 小区开放对周边道路通行影响的评价指标体系

2. 对问题一的求解

模型Ⅰ——层次分析模型(AHP)

根据上述建立的指标体系,我们主要选取小区道路宽度、小区道路通行量、小区出入口数量、小区周边道路车流量、小区周边道路等级、延误时间及交叉口饱和度这七个指标来评价小区开放对周边道路通行的影响。

(1)模型的准备。

建立方案评价的递阶层次模型：

第一层为总目标。T：改善道路通行状况；

第二层为方案评价的准则层。I_1：道路宽度，I_2：通行量，I_3：出入口数量，I_4：车流量，I_5：道路等级，I_6：交叉口通行能力，I_7：交叉口饱和度；

第三层为方案层。D_1：开放式小区，D_2 封闭式小区。其层次结构模型见图4.3。

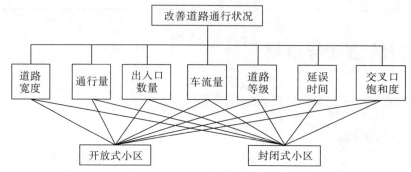

图4.3 改善道路通行的递阶层次结

(2)模型的建立。

设以改善道路通行状况为比较准则，I层各因素的两两比较判断矩阵为T，同样的，以每一个I_i为比较准则，D层次各因素的两两比较判断矩阵为I_i-D，$(i=1,2,3,\cdots,7)$。因此，得到的比较判断矩阵如下：

$$T=\begin{bmatrix} 1 & 2 & 3 & 4 & 5 & 4 & 3 \\ 1/2 & 1 & 4 & 5 & 6 & 5 & 4 \\ 1/3 & 1/4 & 1 & 3 & 2 & 3 & 1 \\ 1/4 & 1/5 & 1/3 & 1 & 6 & 1 & 1/4 \\ 1/5 & 1/6 & 1/2 & 1/6 & 1 & 1/3 & 1/4 \\ 1/4 & 1/5 & 1/3 & 1/5 & 3 & 1 & 1/3 \\ 1/3 & 1/4 & 1 & 4 & 4 & 3 & 1 \end{bmatrix}。$$

综合实际经验和参考文献分析后得到方案层相对准则层的各个比较判断矩阵为：

$$I_1-D=\begin{bmatrix} 1 & 6 \\ 1/6 & 1 \end{bmatrix}, \quad I_2-D=\begin{bmatrix} 1 & 5 \\ 1/5 & 1 \end{bmatrix}, \quad I_3-D=\begin{bmatrix} 1 & 5 \\ 1/5 & 1 \end{bmatrix}, \quad I_4-D=\begin{bmatrix} 1 & 4 \\ 1/4 & 1 \end{bmatrix},$$

$$I_5-D=\begin{bmatrix} 1 & 2 \\ 1/2 & 1 \end{bmatrix}, \quad I_6-D=\begin{bmatrix} 1 & 7 \\ 1/7 & 1 \end{bmatrix}, \quad I_7-D=\begin{bmatrix} 1 & 6 \\ 1/6 & 1 \end{bmatrix}。$$

(3)模型的求解。

①层次单排序及一致性检验。对于上述各比较判断矩阵，用MATLAB软件求出其最大的特征值及其对应的特征向量（程序详见附录），将其归一化后，得到层次单排序的权重向量，以及一致性指标CI和一致性比例CR，见表4.2。

表 4.2　改善道路通行状况

矩阵	层次单排序的权重向量	$\lambda_{\max}$	CI	RI	CR
$T-I$	$(0.3026,0.2964,0.1129,0.0708,0.0347,0.0510,0.1315)^{\mathrm{T}}$	7.5653	0.0942	1.24	0.0760
I_1-D	$(0.8571,0.1429)^{\mathrm{T}}$	2.0000	0	0.58	0
I_2-D	$(0.8333,0.1667)^{\mathrm{T}}$	2.0000	0	1.24	0
I_3-D	$(0.8333,0.1667)^{\mathrm{T}}$	2.0000	0	1.24	0
I_4-D	$(0.8000,0.2000)^{\mathrm{T}}$	2.0000	0	1.12	0
I_5-D	$(0.6667,0.3333)^{\mathrm{T}}$	2.0000	0	0.90	0
I_6-D	$(0.8750,0.1250)^{\mathrm{T}}$	2.0000	0	0.90	0
I_7-D	$(0.8571,0.1429)^{\mathrm{T}}$	2.0000	0	0.90	0

由此可见,所有层次单排序的 CR 值均小于 0.1,符合一致性要求。

②层次总排序。方案层相对于目标层的排序向量为

$$W = (d_1^{(7)},d_2^{(7)},d_3^{(7)},d_4^{(7)},d_5^{(7)},d_6^{(7)},d_7^{(7)}) \cdot \omega^{(2)} =$$

$$\begin{pmatrix} 0.8571 & 0.8333 & 0.8333 & 0.8000 & 0.6667 & 0.8750 & 0.8571 \\ 0.1429 & 0.1667 & 0.1667 & 0.2000 & 0.3333 & 0.1250 & 0.1429 \end{pmatrix} \cdot \begin{pmatrix} 0.3026 \\ 0.2964 \\ 0.1129 \\ 0.0708 \\ 0.0347 \\ 0.0510 \\ 0.1315 \end{pmatrix} =$$

$$(0.8375, 0.1624)^{\mathrm{T}}$$

③层次总排序的一致性检验。

由于 $CI^{(2)} = (CI_1^{(2)}, CI_{21}^{(2)}, CI_3^{(2)}, CI_4^{(2)}, CI_5^{(2)}, CI_6^{(2)}, CI_7^{(2)}) = (0,0,0,0,0,0,0)$

$RI^{(2)} = (RI_1^{(2)}, RI_2^{(2)}, RI_3^{(2)}, RI_4^{(2)}, RI_5^{(2)}, RI_6^{(2)}, RI_7^{(2)}) =$

$(0.58, 1.24, 1.24, 1.12, 0.90, 0.90, 0.90)$

因此 $CI^{(3)} CI^{(2)} \cdot \omega^{(2)} = 0$

$RI^{(3)} = RI^{(2)} \cdot \omega^{(2)} = 0.9578$

$$CR^{(3)} = CR^{(2)} + \frac{CI^{(3)}}{RI^{(3)}} = 0.0760 + \frac{0}{0.9578} = 0.0760 < 0.1$$

层次总排序通过一致性检验。

(4)结论。

为达到改善道路通行状况这一目标,以上两种方案的相对优先排序分别为①开放式小区,权重为 0.8375。②封闭式小区,权重为 0.1624。由此结果可得,开放式小区较封闭式小区能够有效地改善道路交通状况。

结合实际经验以及参考相关资料我们了解到,小区开放后,外部车辆就能充分利用

小区内部原有的道路,一方面减轻了一部分周边道路的交通压力,另一方面相当于从宏观上增加了外部道路的宽度,使道路通行量得到提高,车流量也随之增加,从而减少了延误时间,提高了主路的通行速度,改善了原有的交通状况。

4.5.2 问题二的分析及求解

1. 对问题二的分析

问题二要求我们建立关于车辆通行的数学模型,用以研究小区开放对周边道路通行的影响。在问题一研究的基础上,首先分别建立小区内部因素、小区外部因素及路阻因素三个方面对小区开放后道路通行的影响 Y 的函数模型,然后将其按照权重进行结合,最终建立小区开放对周边道路通行影响的总函数关系模型。

2. 对问题二的求解

模型Ⅱ——函数关系模型(能衡量小区开放对周边道路通行总体影响效果)

(1)模型的准备。

我们知道小区开放对周边道路通行的影响效果是一个抽象的量,为了能从定性到定量地研究小区开放对周边道路通行的影响,基于车辆通行,我们分别建立了小区内部因素、小区外部因素及路阻因素与其相关因素有关的三组函数关系方程,然后按照一定的权重,将其组合成一个关于小区开放后道路通行影响 Y 与各指标因素的影响即 Y_1、Y_2、Y_3 之间的总函数关系方程式,方便问题三的定量计算。

(2)模型的建立。

①总函数关系。

$$Y = f(Y_1, Y_2, Y_3)$$
$$Y = \omega_1 Y_1 + \omega_2 Y_2 + \omega_3 Y_3$$

②各因素与其相应指标的函数关系。

a. 小区内部各因素对道路通行的影响函数

$$Y_1 = f(P_0, \alpha_1, \alpha_2, \varphi_i)$$

在这里,我们选取车流量这个指标,作为小区内部因素对道路通行影响效果 Y_1 的衡量指标。车流量是指在一定的时间内,小区内部道路上某点所通过的车辆数,在这里表示小区内部道路的车流量,计算公式如下。

$$Y_1 = P_0 \cdot \alpha_1 \cdot \alpha_2 \cdot \varphi_i,$$

上式中,φ_i:第 i 类小区内部道路密度系数($i=1,2,3,\cdots,n$),P_0:一条车道上某点的理论车流量,由《城市道路设计规范》得到一条车道理论车流量(见表 4.3)中的建议值。

表 4.3 一条车道理论车流量

v(km/h)	20	30	40
P_0(pcu/h)	310	510	730

α_1:车道折减系数,查阅相关资料,得到不同车道的折减系数表(见表 4.4)。

表 4.4 车道折减系数表

车道	折减系数	车道	折减系数
第一条车道	1.00	第三条车道	0.65—0.78
第二条车道	0.80—0.89	第四条车道	0.50—0.65

(注:小区内部道路多为单车道,故取 $\alpha_1=1$)

α_2:车道宽度折减系数,得车道宽度折减系数表(见表 4.5)。

表 4.5 车道宽度折减系数表

车道宽度 s(m)	通行能力折减系数	车道宽度 s(m)	通行能力折减系数
3.50	1.00	3.00	0.85
3.25	0.94	2.75	0.77

由表 4.5 知将车道宽度折减系数 α_2 表示为车道宽度 s 的一元线性函数(在这里将 s 与实际生活中的数据相联系,扩大其使用范围):

$$\alpha_2 = 0.3120s - 0.0850 \quad (2.75 \leqslant s \leqslant 6)$$

综上 $Y_1 = P_0 \cdot \alpha_1 \cdot \alpha_2 \cdot \varphi_i = P_0 \cdot \varphi_i \cdot (0.3120s - 0.0850)(i=1,2,3,\cdots,n)$

b. 小区外部各因素对道路通行的影响函数

$$Y_2 = f(\theta_i, h, v, L)。$$

在这里,同样选取车流量这个指标,作为小区外部因素 Y_2 对道路通行影响效果的衡量指标。表示小区周围道路的车流量。

$$Y_2 = \theta_i \cdot \frac{h \cdot v}{L} = \theta_i \cdot \frac{h \cdot f(\eta)}{L}$$

θ_i:第 i 类城市分区道路密度系数$(i=1,2,3,\cdots,n)$;

h:单位时间/小时;

L:车辆道路占用长度,表示为车距与车身长度之和,即 $L=l_{车距}+l_{车身长}$(在模型中设定安全车距均为 50 m,车身长均为 5 m),所以 $L=0.055$ km;

v:车速(单位:km/h);

η:道路等级,其中主干路为 8,次干路为 5,支路为 1。

查阅《城市道路设计规范》,得到由道路等级确定的车速表 4.6。

表 4.6 道路等级设计速度表

道路等级	主干路	次干路	支路
设计速度(km/h)	50	35	15

由表 4.6 将车速 v 表示为道路等级 η 的一元线性函数:$v=10+5\eta$。

综上 $Y_2 = \theta_i \cdot \dfrac{h \cdot v}{L} = \theta_i \cdot \dfrac{h \cdot f(\eta)}{l_{车距}+l_{车身}} = \theta_i \cdot \dfrac{10+5\eta}{0.055} \quad (i=1,2,3,\cdots,n)$。

c. 路阻因素对道路通行的影响函数(基于 BPR 路阻函数)

$$Y_3 = f(t, v, L)。$$

同前两问,选取车流量作为路阻因素 Y_3 对道路通行影响效果的衡量指标。当路阻存在时,在一定的时间内,未通过道路上某点的车辆数,车流量的计算公式如下。

$$Y_3 = \frac{t \cdot v}{L} = \frac{t \cdot v}{l_{车距} + l_{车身}},$$

上式中 t:路阻存在情况下,通过某一段道路所需的实际时间

$$t = t_0\left[1 + a\left(\frac{p}{c}\right)^b\right] + c' = \frac{F}{v}\left[1 + a\left(\frac{p}{c}\right)^b\right] + c',$$

(以上函数方程是对 BPR 路阻模型的改进形式)。

t_0:理想状态下,即不存在任何阻碍因素时,通过某一段道路所需的时间;

p/c:是通过某路段的实际交通量与该路段的理论通行能力的比值,即交叉口饱和度;

a、b:模型待定参数,在模型中,设定 a、b 的取值分别为 $a=2.5$,$b=4.0$;

c':模型的修正系数(可以理解为平均交叉口延误时间);

F/v:是单位长度与某一车速的比值,表示以某一车速通过某一单位距离所需的时间(模型中取单位长度为 1 km);

综上 $Y_3 = \dfrac{t \cdot v}{L} = \dfrac{t \cdot v}{l_{车距} + l_{车身}} = \dfrac{\left\{\dfrac{F}{v}\left[1 + a\left(\dfrac{p}{c}\right)^b\right] + c'\right\} \cdot v}{l_{车距} + l_{车身}} = \dfrac{1 + 2.5\left(\dfrac{p}{c}\right)^{4.0} + c'}{0.055}$。

(3)模型的求解。

以上分别建立了小区开放对周边道路通行的影响与小区内部因素、小区外部因素及路阻因素之间的函数模型 Y_1、Y_2、Y_3,现在应将这三个函数模型综合起来,建立小区开放对周边道路通行的总影响 Y 与 Y_1、Y_2、Y_3 之间的函数关系模型。

因此,需要考虑赋予这三方面一定的权重指数,建立起 Y 与 Y_1,Y_2,Y_3 之间的函数模型。通过查阅大量资料,赋予 Y_1,Y_2,Y_3 的权重指数分别为 0.6738,0.2255 和 0.1007。即 $(\omega_1, \omega_2, \omega_3) = (0.6738, 0.2255, 0.1007)$。

综上,得到小区开放对周边道路通行影响的总函数关系方程式,如下:

$Y = 0.6738 Y_1 + 0.2255 Y_2 - 0.1007 Y_3 =$

$\qquad 0.6738 \cdot P_0 \cdot \varphi_i \cdot (0.3120s - 0.0850) + 0.2255 \cdot \theta_i \cdot \dfrac{10 + 5\eta}{0.055} -$

$\qquad 0.1007 \cdot \left[1 + 2.5\left(\dfrac{p}{c}\right)^{4.0} + c'\right]/0.055$

4.5.3 问题三的分析及求解

1. 对问题三的分析

本题要求我们对具有不同结构、周边道路结构及车流量的小区开放前后对道路通行的影响进行定量分析。

经过查阅大量资料,我们将城市分为三个分区,分别是城市核心区(Ⅰ型),中间城区(Ⅱ型),城市郊区(Ⅲ型),道路网密度依次降低。小区位于城市的不同分区,其开放后的周边道路情况会有不同。

小区结构分为四种类型,即 A 型(密网—格网状)、B 型(疏网—格网状)、C 型(密网—环网状)以及 D 型(疏网—环网状)。应用问题二中建立的模型,代入小区开放前后相关数据进行计算,定量比较出位于不同区域的,具有不同结构的小区开放前后对道路通行的影响,并得出对道路通行的最佳影响。

2. 对问题三的求解

(1)小区内部因素。

【小区结构】

我们对现有的小区结构进行总结,将小区路网按照密度分为两类:密网(小区路网长度密度大于等于 12 km/km²),疏网(小区路网长度密度小于 12 km/km²);按照内部连接关系,可分为格网状和环网状。

路网连续是保障小区道路可供穿越和通行功能的基础,也是道路可达性评价的基本要素。小区路网的长度密度决定了住区中可以提供的路径长度,在同样的路网密度条件下,"十"字形交叉口比"T"形交叉口具有更强的连接性和渗透性。对住区街道网络形态连接结构的分析主要从其断头路数量、交叉口形式、路网长度密度来进行分析。

综合路网密度和内部连接状况,可将小区结构分为"密网—格网状""疏网—格网状""密网—环网状"以及"疏网—环网状"四种类型,则小区结构简图及其对应的实例小区地图示意如下:

①格网模式。由若干条贯通式的道路纵横交错组成。这种道路形成的居住区拥有多个出入口,住宅群均匀地分布在网格形状的块状空间内。按照小区路网的长度密度分类如下:

A 型:密网—格网状

实例:上海奥林匹克花园

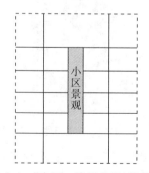 

图 4.4 "密网—格网状"结构简图　　图 4.5 上海奥林匹克花园小区结构图

特点:街道网密度相对较大,网络系统中无断头路或尽端道路,节点"十"字行交叉连接较多,路网连接性相对较好,如上海奥林匹克花园,该小区属于片块式,且小区内路网的长度密度较大。

B型：疏网—格网状

实例：昆明跨世纪经济示范居住区

图 4.6 "疏网—格网状"结构简图

图 4.7 昆明跨世纪经济示范居住区结构图

特点：街道网密度相对较低，街道宽而疏，节点以"十"字形交叉为主，街道相互连接成网格状，如昆明跨世纪经济示范居住区。该小区属于片块式布局，采用了人车混行的格网模式道路。三条城市干道位于周边，东面临河流。共有六个出入口，五条路径将整个小区划分为五个大块，即五个不同的住宅群，住宅群间依靠宅间小路相互连接。整个小区布局完整，道路网结构清晰。

②环网状模式。

其分为内环模式与外环模式。内环模式指由一条环通式和若干尽端式道路组成，环通式的道路穿过居住区中部，尽端道路分布在环通式道路的周边。这种道路形成的居住区通常拥有两个出入口，住宅群分布在尽端式道路附近的空间内。外环模式与其区别在于其环通式的道路分布在居住区边缘。

C型：密网—环网状

实例：兰州市黄河家园住宅区

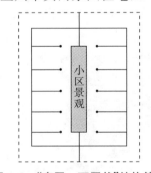

图 4.8 "密网—环网状"结构简图

图 4.9 兰州市黄河家园住宅区结构图

特点：小区路网长度密度较大，"T"行交叉口较多，且存在一定数量的断头路、尽端路等，路径连接性差，街道网络整体渗透性不强。如兰州市黄河家园住宅区，该小区采用了人车混行的外环模式道路。三条城市干道位于周边，小区共有南北两个主要出入口，一条外环路环绕整个小区的外部。从外环路的单侧发散出若干次路径联系各片住宅群。整个小区营造出了向内围合的舒适空间,内部有环形的河流网,河流周围布置景观。整个小区内的住户可以共享小区内的河流。

D型:疏网—环网状

实例:苏州市姑苏雅苑

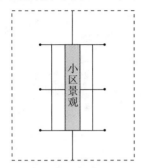

图 4.10 "密网—环网状"结构简图　　图 4.11 苏州市姑苏雅苑小区结构图

特点:小区路网长度密度相对较低,等级衔接以网状和树枝状相结合为主,"T"行交叉较多。如苏州市姑苏雅苑小区,该小区采用了人车混行的内环模式道路。整个小区围绕着中央湖泊水景。小区有两个主要出入口,一条主道路环绕整个小区的内部。从内环路上发散出若干次路径联系各片住宅群。

【道路宽度】

根据"窄路密网""宽路宽网"的理论,我们可以知道 A 型密网—格网状和 C 型密网—环网状的道路宽度较窄,B 型疏网—格网状和 D 型疏网—环网状的小区道路宽度较宽。单车道取值范围定为[3,6],则道路宽度取值范围为[6,12]。则 A、B、C、D 型小区的道路宽度依次为 6 m、9 m、7 m、10 m。

(2)小区外部因素。

【城市分区(表现小区所处的位置)】

接下来我们对小区的外部因素进行分析,即当小区处于所在城市的不同位置时,其周边的道路等级情况不同,这导致车辆的平均行驶速度有所区别。

经过查阅大量的资料,我们按照路网密度高低将城市分为三个分区,第一分区是 I 型城市核心区(路网的长度密度 12～16 km/km²),第二分区是 II 型中间城区(路网密度 7～10 km/km²),第三分区是 III 型城市郊区(路网密度 6～9 km/km²)。这属于小区的位置因素。经整理计算得到这三类城市分区道路密度系数 θ_i 的值分别为 $\theta_1=0.4667$,$\theta_2=0.2833$,$\theta_3=0.2500$。

【道路等级分布情况】

通过大量比较我国各城市的城市道路网规划图,以国内 X 市城市道路网专项规划图(即图 4.12)为例,我们总结出以下结论:

道路等级分布情况 $\begin{cases} 第一分区(城市核心区):主干路,次干路,支路 \\ 第二分区(中间城区):主干路,支路 \\ 第三分区(城市郊区):主干路 \end{cases}$

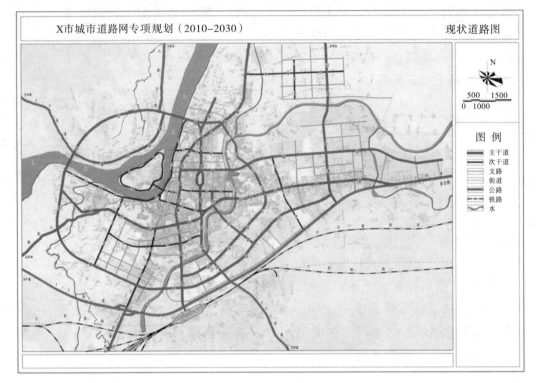

图 4.12　X 市城市道路网专项规划图

各道路等级在每个城市分区中的分布情况并不相等，由问题二可知主干路、次干路、支路的道路等级系数分别为 8、5、1。结合实际生活，并查阅大量文献资料后，得第一分区的综合道路等级系数 $\eta_1=7$，第二分区的 $\eta_2=11$，第三分区的 $\eta_3=14$。

再根据问题二中道路等级与车辆通行平均速度的公式 $v=10+5\eta$，可计算得到第一分区的车辆平均行驶速度为 $v_1=45\ \text{km/h}$，第二分区 $v_2=65\ \text{km/h}$，而在第三分区时 $v_3=80\ \text{km/h}$。

(3) 路阻。

在计算路阻时，需要知道各类型小区在城市不同分区时，其交叉口饱和度不同。这里的交叉口饱和度是指考虑小区结构(主要是出入口数量及交叉口数量)和小区位置的综合交叉口饱和度 Z。计算公式如下：

$$Z = p/c \cdot \zeta$$

这里 P/C 代表小区位置对综合交叉口饱和度 Z 的影响，ζ 为交叉口折算系数，代表小区结构(主要是出入口数量及交叉口数量)对 Z 的影响。

道路饱和度是反映道路服务水平的重要指标之一，值越高，代表道路服务水平越低。由于道路服务水平、拥挤程度受多方面因素的制约，实际中因难以考虑多方面因素，常以饱和度数值作为评价服务水平的主要指标。美国的《通行能力手册》将道路的服务水平根据饱和度等指标的不同分为六级，我国则一般根据饱和度值将道路拥挤程度、服务水平分为四级。

$\begin{cases} 一级服务水平:道路交通顺畅、服务水平好;P/C 介于 0 和 0.6 之间 \\ 二级服务水平:道路稍有拥堵,服务水平较高;P/C 介于 0.6 和 0.8 之间 \\ 三级服务水平:道路拥堵,服务水平较差;P/C 介于 0.8 和 1.0 之间 \\ 四级服务水平:道路严重拥堵,服务水平极差;P/C>1.0 \end{cases}$

我们可以根据各城市分区的道路拥挤程度,得到相应的城区服务水平。则城市核心区、中间城区和城市郊区的服务水平依次为三级、二级、一级。我们这里以均值来表示每个城市分区的综合交叉口饱和度 Z 的其中一个因素。

下面我们考虑小区结构对综合交叉口饱和度 Z 的影响。不妨以我们的小区结构简示图为例进行说明,统计得 A、B、C、D 型小区中的出入口数量和交叉口数量之和,则得 $n_A:n_B:n_C:n_D=12:6:9:5$,对应的交叉口折减系数的 $\zeta_A:\zeta_B:\zeta_C:\zeta_D=15:30:20:36$,按比例可得其在 $[0,1]$ 范围内的数值,则 $\zeta_A=0.3, \zeta_B=0.6, \zeta_C=0.4, \zeta_D=0.72$。

其他小区结构的交叉口折算系数以此类推。

综上,可得综合交叉口饱和度 Z 在考虑小区结构(主要是出入口数量及交叉口数量)、小区位置以及小区是否开放条件下的 15 种取值。如表 4.7 所示。

表 4.7 综合交叉口饱和度 Z 取值汇总表

综合交叉口饱和度 Z		城市核心区	中间城区	城市郊区
小区未开放		0.9	0.7	0.3
小区开放 (考虑四种小区结构)	A 型	0.27	0.21	0.09
	B 型	0.54	0.42	0.18
	C 型	0.36	0.28	0.12
	D 型	0.648	0.504	0.216

初步总结:经过以上分析,我们对本问题进行信息的基本汇总,见表 4.8。

表 4.8 小区类型及相关信息汇总表

小区类型及相关信息	A 型	B 型	C 型	D 型
基本假设	假设各小区外形和面积相同,仅有内部结构不同			
小区结构名称	密网—格网状	疏网—格网状	密网—环网状	疏网—环网状
结构简图				
实例小区	上海奥林匹克花园	昆明跨世纪经济示范居住区	兰州市黄河家园住宅区	苏州市姑苏雅苑
道路网的长度密度	高	低	高	低
小区道路宽度(m)	6	9	7	10
研究目的	在城市的不同分区,定量比较出位于不同区域的,具有不同结构的小区开放前后对道路通行的影响,并得出其对道路通行的最佳影响。			

下面根据问题二构建的总体函数关系模型,代入相关数据进行计算,可定量比较出各类型小区开放前后对道路通行的影响。这里我们针对城市的三种分区,即城市核心区、中心城区和城市郊区,分别进行4种类型小区的定量计算,可知最后得3组数据,每组数据各有5个,分别是小区未开放时对周边道路交通的影响量Y_{i0}($i=1、2、3$)和小区开放后对周边道路交通的影响量Y_{ij}($i=1、2、3$,表示城市的区域类型;$j=1、2、3、4$,表示小区的结构类型),依题意i表示三种城市区位,j表示四种小区结构类型。

问题二中求得的公式为:

小区开放前:

$$Y_{i0} = 0.5089Y_2 + 0.4911Y_3 =$$
$$0.5089 \cdot \theta_i \cdot \frac{10+5\eta}{0.055} + 0.4911 \cdot \left[1 + 2.5\left(\frac{p}{c}\right)^{4.0} + c'\right]/0.055$$

小区开放后:

$$Y_{i0} = 0.5089Y_2 + 0.4911Y_3 =$$
$$0.5089 \cdot \theta_i \cdot \frac{10+5\eta}{0.055} + 0.4911 \cdot \left[1 + 2.5\left(\frac{p}{c}\right)^{4.0} + c'\right]/0.055$$

根据以上公式,使用 MATLAB 可计算得出当考虑小区位置时,小区开放前后对道路通行的影响效果大小表,见表4.9。Y_{ij}的数值越大,说明小区开放前后对道路通行的影响效果越大,促进了周边道路的通行能力。

表4.9 小区开放前后对道路通行的影响效果大小表

小区开放前后对道路通行的影响效果大小 (考虑小区位置)		城市核心区	中间城区	城市郊区
小区开放前		171.2190	156.4959	176.3650
小区开放后 (考虑四种小区结构)	A型	178.3276	167.7363	174.2454
	B型	168.6907	158.3308	164.9690
	C型	192.8906	182.3327	188.8604
	D型	173.2342	163.1392	169.9251

表4.10 小区开放前后对道路通行的影响效果大小的相对改变量

相对改变量	城市核心区	中间城区	城市郊区
A型	4.15%	7.18%	−1.20%
B型	−1.48%	1.17%	−6.46%
C型	12.66%	16.51%	7.08%
D型	1.18%	4.25%	−3.65%

定量比较出位于不同区域的,具有不同结构的小区开放前后对道路通行的影响,并计算得其数值的相对改变量,见表4.10,最后得出其对道路通行的最佳影响。

结论:

(1) 当小区处于城市核心区时,选择开放 C 型最佳。

由数值增加值和相对改变量可以知道, A 型、C 型和 D 型都促进了道路通行能力, 都产生了一定的影响效果。考虑到该小区处于城市核心区,其拥堵程度和生活成本都是较高的,因此选择开放 C 型小区可以大大缓解市中心的拥堵情况,选择开放 C 型小区是最适合城市核心区的决策方案。

(2) 当小区处于中间城区时,选择开放 A 型最佳。

处于中间城区时,四种类型的小区开放均可以使 $Y_{ij}(i=1,2,3,j=1,2,3,4)$ 的值增大,促进了周边道路的通行能力。效果最好的是 C 型,其次是 A 型。

在不考虑建设小区费用等成本的情况下,选择开放 C 型最佳;但在实际生活中,由于 C 型的小区内部道路的长度密度较大,建设与后期维修的成本较高,且将 A 型对中间城区和城市核心区的影响效果值进行比较,发现其对中间城区的影响效果更佳,更能促进中间城区周边道路的通行能力。且中间城区的交通拥挤程度适中,因此,若结合实际生活进行分析,选择开放 A 型小区是较适合中间城区的决策方案。

(3) 当小区处于城市郊区时,可根据实际车流量情况,选择开放 C 型小区或不开放小区。

由表 4.9 和表 4.10 可得,只有当开放 C 型小区时,周边道路通行情况才会得到改善,其他三种会产生负影响。

在此,我们对城市郊区的车流量进行讨论,若其实际车流量较大,则可以选择开放 C 型小区来促进周边道路的通行能力,产生正的影响效果;若其实际车流量较小,开放小区前后对周边道路交通情况并没有明显变化,如表 4.9 中显示 A 型、B 型、D 型小区的数值区别不大,同时考虑 C 型小区不仅建设成本较高,且其小区内部道路的后期维护费用也很昂贵,综合分析得,不开放小区也是城市郊区的一个合理决策方案。

4.5.4 问题四的分析及求解

1. 对问题四的分析

问题四要求我们就有关小区开放问题,向城市规划和交通管理部门提出建议,结合上述研究结果,查阅大量资料,结合生活实际,从交通通行的角度给出合理化的建议。

2. 对问题四的求解

关于小区开放的建议报告

我们通过以上研究了解到传统的封闭式小区阻碍了城市支路穿越,减少了城市道路系统中的支路条数,从而导致主、次干路交通压力过大,引发交通堵塞。而小区开放后将大大增加城市道路密度,有效改善城市交通状况。因此开放的小区住宅形式才能逐步适应发展迅速的当代社会。然而,小区开放后也会为周边带来一系列的影响,如小区开放后内部安保问题等。为解决这些问题,我们向城市规划和交通管理部门提出部分合理化的建议如下。

1. 向城市规划部门提出的建议

(1)城市规划部门应将城市内已建成的封闭住宅小区及单位大院逐步开放。为大力开发建设开放式小区做好铺垫，这有利于增强城市内部系统完整性，充分利用土地资源，减少重复建设，推动城市的可持续发展。

(2)城市规划部门应实施区域开发的开放式住区建设模式。所谓区域开发，是指以和政府部门合作的形式，对城市进行区域整体规划，合理分配周边资源，并正确引导协调各开发商对住区模式的开发建设。这种模式有利于促进开发商建设出更高水平的住宅小区，同时更有利于人民群众对其建设过程进行监督，维护居民的个人权益。

(3)城市规划部门应调整居住区的组织结构，将传统相互孤立的小区结构建设为开放的居住群，建设出开放便捷、尺度适宜、配套完善、邻里和谐的生活街区。

(4)城市规划部门应广招专业技术人员，建立一个完善的小区建设效应衡量体系，比如在城市不同的地理位置建设哪种开放结构的小区能使效应最优化，根据建立的体系就可以迅速对其作出评判抉择。

2. 向交通管理部门提出的建议

(1)加强小区内部道路标准化改造与管理。首先，由于小区内部道路具有非公共性特征，导致其很可能不符合通行道路标准，交通管理部门应严格进行筛选，对符合公共道路通行标准的小区内部道路进行道路标准画线，标志张贴与树立，加强对相应道路的监管，实现小区内部道路管理常态化，将其正式纳入城市路网；建立小区内部道路监控系统、信号系统；对转弯半径、驾驶者视野、道路宽度不合格等问题经改造后仍不能达到公路通行标准的内部道路，不应将其纳入城市路网，仅保留其作为小区内部通行道路的功能。

(2)加强小区内部道路与城市公共道路连接处的管理。对驶入小区内部道路的驾驶者应在路标提醒、信号灯等方面给予充分的提示，以保证驾驶者能提前准备适应进入不同状况条件的道路；

(3)加强在原公共道路的标志提醒。完善相应路标，及时更新道路指示牌所提示的道路结构。妥善解决已开放道路但社会公众知情率低而导致的小区内部道路使用率低问题。

(4)完善小区内部车辆管理。小区内部存在的多岔口问题，会严重影响公共化道路的通行能力，甚至阻碍小区外部道路的正常通行。因此，交通管理部门应相应制定小区内部车辆的通行措施，如设置专供内部车辆使用的道路或设置单独出入口等。加强对小区内部车辆停放的管理，对占据公共道路使用空间的车辆应当及时拖离，实行相当于正常公共道路的管理与处罚标准。

(5)在完善城市路网的同时，也应兼顾小区的职能。交通管理部门可统一设置区域内小区内部道路使用时间段，借助建立的路网监控系统进行实时监控，在非通行时间段禁止外来车辆的进入。

(6)设立完备的执勤体系,包括在通行能力与实际车流量较大的公共化后的内部道路设立固定执勤站点和建立执勤流动小队,对小区内部道路实行动态化管理。

(7)设置隔离带。在向外界开放的小区内部道路与小区内部公共生活休闲区域,建立软性隔离带,在保证小区环境美观性的情况下,一方面既能保护小区住户的安全,另一方面也能给驾驶者营造一个舒适又安全的驾驶条件。同时为保证小区住户的通行需求,交通管理部门应尽量将人行道与汽车行驶道路分离。在不能分离的情况下,应在人行道上给予驾驶者醒目的标志提醒。

4.6 误差分析与灵敏度分析

4.6.1 误差分析

在问题一我们建立的小区开放对周边道路通行影响的评价指标体系中,我们只选取了一些主要的影响因素指标,实际对小区开放的周边道路通行有影响的因素还有很多,这可能对问题二、三中要求建立车辆通行的模型,对各类型小区开放前后对道路通行的定量比较有一定的影响,存在着不可避免的误差。

在问题二建立路阻因素对道路通行的影响效果函数时,我们使用了 BPR 路阻函数模型,在这里我们忽略了非机动车和行人对机动车行驶的干扰,存在一定的误差。我们在模型的改进中考虑了非机动车和行人对机动车行驶造成的干扰以及交叉口延误时间,对 BRP 路阻函数模型进行了改进。

4.6.2 灵敏度分析

在问题三小区开放对周边道路通行影响的函数关系模型中,计算总函数关系式 $Y = 0.6738Y_1 + 0.2255Y_2 - 0.1007Y_3$,其中 $Y_1 = P_0 \cdot \varphi_i \cdot (0.3120s - 0.0850)$,$Y_2 = \theta_i \cdot \frac{10+5\eta}{L} \eta$,$Y_3 = [1 + 2.5(p/c)^{4.0} + c']/L$ 时,为了计算简便,我们直接设定所有小区内部道路上某点的车流量 P_0 为 510 pcu/h,并且设定道路上所有行驶车辆之间的车距 $l_{车距}$ 为 50 m,车身长 $l_{车身长}$ 为 5 m,即车辆道路占用总长为 $L = 0.055$ km,但在实际生活中,不同车型车身长的值不同,而且实际道路行驶过程中车辆间车距也不会保持某个值不变,因此针对车流量 P_0 及车距 $l_{车距}$,车身长 $l_{车身长}$ 的不同取值我们运用 MATLAB 软件进行灵敏度分析,结果见表 4.11。

表 4.11　不同 P_0 值和 L 值下对于的 Y 值

车流量	0.035	0.045	0.055	0.065	0.075
310	192.4690	163.1934	144.5635	131.6659	122.2077
410	212.0590	182.7834	164.1535	151.2559	141.7977
510	231.6490	202.3734	183.7435	170.8459	161.3877
610	251.2390	221.9634	203.3335	190.4359	180.9777
710	270.8290	241.5534	222.9235	210.0259	200.5677

根据表 4.11 中的数据作图,可得 Y 对 P_0 和 L 的灵敏度分析图,见图 4.13。由灵敏度分析可知,当 P_0 值一定时,Y 值随着车辆道路占用长度的增大而减小,且减少量越来越小;当 P_0 值不同时,车辆道路占用长度在 0.035～0.055 km 时减少较快,0.055 km 之后,减少较慢。

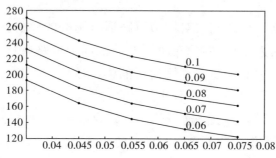

图 4.13　灵敏度分析图

4.7　模型的评价与推广

4.7.1　模型的评价

1. 优点

(1) 本文巧妙地运用流程图,将建模思路完整清晰地展现出来;

(2) 利用 EXCEL 软件对数据进行处理并作出图表,快捷、简便、直观;

(3) 采用层次分析法将研究对象作为一个系统,不隔断各个因素对结果的影响,同时将定性法和定量法有机地结合起来,量化每个因素对结果的影响程度,结果明确;

(4) 采用 BPR 路阻函数,建立道路通行时间与道路负荷之间的函数关系,转化为因路阻而导致的时间延误,解决路阻因素对周边道路通行能力影响的问题;

(5) 采用函数关系模型,建立不同道路状况指标与车流量之间的函数关系,方法独特,并且能有效衡量道路通行能力。

(6) 一方面将小区结构分为 A 型(密网—格网状)、B 型(疏网—格网状)、C 型(密网—环网状)以及 D 型(疏网—环网状)四种类型,另一方面按照城市道路网密度由高到低依次将城市各区域分为 Ⅰ 型城市核心区、Ⅱ 型中间城区、Ⅲ 型城市郊区,再定量比较得位于城市不同区域,具有不同结构的小区开放前后对道路通行的影响。

2. 缺点

(1)层次分析法定性成分较多,比较容易受主观判断影响;

(2)BPR 函数计算路段阻抗时,不能反映出交通状况由顺畅到堵塞的过程中交通量先增后减的事实,这与实际交通情况不相符;

(3)模型中为使计算简便,使所得结果更理想化,忽略了一些次要的影响因素,并对一些数据进行了近似处理。

4.7.2 模型的推广

层次分析法在经济、科技、文化、军事、环境乃至社会发展等方面的管理决策中都有广泛的应用。常用来解决诸如综合评价、选择决策方案、估计和预测、投入量的分配等问题,例如洗车方式、农业技术推广等实际生活问题。

4.8 模型的改进

在问题二中,我们建立了路阻因素对道路通行的影响函数,路阻最直接的表现效果就是会产生延误时间,延误时间为道路行驶延误时间和交叉口延误时间之和。

下面给出我们在问题二中用到的最基本的 BPR 路阻函数模型,公式如下:

$$t = t_0 \left[1 + a\left(\frac{p}{c}\right)^b\right]'$$

但是该模型只考虑了机动车交通负荷的影响,使用比较方便,在国内广泛使用于公路网规划,但由于国内城市道路上,除了机动车的交通负荷外还有非机动车的交通负荷,所以我们需要对其进行改进,来减少误差。

1. 考虑非机动车及行人对机动车通行的干扰

在实际生活中道路的通行能力不仅受自行车干扰,在小区内部道路没有设置实物分隔带时,对向车辆和行人横流也会对其进行干扰。我们参考了大量文献资料,总结出行人干扰系数表(见表 4.12)。

表 4.12　行人干扰系数表

干扰度	严重	较严重	一般	较少	无
k	0.3	0.5	0.7	0.9	1

当自行车交通量没有超过通行能力时系数取 0.7,而当超过通行能力时干扰系数可由下面公式计算得到。

$$k_b = 0.7 - \left(\frac{q_b}{Q_b} + 0.5 - W_j\right)/W_f$$

其中:K_b 为自行车对机动车通行能力的干扰系数,q_b 为某时刻该道路上实际的自行车(以及电动车)数量,Q_b 为自行车在非机动车道上的设计通行能力,W_f 和 W_j 分别为单向非机动车道的宽度以及单向机动车宽度。

根据德国和日本的线性函数,有 $t = t_0 + X_i h_i$,其中: h_i 为路段 i 上的流量,单位为 pcu/h。路段延误参数:指在路段 i 上的延误系数,有 $X_i = a \left(\dfrac{p}{c}\right)^b t_0$。

2. 考虑增添交叉口延误时间因素

关于交叉口的延误时间,我们可以计算其平均延误时间。其中:T 为交叉口的信号周期长度,单位 s;是 t_g 信号灯的有效绿灯时间,单位 s。综上,我们可以将其改进为 BPR 综合路阻函数模型,则有如下的计算公式:

$$y_A = \dfrac{0.5T\left(1 - \dfrac{t_g}{T}\right)}{1 - \left[\min\left(1, \dfrac{p}{c}\right) \cdot \dfrac{t_g}{T}\right]}, \quad T = \begin{cases} t_0 + a \left(\dfrac{p}{k \cdot k_b \cdot c}\right)^b \cdot h_i + y_A, & 0 \leq k_b \leq 1, \\ t_0 + a \left(\dfrac{k_b \cdot p}{k \cdot c}\right)^b \cdot h_i + y_A, & k_b > 1。 \end{cases}$$

参考文献

[1] 杨桂元. 数学建模[M]. 上海:上海财经大学,2015.157-181.

[2] 杨桂元,朱家明. 数学建模竞赛优秀论文评析[M]. 合肥:中国科学技术大学出版社,2013.1-12.

[3] 吴礼斌,李柏年,闫云侠. 经济数学实验与建模[M]. 北京:国防工业出版社,2013.202-212.

[4] 高亚楠. 基于交通微循环的住区街道空间优化策略研究[D]. 成都:西南交通大学,2015,(12).34-72.

[5] 李向朋. 城市交通拥堵对策——封闭型小区交通开放研究[D]. 长沙:长沙理工大学,2014.32-55.

[6] 刘远才,朱德滨. 城市小区道路交通与管理[J]. 林业建设,2001,03:29-33.

[7] 李飞. 对《城市居住区规划设计规范》(2002)中居住小区理论概念的再审视与调整[J]. 城市规划学刊,2011,03:96-102.

[8] 钟慧荣,顾雪平. 基于模糊层次分析法的黑启动方案评估及灵敏度分析[J]. 电力系统自动化,2010,16:34-49.

[9] 王燕,康睿,张卫东. 开放式社区交通微循环体系规划与运营[J]. 城市发展研究. 2012,19(08):102~106.

[10] 熊烈强. 路段通行能力及其服务水平指标的研究[J]. 武汉理工大学学报:交通科学与工程版,2014,28(4):511-514.

建模特色点评

❀**论文特色**❀

◆标题定位:"小区开放对道路通行的影响"标题即是赛题,没有突出特色,标题定位一般。

◆方法鉴赏:首先是层次分析法构建评价指标体系,其次权重组合法构建小区开放对周边道路通行总体影响的总函数关系(BPR 路阻函数)。最后针对不同小区,以案例式结合进行建模分析,这也是文章最大特色所在。

◆写作评析:内容结构合理,内容完整,思路清晰。

❀**不足之处**❀

摘要里面出现公式,细节方面存在问题,比如标点符号及格式首行缩进等方面小问题较多。

第 5 篇
"拍照赚钱"的任务定价

◆ **竞赛原题再现**

2017 年 B 题 "拍照赚钱"的任务定价

"拍照赚钱"是移动互联网下的一种自助式服务模式。用户下载 APP，注册成为 APP 的会员，然后从 APP 上领取需要拍照的任务(比如上超市去检查某种商品的上架情况)，赚取 APP 对任务所标定的酬金。这种基于移动互联网的自助式劳务众包平台，为企业提供各种商业检查和信息搜集，相比传统的市场调查方式可以大大节省调查成本，而且有效地保证了调查数据真实性，缩短了调查的周期。因此 APP 成为该平台运行的核心，而 APP 中的任务定价又是其核心要素。如果定价不合理，有的任务就会无人问津，而导致商品检查的失败。

附件一是一个已结束项目的任务数据，包含了每个任务的位置、定价和完成情况（"1"表示完成，"0"表示未完成）；附件二是会员信息数据，包含了会员的位置、信誉值、参考其信誉给出的任务开始预订时间和预订限额，原则上会员信誉越高，越优先开始挑选任务，其配额也就越大（任务分配时实际上是根据预订限额所占比例进行配发）；附件三是一个新的检查项目任务数据，只有任务的位置信息。请完成下面的问题：

1. 研究附件一中项目的任务定价规律，分析任务未完成的原因。

2. 为附件一中的项目设计新的任务定价方案，并和原方案进行比较。

3. 实际情况下，多个任务可能因为位置比较集中，导致用户会争相选择，一种考虑是将这些任务联合在一起打包发布。在这种考虑下，如何修改前面的定价模型，对最终的任务完成情况又有什么影响？

4. 对附件三中的新项目给出你的任务定价方案，并评价该方案的实施效果。

附件一：已结束项目任务数据；

附件二：会员信息数据；

附件三：新项目任务数据；

原题详见 2017 年全国大学生数学建模竞赛 B 题。

获奖论文精选

基于多元拟合和 Logistic 回归方法对任务定价的研究[①]

摘要: 本文针对移动互联网的劳务众包任务的定价问题,运用 K-Means 聚类、多元线性回归及 Logistic 二分类等方法,构建了基于会员位置的中心城区点确定模型,基于经验的任务定价模型及基于二分类回归的完成度测度模型,综合运用了 MATLAB、SPSS Modeler 及 STATA 等软件编程求解,得出了不同城区的任务定价规律、完成度预测结果及任务打包对定价的影响等结论,最后结合模型为新项目制定了任务定价方案并进行预测评估。

本文的特色是对经纬度数据的挖掘和可视化方法,使得信息更加直观、易于理解。

针对问题一,要求研究附件一中任务的定价规律,分析任务未完成的原因。首先,运用了数据挖掘的方法,对任务点的位置数据进行统计量化和可视化分析;其次,运用了快速聚类、多元线性拟合等理论,构建了三个不同城区的任务定价模型,运用 SPSS 和 STATA 等软件编程求解,得到了不同城区任务定价的规律及任务未完成的原因。

针对问题二,要求在问题一的基础上为附件一的项目设计新的定价方案,并与原方案进行比较。首先,分析新的定价方案应该满足的条件是提高任务完成度和控制单位成本;其次,在对完成任务数据的挖掘基础上,建立了基于经验的任务定价模型,并为未完成的任务制定了新的定价方案;最后,基于 Logistic 二分类理论建立了任务完成度的测度模型,并对优化后的定价方案的完成情况进行了预测,发现任务完成率由原来的 62.5% 提高到 86.7%。

针对问题三,要求考虑将位置比较集中的多个任务联合在一起打包分布,据此修改问题二中的定价模型,并研究对最终任务完成情况的影响。首先,根据任务点的分布,我们分三种情况来考虑:①不参与打包的任务点;②参与打包的任务点之间距离较近;③参与打包的任务点之间距离较远;其次,从不同方面分别分析这三种情况对定价模型的影响,对模型进行修改;最后,总体分析将任务打包分配对任务完成情况的影响。

针对问题四,要求为附件三中的新任务制定定价方案并对实施效果进行评价。本问题建立在问题一和问题二的基础之前,应用 K-Means 聚类模型和基于经验的多元拟合模型,运用 MATLAB 等软件编程求解,得到新项目的定价方案的实施效果,其中任务完成度比例为 87.5%,平均成本价格为 68.37 元。

本文最后还对模型进行了误差分析,对模型的优点和缺点进行了客观评价,对模型中不确定的量,如平均距离,进行了灵敏度分析。

[①] 本文获 2017 年全国一等奖。队员:倪梦莹、彭晓曼、李姗姗;指导教师:朱艳玲。

关键词：任务定价；Logistic 二分类；K-Means 聚类；多元线性回归；SPSS Modeler

5.1 问题的重述

5.1.1 背景知识

1. 引言部分

"拍照赚钱"是伴随着移动互联网的发展，出现的一种自助式服务模式。它基于移动互联网的自助式劳务众包平台，为企业各种商业检查和信息搜集提供服务，助力于 O2O 大潮中的数据采集和产品推广。相比传统的市场调查方式，这种自助式调查可以大大节省调查成本，提高效率的同时有效保证了调查数据的真实性，缩短了调查的周期，并且可以时刻呈现项目动态。

用户通过 APP 领取拍照任务，因此 APP 成为该平台运行的核心，而 APP 中的任务定价又决定了任务的完成情况。如果定价不合理，有的任务就会无人问津，从而导致商品检查的失败；如果定价太高，就会导致任务花费代价偏大。因此研究任务的合理定价，提高任务完成度，对于企业来说意义重大。

2. 任务定价

相同情况下，用户会优先选择完成定价相对较高的任务，如果定价不合理，会导致定价相对较低的任务执行失败。合理的任务定价会带动用户的积极性，即使有的任务距离较远，但由于定价高，用户也会选择完成，从而提高任务的完成度。通过各种商业检查，及时发现问题并改进，促进企业的发展。

3. 研究意义

首先，传统的调查方法存在很多弊端，尤其是人为因素。无法对调查员的行为进行标准化评估，其搜集信息的准确性也难以保证，且调查周期长，不能及时反映现实情况，调查效率低下。而"拍照赚钱"这种自助式服务模式，可以去中介化，重塑价值链，大幅度缩减了中间环节的人力成本，同时提高调查效率。

其次，自助式劳务众包平台可以发动全国各地的用户，去周边任务点采集商品的各类数据，为企业进行各种商业检查提供服务，对促进企业发展、提高企业的竞争力有着重要意义。

5.1.2 相关数据

(1) 已结束项目任务数据（见题目附件 1）；
(2) 会员信息数据（见题目附件 2）；
(3) 新项目任务数据（见题目附件 3）。

5.1.3 具体问题

1. 问题一

研究题目附件一中项目的任务定价规律,分析任务未完成的原因。

2. 问题二

为题目附件一中的项目设计新的任务定价方案,并和原方案进行比较。

3. 问题三

实际情况下,多个任务可能因为位置比较集中,导致用户会争相选择,一种考虑是将这些任务联合在一起打包发布。在这种考虑下,修改前面的定价模型,并研究对最终任务完成情况的影响。

4. 问题四

对题目附件三中的新项目给出任务定价方案,并评价该方案的实施效果。

5.2 问题的分析

5.2.1 研究现状综述

张鹏[1]对众包激励机制做了较深入的研究,通过考虑众包平台企业自营和第三方运营以及创新方案单独产出和联合产出两个维度,分四个象限进行分析,提出线性定价的定价机制,并推导出线性奖金的激励机制优于固定奖金的激励机制的结论。但由于是基于委托代理理论建模,未进行实证分析,不能证明模型的可行性。

Y. Singer and M. Mittal[2]曾提出一种在线定价机制,这种定价机制综合考虑了企业与会员两方面因素,一方面,企业不需要提前为每个任务设定价格,只需要提供任务的总体预算或任务的总数量;另一方面,当会员申请任务的时候,需要其给出期望的任务定价及期望完成的任务数量。这种在线定价策略虽然价格更加合理,但要对任务定价进行多次调整,增加了任务选择时间。

刘晓钢[3]利用中国最大的威客网的实际数据分析了任务自身属性以及市场竞争状况对任务标价的影响。研究得出,任务期限越长,任务期望获得的作品越多,近期相似任务的出价越高,任务的最终标价越高。想要降低劳务成本并节省时间、保证质量,在定价时应该从相似任务的标价、任务难度、任务期限、外包报价等方面来综合考虑悬赏金额以激励用户参与任务的解决。但没有对其他典型的众包网站进行对比分析,因此研究具有片面性,不足以保证研究结论的普适性。

总而言之,现有的文献或多或少都有其不足之处,相关研究大都基于理论层面,缺乏经验及数据支持,需要加以完善。

5.2.2 对问题的总体分析和解题思路

本文是研究自助式劳务众包平台任务定价的问题,针对此问题,我们分为四个小问题来研究。第一:选取合适的指标,研究题目附件一中项目的任务定价规律,并分析任务未完成的原因;第二:基于 Logistic 二分类理论建立了任务完成度的测度模型,并对优化后的定价方案的完成情况进行了预测;第三:考虑三种不同的打包类型,并分别在问题二建立的定价模型的基础上进行修改,然后研究其对最终任务完成情况的影响;第四:根据问题二的定价模型,为附件三中的新项目制定任务定价方案,并评价方案的实施效果。

5.2.3 对具体问题的分析和对策

1. 对问题一的分析和对策

问题一要求研究题目附件一中项目的任务定价规律,并分析任务未完成的原因。针对这一问题,我们分为两个步骤来解决。首先,我们对附件一中的数据进行数据挖掘,使用 SPSS Modler14.1 软件和百度地图对任务点坐标数据进行简单的统计和量化,分析其基于定价和完成度的位置分布图;其次我们选取合适的会影响定价的指标,分别为单位范围内与任务点最近的会员的距离以及会员的密度,进行相关性检验和多元线性拟合,得到附件一中的任务定价规律;最后,我们筛选出未完成任务的数据,对其相应指标进行分析,总结任务未完成的原因。

2. 对问题二的分析和对策

问题二要求为题目附件一中的项目设计新的任务定价方案,并和原方案进行比较。针对这一问题,我们分为两个方面来解决。首先,统计 522 项已完成任务的数据信息,对其进行数据挖掘,得到归于函数的定价规律,在此基础上添加可能对定价产生显著影响的指标,建立新的定价规律模型;其次,将新的定价规律模型应用到附件一的项目中,将其结果与原方案进行比较。

3. 对问题三的分析和对策

问题三要求考虑将位置比较集中的多个任务联合在一起打包分布,据此修改问题二中的定价模型,并研究对最终任务完成情况的影响。针对此问题,首先,根据任务点的分布,我们分三种情况来考虑:①不参与打包的任务点;②参与打包的任务点之间距离较近;③参与打包的任务点之间距离较远;其次,分别分析这三种情况对定价模型不同方面的影响,对模型进行修改;最后,总体分析将任务打包分配对任务完成情况的影响。

4. 对问题四的分析和对策

问题四要求对题目附件三中的新项目给出任务定价方案,并评价该方案的实施效果。针对这一问题,我们选择在前面三个问题的基础上求解。首先,对附件三中的任务数据进行简单的统计和量化,分析可视化结果;其次,根据经纬度指标对任务进行聚类,与问题一中的区域进行对比,发现位置规律,根据问题二中优化的定价模型求出定价方案;最后,根据问题二中的 Logistic 回归模型对方案实施效果进行评估。

5.3 模型的假设

(1)假设没有恶性预定任务的情况,即所有未完成的任务均是偶然因素造成的;
(2)假设发布的任务难易程度相同,不存在差别;
(3)假设会员不能在不属于自己的区域内完成任务;
(4)假设各区域的气候、交通状况等都相同,不会影响会员对任务的选择;
(5)假设会员完成任务的成本只与距离相关。

5.4 名词解释与符号说明

5.4.1 名词解释

1. 拍照赚钱

"拍照赚钱"是移动互联网下的一种自助式服务模式。用户可以通过 APP 领取需要拍照的任务,赚取 APP 对任务所标定的酬金。

2. 众包

众包指的是一个公司或机构把过去由员工执行的工作任务,以自由自愿的形式外包给非特定的(而且通常是大型的)大众网络的做法。众包的任务通常由个人来承担,但如果涉及需要多人协作完成的任务,也有可能以依靠开源的个体生产的形式出现。

3. 多元回归

研究一个因变量,与两个或两个以上自变量的回归。亦称为多元线性回归,是反映一种现象或事物的数量依多种现象或事物的数量的变动而相应地变动的规律。

4. Logistic 回归

又称 Logistic 回归分析,是一种广义的线性回归分析模型,常用于数据挖掘,疾病自动诊断,经济预测等领域。

5.4.2 符号说明

表 5.1 符号说明

序号	符号	符号说明
1	$d(x_k, x_i^{(i)})$	第 k 个任务点与第 i 个聚核的距离
2	d_i	第 i 个任务点到第 j 个会员的距离
3	k	任务完成度提高率

续表

序号	符号	符号说明
4	W_i	任务完成度比例
5	W_2	平均成本价格
6	x_1	平均距离
7	x_2	任务密度
8	x_3	每个任务点的会员密度
9	x_4	每个任务点会员平均信誉值

5.5 模型的建立与求解

5.5.1 问题一的分析与求解

1. 对问题一的分析

本问题要求我们研究附件一中的任务定价规律,并分析任务未完成的原因。针对这一问题,我们分两个步骤来解决。首先,我们对附件一中的数据进行数据挖掘,使用 SPSS Modler14.1 软件和百度地图对任务点坐标数据进行简单的统计和量化,分析其基于定价和完成度的位置分布图;其次我们选取合适的会影响定价的指标,分别为单位范围内与任务点最近的会员的距离以及会员的密度,进行相关性检验和多元线性拟合,得到附件一中的任务定价规律;最后,我们筛选出未完成任务的数据,对其相应指标进行分析,总结任务未完成的原因。

定义 1 平均距离:指在单位距离范围内与某一任务点距离最近的 n 个会员到该任务点的平均直线距离,单位是 km。

定义 2 任务点密度:指在某一任务点一定距离范围内任务点的个数。

2. 对问题一的求解

模型Ⅰ——基于 K-Means 聚类的中心城区点确定模型

(1)模型的准备。

数据的可视化分析

我们把附件一的数据导入 SPSS Modler14.1 软件中,可以得到基于定价的位置分布图,如图 5.1 所示。

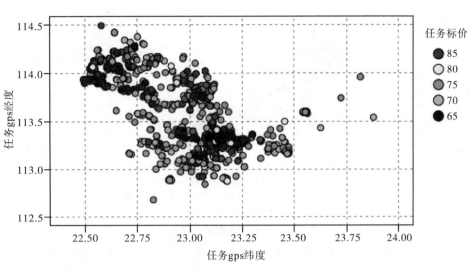

图 5.1　基于定价的位置分布

分析图 5.2,可简单看出任务在地理上的具体分布情况和它们各种的定价数据,大致可以看出三个价格族群,其价格从小到大向外发散,说明任务定价与任务的集中程度是有相关关系的。接着我们统计不同定价的任务数量,见表 5.2。

表 5.2　不同定价的任务数量表

定价	65	65.5	66	66.5	67	67.5	68	68.5	69	69.5	70	70.5
任务数量	65	150	103	63	38	23	30	11	19	8	96	11
定价	71	71.5	72	72.5	73	73.5	74	74.5	75	80	85	—
任务数量	4	5	60	9	10	5	5	2	78	13	27	—

为了更加直观地看出任务数量分布的规律,我们作出图 5.2,通过分析该图可知任务定价和数量直接存在复杂的相关关系,像某些特定价格的任务数量偏多或偏少,说明某些位置上的任务跟定价具有相关性。

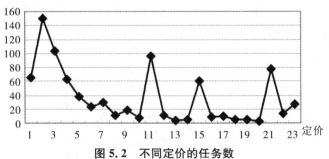

图 5.2　不同定价的任务数

我们统计 835 个任务的完成度情况,如图 5.3。

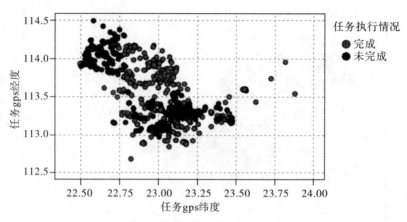

图 5.3 基于完成度的位置分布图

分析图 5.3，我们可知任务的完成情况与位置具有很大的相关性，其中位于中心区域的任务基本都完成了，而处于左上角区域的任务完成度最低，我们认为这可能与城市的特性相关，如该地区工资较高，但任务标价相对来说较低，会员没有意愿去完成任务。实际上还有些偶然因素，如店铺拒访、恶劣天气、道路施工等。

聚类的理论基础

根据可视化分析结果，我们认为可将 835 个任务点划分为 3 个不同的区域来研究他们各自的定价规律。故我们选择通过附件二会员的位置数据来进行 K-Means 聚类求得三个区域的中心城区点位置。

我们采用的算法是 K-Means 算法，该算法采用距离的远近程度作为聚类的标准，其聚类的具体步骤如下：

Ⅰ. 随机选取两个点 $x_1^{(1)}$ 和 $x_1^{(2)}$ 作为聚核；

Ⅱ. 对于任何点 x_k，分别计算 $d(x_k, x_1^{(1)})$ 和 $d(x_k, x_2^{(1)})$；

Ⅲ. 若 $d(x_k, x_1^{(1)}) < d(x_k, x_2^{(1)})$，则将 x_k 划为第一类，否则划给第二类，得到两个分类；

Ⅳ. 分别计算两个类的重心，则得 $x_1^{(1)}$ 和 $x_1^{(2)}$，以其为新的聚核，对空间中的点进行重新分类，得到新分类。

其算法框图如图 5.4。

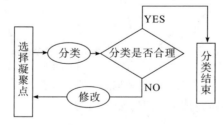

图 5.4 K-Mean 聚类程序框图

(2)结果的求解。

我们将 1877 个会员的位置数据导入 SPSS 软件中，可得到 K-Means 聚类的结果，三个城区分别命名为区域 A，区域 B 和区域 C，三个区域的中心城区点的位置坐标分别为

$aa=(22.64,114.07), bb=(22.92,113.83), cc=(23.12,113.28)$

我们把聚类结果进行可视化,如图 5.5。

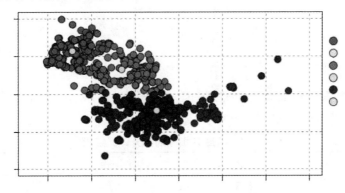

图 5.5 会员位置的聚类结果

从图 5.5,可以直观看出任务点在地理上的分布情况。

模型 Ⅱ——基于多元回归拟合的定价规律模型

根据模型 Ⅰ 中对任务点的分区结果,我们分别对三个区域的定价规律进行研究。

(1) 数据的预处理。

首先我们需要计算区域 B 内所有位置点(包括所有任务点和会员点)两两之间的距离,写 MATLAB 程序,计算结果部分见表 5.3,其中 A_m 表示任务点,单位:km。题目中附件给出的是任务点以及会员的 GPS 定位,即纬度与经度数据,运用下面公式

$$L=\frac{\arccos(\cos(90-AW)\times\cos(90-BW)+\sin(90-AW)\times\sin(90-BW)\times\cos(AJ-BJ))}{180}\times\pi\times R$$

其中,AW、BW 分别为两点的纬度,AJ、BJ 分别为两点的经度,R 为地球半径。

将给出的数据使用 MATLAB 编程进行处理,得到任务与任务、任务与会员两两之间的球面距离。

表 5.3 区域 B 内部分任务点两两之间的距离

	A0070	A0078	A0085	A0087	A0089	A0093	A0094
A0070	0.00	0.00	5.04	1.44	3.78	6.72	3.33
A0078	0.00	0.00	5.04	1.44	3.78	6.72	3.33
A0085	5.04	5.04	0.00	4.68	1.69	2.34	1.72
A0087	1.44	1.44	4.68	0.00	3.87	5.93	3.15
A0089	3.78	3.78	1.69	3.87	0.00	3.98	0.88
A0093	6.72	6.72	2.34	5.93	3.98	0.00	3.72
A0094	3.33	3.33	1.72	3.15	0.88	3.72	0.00
A0097	6.73	6.73	2.23	5.98	3.90	0.20	3.68

(2) 指标的计算。

我们选取的影响指标为平均距离和任务密度,分别表示为 x_1 和 x_2。

首先,我们对任务点到会员之间的距离进行筛选,得到距离每个任务点最近的十个

会员的位置信息，计算这十个会员到该任务点的平均距离。

设 d_i^j 为第 i 个任务点到第 j 个会员的距离，其中 $i=1,2,\cdots,193,j=1,2,\cdots,10$，则

$$x_1 = \frac{\sum_{j=1}^{10} d_i^j}{10}, i=1,2,\cdots,193$$

任务密度的数据可以直接通过 EXCEL 筛选得到，故不用进行计算。以区域 B 为例，其各个指标的计算结果见表 5.4。

表 5.4 区域 B 指标的部分计算结果

任务编号	标价	平均距离	任务密度
A0070	66.5	2.84	34
A0078	66.5	2.84	34
A0085	66	2.35	18
A0087	66	2.62	30
A0089	66.5	3.17	21
A0093	70	2.37	16
⋮	⋮	⋮	⋮
A0385	75	14.18	1
A0388	72	3.41	12
A0390	74	6.7	8

(3) 求解结果。

建立的模型如下：

区域 A： $y_1 = \alpha_1 x_1 + \beta_1 x_2 + \varepsilon_1$

区域 B： $y_2 = \alpha_2 x_1 + \beta_2 x_2 + \gamma_2 \ln x_1 x_2 + \varepsilon_2$

区域 C： $y_3 = \alpha_3 x_1 + \beta_3 x_2 + \gamma_3 \ln x_1 x_2 + \varepsilon_3$

其中，α,β,γ 为拟合参数，将三个区域的指标数值导入 STATA 软件中，模型的参数检验结果见表 5.5。

表 5.5 模型参数检验表

	区域 A		区域 B			区域 C		
参数	α_1	β_1	α_2	β_2	γ_2	α_3	β_3	γ_3
估计值	0.6769	−0.0699	0.705	−0.1094	3.113	0.2464	−0.0847	1.0493
t 检验	4.93	2.93	4.25	−3.28	3.45	4.12	−10.61	3.57
P 值	0	0.004	0	0.001	0.001	0	0	0
F 检验的 P 值	0		0			0		
拟合优度	0.3412		0.2927			0.3024		

从参数检验的结果来看，三个模型的 8 个参数均通过了 t 检验，模型整体也通过了 F 检验，但是我们发现模型的拟合优度都偏低，分析可知这是由于该模型还存在显著影响的因素未参与拟合所导致的，说明在实际定价中还存在其他影响指标，例如当地的居民消费水平，到任务点的交通状况等等。在本文中，我们不对这些问题进行考虑。

故得到三个地区的定价规律模型如下：
$$y_1 = 0.6769x_1 - 0.0699x_2 + 67.9795$$
$$y_2 = 0.705x_1 - 0.1094x_2 + 3.113x_1x_2 + 56.1512$$
$$y_3 = 0.2464x_1 - 0.0847x_2 + 1.0493\ln x_1x_2 + 67.8522$$

未完成任务的分布情况——基于三个地区的定价规律模型

(1)完成任务的描述性统计。

统计可得 835 项任务中，共有 313 项未完成的任务，其中大部分都分布在区域 A 和区域 B，故我们作出区域 A 未完成任务基于价格的位置分布图，见图 5.6。

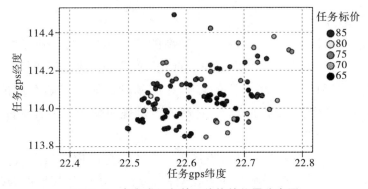

图 5.6　未完成任务基于价格的位置分布图

分析图 5.6，可以看出各种价格的任务都有未完成的情况，可见完成度不仅与价格相关，还存在其他影响定价的偶然因素。但是其中价格低的任务占比较高，达到 64%。故我们可以将某些点视为特殊点，不考虑它们的情况，并对具有明显规律的任务点做下一步分析。

(1)基于区域 B 的定价规律模型对区域 A 未完成任务的分析。

由模型 II 可知区域 B 的定价规律模型，为
$$y_2 = 0.705x_1 - 0.1094x_2 + 3.113x_1x_2 + 56.1512$$

已知区域 A 各项指标的数值，将该定价规律应用于区域 A 任务点的定价中，可以得到基于区域 B 定价模型的定价方案，将其与原定价数据进行比较，得到图 5.7。可以看出，新的方案定价在整体上比原来的高，说明区域任务未完成的原因主要是价格太低。

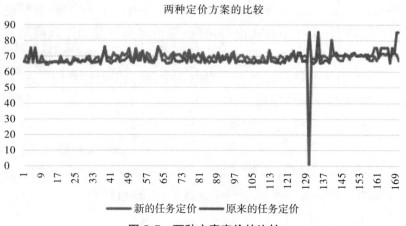

图 5.7　两种方案定价的比较

5.5.2　问题二的分析与求解

1. 对问题二的分析

本问题要求我们为附件一中的项目设计新的任务定价方案，并和原方案进行比较。针对这一问题，我们分为两个方面来解决。首先，统计 522 项已完成任务的数据信息，对其进行数据挖掘，得到归于函数的定价规律，在此基础上添加可能对定价产生显著影响的指标，建立新的定价规律模型。其次，将新的定价规律模型应用到附件一的项目中，将其结果与原方案进行比较。

定义 3　会员密度：指在某一任务点周围一定距离范围内会员的人数。

定义 4　最短距离：指对于某一任务点而言，所有会员中与其最近的距离。

定义 5　平均信誉度：指对某一任务点而言，一定距离范围内所有会员信誉值的平均值。

2. 对问题二的求解

模型Ⅲ——基于多元回归的任务定价模型

(1)模型的准备。

本问题旨在通过挖掘附件一和附件二的数据得到基于经验的定价模型，我们将优化目标定为提高完成度同时控制成本价格。首先在附件一中筛选出 522 项已完成的任务，对其进行数据挖掘，接着分析其指标特征及找出具有显著影响的因子，最后以这些定价因子作为解释变量做多元回归。

该定价方法的思路框图见图 5.8。

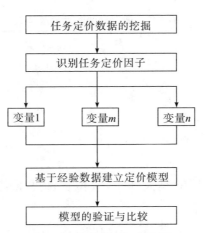

图 5.8　定价模型思路框图

(2) 多元回归定价模型的建立。

我们一共选取五个指标作为自变量，分别为平均距离、任务点密度、会员密度、最短距离和平均信誉度。

以定价为因变量建立以下模型：

$$y = \beta_0 + \beta_1 x_1 + \beta_2 x_2 + \beta_3 x_3 + \beta_4 x_4 + \beta_5 x_5 + \beta$$

其中，β_i 为第 i 个定价因子的拟合参数，β_0 为常数项 ε 为残差项。

(3) 模型的求解。

将变量数据导入 STATA14.0 软件中，得到如下的回归结果。

$$y_1 = 71.63 + 0.318 x_1 - 0.02 x_2 - 0.69 x_3 - 0.2627 x_4 - 0.0022 x_5$$

其参数的检验值和方程整体的检验结果见表 5.6。

表 5.6　第一次参数检验结果

参数	β_0	β_1	β_2	β_3	β_4	β_5
估计值	71.63	0.318	−0.2	−0.069	−0.262	0.0002
t 检验	131.95	2.68	−1.5	−4.88	−1.62	−2.17
P 值	0	0.008	0.134	0	0.106	0.031
F 检验的 P 值	0					
拟合优度	0.2696					

分析该表，发现方程整体通过了 F 检验，即模型整体显著，但是 x_2 和 x_4 的参数没有通 t 检验，并且 x_4 的参数估计值不符合经济意义。我们认为该模型是不合理的。故我们选择去掉变量 x_4，重新进行回归。

其参数的检验值和方程整体的检验结果见表 5.7。

表 5.7 调整后的参数检验结果

参数	β_0	β_1	β_2	β_3	β_4
估计值	72.04	0.152	−0.022	−0.071	−0.0002
t 检验	149.9	2.51	−1.72	−5.08	−2.25
P 值	0	0.013	0.085	0	0.025
F 检验的 P 值			0		
拟合优度			0.2659		

分析该表,由于本题采用的是大样本,故拟合优度不高的问题可以不考虑,则该模型通过了检验,回归模型如下。

$$y = 72.04 + 0.152x_1 - 0.022x_2 - 0.071x_3 - 0.0002x_5$$

模型Ⅳ——基于 Logistic 回归的完成度测度模型

(1) 模型的准备。

Logistic 回归又称 Logistic 回归分析,是一种典型的概率型非线性回归模型,是研究二分类观察结果与一些影响因素之间关系的一种多变量分析方法。

本题中,要求我们将新的任务定价方案与原方案进行比较,我们选取任务的完成度和成本价格为评价指标,通过 Logistic 模型训练附件一中已完成任务的主要特征,将之前未完成的任务输入模型,可以得到任务完成度和成本价格的具体变化。

(2) 模型的建立。

考虑具有 n 个独立变量的向量 $x = (x_1, x_2, x_3, \cdots, x_n)$,设条件概率 $P(y=1|x) = p$,为根据观测量相对于某事件 x 发生的概率。那么 Logistic 回归模型可以表示为:

$$P(y = 1|x) = \pi(x) = 1/(1 + e^{-g(x)})。$$

这里 $f(x) = 1/(1 + e^{-x})$ 称为 Logistic 函数。其中,$g(x) = w_0 + w_1 x_1 + \cdots + w_n x_n$。还可知在 x 条件下 y 不发生的概率

$$P(y = 0|x) = 1 - P(y = 1|x) = 1 - 1/(1 + e^{-g(x)}) = 1/(1 + e^{g(x)})$$

事件发生与不发生的概率之比为

$$\frac{P(y=1|x)}{P(y=0|x)} = \frac{p}{1-p} = e^{g(x)}$$

(3) 模型的求解。

① 数据的准备。本问题采用的目标函数是任务的完成情况,将完成的任务定义为"1",未完成的任务定义为"0",影响任务完成的指标选取有任务标价、平均距离、任务密度、会员密度和平均信誉值共五个指标,设为 x_1, x_2, x_3, x_4, x_5。

② 求解的步骤。本文使用 SPSS Modler14.0 来对模型进行求解,其运行步骤见图 5.9。

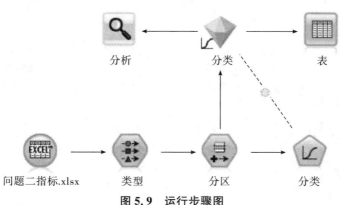

图 5.9 运行步骤图

③求解结果。

ⅰ.参数检验

Logistic 回归结果见表 5.8。

表 5.8 回归结果表

	Intercept	任务标价	平均距离	任务密度	会员密度	平均信誉值
估计值	−9.419	0.281	−1.824	0.953	0.081	−0.001
标准误差	18.915	0.28	0.715	0.38	0.112	0.001
Wald 统计量	0.248	1.006	6.502	6.31	0.518	0.796
显著性	1	1	1	1	1	1
t 检验显著性	0.618	0.016	0.011	0.012	0.472	0.372
优势比		1.324	0.161	0.385	1.084	0.999

对结果的分析：首先是参数的显著性检验，在五个指标中，任务标价、平均距离和任务密度通过了检验，说明这三个指标对任务完成具有显著影响。

故回归方程为

$$\mathrm{logit}(P) = 0.281x_1 - 1.824x_2 + 0.953x_3$$

模型的经济意义为：任务标价越高完成概率越高；平均距离越大完成概率越低；任务密度越大完成概率越大。符合实际情况，可认为模型合理。

ⅱ.模型效果的判断

模型效果的判断见表 5.9。

表 5.9 决定系数检验结果及模型预测效果

	决定系数检验结果			模型的预测效果		
评价	Cox and Snell	Nagelkerke	Nagelkerke	正确	825	98.80%
数值	0.718	0.986	0.972	错误	10	1.20%

分析该表，可知模型整体通过了拟合检验，并且就测试集的预测结果来看，正确率达到了 98.8%，说明该模型预测效果比较好，可用于评价定价方案的实施效果。

①模型的应用。将未完成任务的数据导入 Logistic 模型中，可以得到完成情况的预

测结果,见图 5.10。

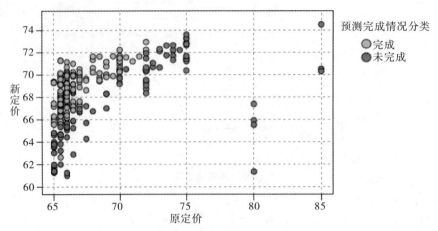

图 5.10 预测完成情况分布

设总的任务个数为 m,原完成任务的个数为 a,可知 $m=835, a=522$。定义衡量定价方案效果的指标任务完成度提高率,设为 k,实施新的定价方案后增加的完成任务数设为 c。则

$$k = c/m \times 100\% = 24.7\%$$

5.5.3 问题三的分析与求解

1. 对问题三的分析

本问题要求我们考虑实际情况,将位置比较集中的多个任务联合在一起打包发布,修改问题二中的定价模型,并研究其对最终任务完成情况的影响。首先,根据任务点的分布,我们分三种情况来考虑:①不参与打包的任务点;②参与打包的任务点之间距离较近;③参与打包的任务点之间距离较远;其次,分别分析这三种情况对定价模型不同方面的影响,对模型进行修改;最后,总体分析将任务打包分配对任务完成情况的影响。

2. 对问题三的求解

模型Ⅴ——基于会员参与度的任务定价模型

(1)模型的准备。

我们可以从多个方面对问题二中的任务定价模型进行改进,但最基本的在于会员参与度,制定合理的任务标价正是为了使会员的参与度最大,从而提高任务的总体完成情况,因此我们可以从会员参与度出发,对模型做进一步修改。

通过分析,我们发现,打包并不会影响问题二定价模型中的变量,即平均距离、任务点密度、会员密度和平均信誉度并不会因为打包方式的存在而发生变化。

(2)模型的建立。

①第 1 种情况对定价模型的影响。如果某个任务离周围任务点的距离都较远,我们可以不将其打包,此种情况下,会员可能因距离较远而不愿去完成此任务,导致任务执行失败。我们可以通过提高任务价格吸引会员,提高其参与度并愿意完成任务。修改后

的模型如下：
$$y = 72.04 + 0.152x_1 - 0.022x_2 - 0.071x_3 + (c - 0.0002)x_5$$

其中，c 是会员完成任务时会获得的奖励金系数。奖励金[4]的设置会使得会员完成任务时获得的报酬增加，且平均信誉度越高，奖励金越大，从而调动其积极性，提高参与度。

②第 2 种情况对定价模型的影响。这种情况下，参与打包的任务点之间距离较近，且每个任务点的标价相同。对于标价的制定，一方面由于任务点较为集中，会员会更愿意参与进来，另一方面，为企业考虑，使其尽可能以最小的成本获得最大的效益，我们定义参与打包的任务点中的最低标价为打包任务点的标价。修改后的模型如下：

$$\begin{cases} y = \min y_i, \\ y = 68.24 + 0.152x_1 - 0.022x_2 - 0.071x_3 - 0.0002x_5。\end{cases}$$

③第 3 种情况对定价模型的影响。这种情况下，参与打包的任务点之间距离较远，且每个任务点的标价相同。对于标价的制定，考虑到由于距离较远，会员参与度不高，因此我们定义参与打包的任务点中的最高标价为打包任务点的标价。修改后的模型如下：

$$\begin{cases} y = \max y_i, \\ y = 78.56 + 0.152x_1 - 0.022x_2 - 0.071x_4 - 0.0002x_5。\end{cases}$$

(3) 模型的意义。

定性分析采取打包发布任务的方式对任务执行情况的影响。将位置比较集中的多个任务联合在一起打包发布，首先通过合理的标价吸引会员，提高其参与度，有效地保证和提高了每个任务的完成度，同时任务点相对密集的地方标价变低，任务点相对稀疏的地方标价变高，两者之间互补，做到效用最大化，且在一定程度上可以缩短任务完成周期，有利于企业及时做出相关改进决策，促进企业长期发展。

5.5.4 问题四的分析与求解

1. 对问题四的分析

本问题要求我们为附件三中的新项目制定任务定价方案，并评价方案的实施效果。针对这一问题，我们选择在前面三个问题的基础上求解。首先，对附件三中的任务数据进行简单的统计和量化，分析可视化结果；其次，根据经纬度指标对任务进行聚类，与问题一中的区域进行对比，发现位置规律，根据问题二中优化的定价模型求出定价方案；最后，根据问题二中的 Logistic 回归模型对方案实施效果进行评估。

2. 对问题四的求解

新项目的可视化——基于 K-Mean 聚类模型。

(1) 聚类结果。

将聚类结果进行可视化，见图 5.11。

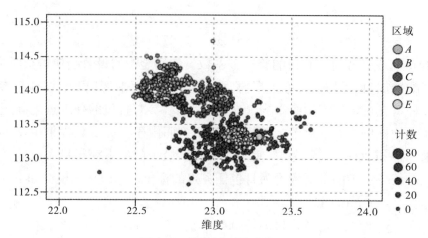

图 5.11 新项目的聚类结果

其中,区域 A、B、C 是模型 I 中聚类划分的结果,区域 D、E 是新项目的聚类划分的结果,从该图可以看出新项目的任务主要分布在区域 A 和区域 C。本文选择对区域 D 的项目进行研究,制定定价方案。

新项目的定价结果——基于多元回归的经验定价模型

(2)定价结果。

由问题二得到的定价模型如下:

$$y = 72.04 + 0.152x_1 - 0.022x_2 - 0.071x_3 - 0.0002x_5$$

其中,x_1 是指变量平均距离,x_2 是指任务密度,x_3 是指会员密度,x_5 是指平均信誉值。将该模型应用到区域 D 的任务数据中,可以得到新项目的定价结果。基于位置的定价分布情况见图 5.12。

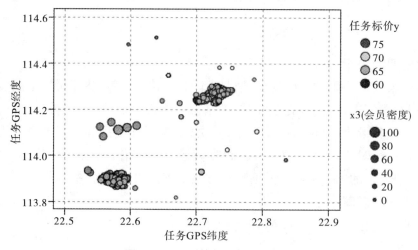

图 5.12 预测的新任务价格分布

实施效果的评估——基于 Logistic 二分类的完成概率模型

(3)实现效果的预测结果。

新任务的完成情况预测模型与问题二相同。设预测结果的评价指标为完成度比例和平均成本价格,为 W_1 和 W_2。

计算可得 $W_1=87.5\%$,$W_2=68.37$。从完成度比例可看出,该定价方案的实施效果较好,任务完成比例较高。平均成本价格为 68.37,也在可以接受的范围之内。

5.6 误差分析与灵敏度分析

5.6.1 误差分析

(1)问题一中根据任务点及其标价的分布情况,我们对其进行分区讨论,由聚类得出的中心点确定区域中心,可能会对后面的计算造成误差;

(2)在指标选取中,我们选择了 10 km 范围内任务点个数,实际上,选定不同范围,任务点个数也会因此而改变,从而对模型的拟合产生不同程度的影响,见表 5.10。

表 5.10 不同单位范围的选取对模型拟合优度的影响

不同单位范围	1 km	2 km	3 km	4 km	5 km	6 km	7 km	8 km	9 km	10 km
模型拟合优度	0.171	0.194	0.217	0.225	0.239	0.256	0.279	0.301	0.326	0.341

5.6.2 灵敏度分析

在问题二中,我们对因变量任务标价和自变量平均距离、任务点密度、会员密度、平均信誉度进行回归拟合,得到如下定价模型:

$$Y = 72.04 + 0.152X_1 - 0.022X_2 - 0.071X_3 - 0.0002X_5$$

在问题四中,根据前面得到的任务定价规律,对新项目的任务标价进行讨论,当其他变量值一定时,任务标价随平均距离的不同增加值而变化。运用 MATLAB 软件对其进行灵敏度分析。

表 5.11 不同 a 值和 x_1 值下对应的 y 值($X_2=260$,$X_3=25$,$X_5=75.5$)

x_1	0.6	0.8	1.0	1.2	1.4	1.6	1.8	2.0
$y_1(a=0.152)$	64.6211	64.6515	64.6819	64.7123	64.7427	64.7731	64.8035	64.8339
$y_2(a=0.172)$	64.6331	64.6675	64.7019	64.7363	64.7707	64.8051	64.8395	64.8739
$y_3(a=0.192)$	64.6451	64.6835	64.7219	64.7603	64.7987	64.8371	64.8755	64.9139
$y_4(a=0.212)$	64.6571	64.6995	64.7419	64.7843	64.8267	64.8691	64.9115	64.9539
$y_5(a=0.232)$	64.6691	64.7155	64.7619	64.8083	64.8547	64.9011	64.9475	64.9939
$y_6(a=0.252)$	64.6811	64.7315	64.7819	64.8323	64.8827	64.9331	64.9835	65.0339

根据表 5.11 作图 5.13,如下。

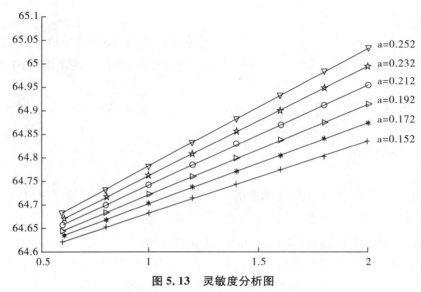

图 5.13 灵敏度分析图

由图 5.13 可知,当 x_1 确定下来时,参变量 a 值变化会引起 y 值变化,且变化量随着任务平均距离的增加而增加;当 a 值一定时,y 值随着 x_1 值的增加而增加。

5.7 模型的评价

5.7.1 模型的优点

(1)本文建立的模型能与实际紧密结合,结合实际情况对问题进行求解,具有现实意义,且通用性较强;

(2)巧妙地运用流程图,将建模思路完整清晰地展现出来;

(3)利用 SPSS、EXCEL 等软件对数据进行处理并绘制图表,不仅方便快捷,而且使结果更加形象直观,如根据经纬度所作出的任务点分布图,可以看出分布的大致规律,有利于对问题的进一步分析;

(4)在建立定价模型时,充分考虑了企业的成本和会员的效益,使得价格不至于太高使公司成本增加,也不会太低使任务完成情况不好。

5.7.2 模型的缺点

(1)为了使结果更加理想化,对于一些数据,我们对其进行了必要的处理,如对异常数据的舍弃,这会带来一定的误差;

(2)为使计算简便,易于实现,模型中忽略了一些次要影响因素,如任务难度、任务期限等,会对结果造成一定程度的影响。

参考文献

[1] 张鹏. 基于委托代理理论的众包式创新激励机制研究[D]. 成都：电子科技大学，2012：47—56.

[2] Y. Singer and M. Mittal，Pricing Mechanisms for Crowdsourcing Markets，in Proceedings of the 22nd International Conference on World Wide Web[C]. Waco：ACM，2013，1157—1166.

[3] 刘晓钢. 众包中任务发布者出价行为的影响因素研究[D]. 重庆：重庆大学，2012.

[4] 芮兰兰，张攀，黄豪球，等. 一种面向众包的基于信誉值的激励机制[J]. 电子与信息学报，2016(07)：1808—1815.

◆ 建模特色点评

❀论文特色❀

◆标题定位：原赛题为"拍照赚钱"的任务定价，竞赛论文为"基于多元拟合和Logistic回归方法对任务定价的研究"，集方法、问题及特色于一体，标题定位准确、简洁、有特色。

◆方法鉴赏：模型能与实际结合进行求解，具有现实意义，且通用性较强。巧妙地运用流程图，将建模思路完整清晰地展现出来，利用SPSS、EXCEL等软件对数据进行处理并绘制图表，不仅方便快捷，而且使结果更加形象直观，如根据经纬度所作出的任务点分布图，可以看出分布的大致规律，有利于对问题的进一步分析；在建立定价模型时，充分考虑了企业的成本和会员的效益，使得价格不至于太高使公司成本增加，也不会太低使任务完成情况不好。

◆写作评析：文章写作格式规范，内容较贴切，文献综述是2017建模论文中的特色之处，误差分析与灵敏度分析为本文突出的亮点之一。

❀不足之处❀

对问题总体分析缺少框图，缺少模型的推广与改进。

第 6 篇 机场的出租车问题

◆ 竞赛原题再现

2019 年 C 题 机场的出租车问题

大多数乘客下飞机后要去市区(或周边)的目的地,出租车是主要的交通工具之一。国内多数机场都是将送客(出发)与接客(到达)通道分开的。送客到机场的出租车司机都将会面临两个选择:

(A)前往到达区排队等待载客返回市区。出租车必须到指定的"蓄车池"排队等候,依"先来后到"排队进场载客,等待时间长短取决于排队出租车和乘客的数量多少,需要付出一定的时间成本。

(B)直接放空返回市区拉客。出租车司机会付出空载费用和可能损失潜在的载客收益。

在某时间段抵达的航班数量和"蓄车池"里已有的车辆数是司机可观测到的确定信息。通常司机的决策与其个人的经验判断有关,比如在某个季节与某时间段抵达航班的多少和可能乘客数量的多寡等。如果乘客在下飞机后想"打车",就要到指定的"乘车区"排队,按先后顺序乘车。机场出租车管理人员负责"分批定量"放行出租车进入"乘车区",同时安排一定数量的乘客上车。在实际中,还有很多影响出租车司机决策的确定和不确定因素,其关联关系各异,影响效果也不尽相同。

请你们团队结合实际情况,建立数学模型研究下列问题:

(1)分析研究与出租车司机决策相关因素的影响机理,综合考虑机场乘客数量的变化规律和出租车司机的收益,建立出租车司机选择决策模型,并给出司机的选择策略。

(2)收集国内某一机场及其所在城市出租车的相关数据,给出该机场出租车司机的选择方案,并分析模型的合理性和对相关因素的依赖性。

(3)在某些时候,经常会出现出租车排队载客和乘客排队乘车的情况。某机场"乘车区"现有两条并行车道,管理部门应如何设置"上车点",并合理安排出租车和乘客,在保证车辆和乘客安全的条件下,使得总的乘车效率最高。

(4) 机场的出租车载客收益与载客的行驶里程有关,乘客的目的地有远有近,出租车司机不能选择乘客和拒载,但允许出租车多次往返载客。管理部门拟对某些短途载客再次返回的出租车给予一定的"优先权",使得这些出租车的收益尽量均衡,试给出一个可行的"优先"安排方案。

原题详见 2019 年全国大学生数学建模竞赛 C 题。

◆ 获奖论文精选

机场出租车返程载客综合研究[①]

摘要:为了提升机场出租车离开机场的载客效率和机场周边交通管理水平,本文通过研究影响出租车司机是否选择载客离开机场的决策因素,构建司机决策模型,在期望收益一致的前提下,对比模型得出司机容忍的最大待客时间和观察到的司机实际待客时间,说明司机对待客时间的心理预期是其做出决策的重要因素;收集了上海浦东国际机场及上海市出租车的相关数据,通过到港航班测算对出租车的需求量,计算此时出租车的等待时间小于前述模型的极限值,从而验证了模型在一般情况下的可靠性;采用假期旅客出行高峰、恶劣天气等极端场景,讨论了非常态情况对模型的影响;研究了国内机场到达区出租车上客的传统单车道路边模式,给出了一种新的双车道上客方案,并通过模拟计算得出了方案中最优的上客点(乘车口)个数,从而尽可能地提升该上客方案的效率;为了支持有关部门改善搭载短途乘客离场的出租车司机的收入和体验,给出了一些具体的支持措施,并分析了其合理性和可行性。

针对问题一,在考虑机场乘客的变化规律和出租车收益的条件下,运用 Laplace 准则构建出租车司机的决策模型,得到出租车等待时长上限的描述函数,在机场待客和直接返回二者收益相同情况下,利用该描述函数计算出司机最大等待容忍时间,以先前研究者现场观察到的实际待客时间作为实际可接受的等待时间,在模型计算出的最大等待容忍时间大于实际可接受的等待时间时,司机将做出在机场待客的决策。

针对问题二,要求找到实际相关数据,对出租车选择方案的模型进行可靠性分析,并探究对影响因素的依赖性。根据上海浦东机场及该市的相关数据,对出租车等待时长的理论值和实际值进行比较,验证模型可靠性,再通过对上海假期乘客数量及恶劣天气的量化分析,说明极大客流吞吐量、极端天气状况对出租车等待时长具有一定的影响。

针对问题三,在机场现有两条并行车道的前提下,设计了一种新的双车道边上车方案,并通过 Monte Carlo(蒙特卡罗)算法计算出每条车道设置乘车口的最优数量。

---

① 本文获 2019 年全国二等奖。队员:凌学轩,周骅莉,张秦;指导教师:朱江乐。

针对问题四,考虑到出租车司机不能拒载短途乘客,管理部门拟对某些短途载客再次返回的出租车给予一定的"优先权",通过分析为此类出租车给出了返回后优先载客的支持措施,并分析了措施的合理性和可行性。

针对机场出租车返程载客的司机决策、乘客搭乘便利、管理部门的人性化管理等一系列问题,本文综合运用收益均衡理论、Monte Carlo 计算方法、排队理论模型等数学工具,对出租车送客问题进行量化、建模处理,并结合上海浦东机场的实际数据进行了模型验证,为了减少司机和乘客的等待时间,给出了一种新的机场双车道出租车上客方案,最后为管理部门支持搭载短途乘客的司机提出了方案。

关键词:出租车;决策优化;Laplace;Monte Carlo 算法;排队论模型

6.1 问题的重述

6.1.1 背景知识

1. 行业现状

改革开放至今,我国民用航空运输发展迅速,整体经济的发展和居民消费水平的提升激发了航空客运需求的急速增长。机场旅客吞吐量迅速增加为大型机场的建设带来了蓬勃发展,也给机场周边多元交通方式的载客能力带来巨大的压力,周边道路的交通通行能力承受较大的考验。

在转乘的多样化交通工具中,出租车作为一种灵活、快捷、无时间限制的交通工具,对机场乘客的流动起到重要作用。我国不同机场针对出租车的运行模式,有不同的硬件、政策的规划。

- 北京首都国际机场:出租车蓄车池采取双车道模式,短途运出的出租车可领取免排队票,再次返回机场时优先载客,并利用网络对出租车配给进行调控。
- 武汉天河国际机场:天河机场最新启用智慧停车系统,采用系统匹配出租车短途返程优先配客措施,在站点按长短途划分排队,提高效率。

2. 提出问题

机场返港乘客中,转乘出租车的人数占较高比例,但随着乘坐出租的人数不断增加,出租车的需求量较大,容易造成人车供给不匹配、车辆拥堵引发连锁交通瘫痪、机场乘客滞留等问题,影响机场这一大型枢纽的整体运转。

为更好地优化出租车系统,本文首先站在出租车司机的角度,在出租车收益最大化的条件下探讨出租车的决策,并结合实际数据对影响因素进行探究;其次,对机场出租车的蓄车区域进行最优规划,提高出租车和乘客的运载效率;最后在机场的角度,为解决长短途司机收益不均问题,给出有效的鼓励策略,提高出租车的使用效率。

3. 研究意义

机场作为近年来飞速发展的大型交通枢纽,随着经济的迅猛发展它有着巨大的发展潜力,但随之而来的是较大的服务压力。推动机场出租车系统的完善,有利于缓解机场及其周边整个交通系统的拥堵状况;同时,由出租车的系统优化模式可以辐射到其他多元交通工具,提高整体交通运输效率;最后,便捷的交通有利于增强乘客的服务体验感,增强机场的口碑和好评度。

6.1.2 相关数据

(1)2018年12月上海机场发展报告;
(2)文献基于天气影响的离场航班延误分析及预测。

6.1.3 具体问题

1. 问题一

本题要求在考虑飞机航班、司机最大收益等因素的情况下,对司机的决策进行推断。司机有留在机场载客和返回市中心载客的两种决策,在相同收益的情况下,考虑司机的极限等待时间,并根据实际进行对比做出决策。

2. 问题二

本题要求在找到某机场的实际数据情况下,对问题一中的模型进行验证,并对影响因素的可靠性进行探究。在找到上海浦东机场的数据后对模型进行验证,并找到有关天气、人流量等极端情况下的数据,对影响因素进行验证。

3. 问题三

本问题要求在机场陆侧交通实际情况下,经常会出现出租车排队载客和乘客排队乘车的情况。假定某机场乘车区现有两条并行车道,在保证车辆和乘客安全的条件下,给管理部门设置上客点的建议,合理安排出租车和乘客,使得总的乘车效率最高。

4. 问题四

本问题要求对于在机场短途行驶的短途出租车司机给出可行的"优先"安排方案。根据相关文献、数据和报告,对于司机返回机场继续载客进行可行性的判断,依据结果对于现有机场方案进行改进得出最终优化方案。

6.2 问题的分析

6.2.1 研究现状综述

李嘉靖[1]通过构建出租车需求量实时动态预估模型,出租车保有水平预测模型构建预警系统的系统框架以实现出租车合理调配,使用上海市数据进行仿真实验。文章仅考虑某日数据,未考虑节假日、晴雨天气等其他因素。

柳伍生[2]对于出租车车道边通行能力进行仿真研究,利用统计分析对于数据进行初步处理后设计仿真流程和模块,利用MATLAB进行仿真模拟,说明上车点的最佳选择。不足之处是仅考虑单车道而未考虑到双车道或是多车道。

黄业文[3]通过对于系统理论分析得出系统队长、等待队长、等待时间的公式,并对于第1级顾客、第2级顾客进行分类讨论,最终得出总公式,利用MATLAB进行模拟实验。其未考虑可能存在的顾客损失率等方面。

总而言之,现有的文献或多或少的都有其不足之处,需要加以完善。

6.2.2 本文的研究思路和步骤

1. 对问题的总体分析和解题思路

本文是研究机场出租车载客的问题,针对此问题,我们分为四个小问题来研究。

第一:建立模型探究影响司机决策的因素,并给出选择策略;

第二:根据实际数据对模型进行验证,并对影响司机载客的因素进行探究;

第三:先估计出高峰时客流量,分析车道边的通行能力,得出最优上客点个数;

第四:先得出供求关系以及收益函数,分析回机场是否具有可行性,通过排队论得出返回后所需排队时间判断合理性最终得出设计方案。

根据本文的研究思路,做出整体的思路流程图,如图6.1所示。

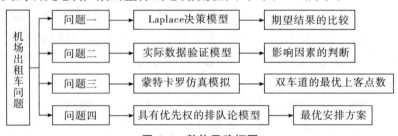

图6.1 整体思路框图

2. 对问题的具体分析和对策

(1)对问题一的分析。

问题一要求我们在考虑四季收益最大化和飞机航班的情况下,找出司机的决策影响因素,并根据模型对司机的决策策略进行设计。运用Laplace准则对司机的选择进行判断,在两种决策结果收益相等的情况下对极限时间进行求解,并与实际时间进行比较。

(2)对问题二的分析。

问题二要求我们在考虑实际的数据情况下对模型进行测定。我们找到了2018年上海浦东机场的数据,并找到了春节高客流量情况下的航班数据、极端环境的天气量化,对影响因素进行判断。

(3)对问题三的分析。

问题三要求我们在机场有两条并行车道时,管理部门应如何设置"上车点",传统的机场车道设置的是单侧上车模式,本文采用双侧车道上车模式,运用蒙特卡罗仿真模拟出不断增加上客点,对其通行能力进行对比分析。得出两条并行车道的最优上客点个数。

(4) 对问题四的分析。

本问题要求对于在机场短途行驶的短途出租车司机给出可行的"优先"安排方案。根据相关文献、数据和报告,了解上海市出租车定价方案,根据供求模型得出机场出租车的供求关系,通过非强占型排队论模型对于司机返回机场继续载客进行排队时间、排队效率等方面的分析,依据结果对于现有机场方案进行改进得出最终优化方案。

6.3 模型的假设

(1) 假设机场吞吐量较大,不存在出租车乘客断流,即有出租车在机场等待而无乘客乘车的情况。

(2) 假设司机选择空载返回市区,不考虑机场至市区途中载客的可能性,且出租车一直处于行驶状态不停止驾驶。

(3) 蒙特卡罗仿真时,假设 1 辆出租车只搭载 1 位乘客。

(4) 问题三中乘客为两条双侧队列排队候车,先到先服务,且两条队列的乘车概率相同。

(5) 假设出租车行驶过程中未遭遇堵车等情况。

(6) 假设出租车司机均为"理性人"。

6.4 名词解释与符号说明

6.4.1 名词解释

1. 乘车效率

乘车效率代表整个机场出租车及乘客构成的系统总的运行效率,在乘客和出租车足够多的情况下,可以表现为在相同的时间内,所能通过的最大车辆数(*辆/小时);计算进入上客点的(1 000 辆)出租车的平均载客时间(*秒)。乘车效率越大,乘客和出租车的排队时间越小。

2. 上客点

上客点即出租车开始载客的泊位,之前出租车需要在蓄车位等待,有客流时就驶进上客点。上客点的设置是影响乘车效率的主要因素之一。

3. 车道边

车道边是机场陆侧交通中至关重要的节点,是人车换乘的关键场所。由于各种类型的车流交织以及车流与人流在车道边相互交错,车道边向来是机场的陆侧交通体系中最容易形成"瓶颈"的地方之一,也是机场容量达到饱和时容易出问题的敏感地区[5]。

4. Laplace 准则

它是指决策者面临事件集合时,赋予每个事件等可能性,并计算每个策略的期望值,以进行最优决策[1]。

6.4.2 符号说明

序号	符号	符号说明
1	Q	时刻到达站台的乘客流出量
2	α	在站台换乘出租车的乘客换乘率
3	ω	出租车平均载客量
4	S	距离
5	G	机场等客出租车收益
6	v	速度
7	t	时间
8	c	单车道单通道车道边通行能力
9	β	短途选择比
10	M	出租车保有量
11	μ	每分钟可搭载乘客数

6.5 模型的建立与求解

6.5.1 问题一的分析与求解

1. 对问题一的分析

本问题要求我们对出租车行为决策影响因素进行探究,并建立模型,找出出租车司机的最优决策。到达机场的出租车司机面临两种选择:

A:留在机场等待顾客,但可能会因为等待时间过长而遭受损失。

B:返回市区载客,可能会因为其他成本或减少隐含的时间收益而减少总收益。

在面临这样的抉择环境下,建立基于 Laplace 准则的出租车选择决策模型,分别对方案 A、B 的期望结果进行预测,并对面临的实际情况进行比较,做出收益最大化的最优决策。

2. 对问题一的求解

模型 I——基于 Laplace 准则的出租车选择决策模型

(1)模型的准备。

模型原理

在不确定决策问题中,悲观主义者通常采取保守型的决策准则,处理问题小心、谨慎,从最坏的结果出发选择最优方案;另一种乐观主义者乐于冒险,是基于所有最优结果的选取最优方案。Laplace 准则(等可能性决策准则)在无特殊理由时认为最好结果和最坏结果是等价的,即依据:

$$\max\left\{\frac{1}{2}\max(A)+\frac{1}{2}\min(A)\right\}$$

选择事件 A 的最优决策结果。

建模思路

在空载回市区、留在机场载客的选择中,影响司机决策的直接因素为机场等待时间。若等待时间过长,等待时间的机会成本高于留在机场载客获得的收益,"车等人"的情况会驱使司机选择返回市区载客。分析等待时间的因素影响,并计算当返回市区与等待的收入支出平衡时,司机所能接受的等待时间上限。

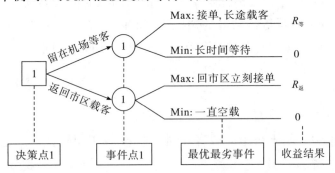

图 6.2 基于 Laplace 准则的决策分析图

基于上述决策分析流程,不计算同决策下的最优事件结果,进行比较得出结论。

(2)模型的建立。

若司机在机场等待,等待时间受到已排队出租车数量、飞机航班的抵达时间、乘客上车的时间等因素影响。其中已排队出租车数量 a 可直接观察,通过对该时段乘客数量进行估算,利用模型结果做出判断是否可以成功载到乘客,做出抉择。具体的思路如图 6.3 所示。

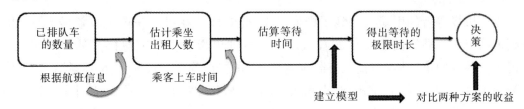

图 6.3 决策思维导图

根据表 6.1,分别对方案 A、B 进行量化分析和比较。

表 6.1 参数表

f	出租车需求量	$G_{等}$	机场等客出租车的净收益
Q_t	t 时刻到达站台的乘客流出量	r_0	起步价格
α	在站台换乘出租车的乘客换乘率	S_0	起步距离
ω	出租车平均载客量	r_1	超出起步距离每千米的价格
a	排队出租车	h_0	机场—市中心的油费
S	机场—市中心的距离	$G_{返}$	返回市中心接客的净收益
t_0	机场—市中心的行驶时间	h_1	市中心内的载客油费
t_1	机场等待载客的时间	v_2	市内行驶的低速度
v_1	市中心外的高速度		

- A 决策：机场等待接客的最优事件结果

已知出租车需求量为：$f = Q_t \alpha / \omega$。

根据文献可得现场调研参数值，出租车入口排队区放行时间为 35 s，待客区载客出租车离去间隔为 10 s，则站台处一组乘客上出租车的总时间参数标定为 45 s[2]，排队出租车为 a 辆，则选择在机场等待载客的司机等待时间 t_1 为：$t_1 = 45a$，且 $1.6(a+1) \leqslant Q_t$。

设出租车从机场载客后行驶最远至市中心，则耗费的总时间和净收益为：

$$t = t_1 + t_0,$$
$$t_0 = S/v_1,$$
$$G_{等} = r_0 + (s - s_0) * r_1 - h_0。$$

由于机场至市中心属于长距离载客，净收益 $G_{等}$ 等于司机获得的起步价格和额外价格减去这段路的油费。

- B 决策：返回市区载客的最优事件结果

若司机选择返回市区载客，假设机场至市区之间无载客，且一到达市区立刻接客，之后的时间一直保持短途接客的理想状态，则选择 B 方案得到的收益为：

$$G_{返} = (t - t_0) * v_2 * r_0 / s_0 - h_0 - h_1 = t_1 * v_2 * r_0 / s_0 - h_0 - h_1。$$

在 t 时间内，B 方案的净收益 $G_{返}$ 等于除去赶回市中心的时间外，按照单位起步价获得的收益，再减去机场—市中心的油费（h_0）、市中心内载客油费（h_1）。

(3) 模型的求解。

当在机场等待的甲司机等到了极限时长，此时即使在机场获得了收益最大化，也无法超过回市中心载客的乙司机的收入。为了得到出租车司机愿意接受的极限时长，设乙司机空载返回市区后，一直保持载客状态；甲司机在等待一段时间 t_1 后载客返回市区时刻为止，计算此时使甲、乙两司机收入相同的时间，即在 $G_{等} = G_{返}$ 的条件下，计算极限值 t：$[r_0 + (s - s_0) * r_1 - h_0] = t_1 * v_2 * \dfrac{r_0}{s_0} - h_0 - h_1$，

得到：$t_{极限} = [r_0 + (s - s_0) * r_1 - h_1] * \dfrac{s_0}{t_1 * v_2 * r_0} + t_0$，

上面的 $t_{极限}$ 就是司机愿意等待的最长时间。

当司机预估的等待时间小于极限值,较大可能性会选择等待在机场载客;当预估时间大于极限值,等待在机场的机会成本更高,司机会选择返回市区再载客。

6.5.2 问题二的分析与求解

1. 对问题二的分析

在问题一的基础上,首先结合上海浦东机场、北京首都国际机场、乌鲁木齐机场等数据,计算模型Ⅰ的运行结果,即为出租车理论等待时长极限;再根据各机场的航班时刻表、出租车换乘率等参数,计算与时间相关的每时刻出租车需求量,假设在机场已等待的出租车数量等于该需求量,从而求出实际等待时长极限。

图6.4　问题二流程图

定义1　理论等待时长极限:指根据问题一的模型结论,将参数带入得到一个平的出租车等待时间的理论预期结果。

定义2　实际等待时长极限:指基于城市具体的航班表和其他参数推出的与时间相关的实际最长等待时间。

2. 对问题二的求解

(1)数据的预处理。

根据2018年12月上海市机场调查报告里上海浦东国际机场的相关资料,得到2018年11月日均的回航数量,并找出乘客换乘率、出租载客率等参数数据,并对其他必要参数值进行推算,如出租车需求量等。然后利用这些数据求出司机实际等待的时间上限,再利用问题一结论得出理论极限时长结果,与实际情况进行比对。

(2)理论值对比。

• 实际等待时长极限

根据2018年12月上海市机场发展报告,上海浦东机场11月平均每小时进港的班次如图6.5所示。浦东机场主要运营时段为7点至23点,9点至23点入港数均大于30架,机场进港高峰为15点和19点时段。

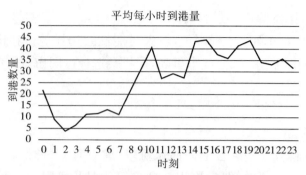

图 6.5　2018 年 11 月日均时刻返航数

经查阅资料,对浦东机场航空换乘出租车的参数进行标定:$\alpha=0.05, \omega=1.6$,即每架飞机约有 5% 的乘客选择乘坐出租离开站台,从站台载客的出租车平均量载 1.6 人[3]。则由问题一中出租车需求量的表达式可知,出租车在不同时刻的实际等待最长时间为:$t_{实际}=f_t*0.0125=0.0125*Q_t\alpha/\omega$,其中 f_t 表示 t 时刻的出租车需求量,绘制成折线图,如图 6.6 所示。

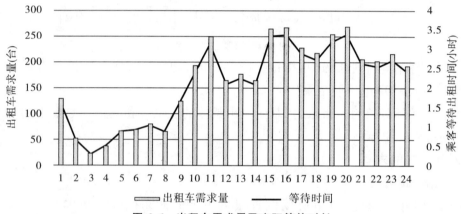

图 6.6　出租车需求量及实际等待时长

图 6.6 表明了不同时刻的出租车需求量,及在此需求量后出租车继续排队,需要的最长等待时间。

• 理论等待时长极限

选取上海陆家嘴作为市中心,陆家嘴至上海浦东机场的距离为 48 km。根据上海市出租车公交规定,查询到出租车 15 km 的白天、晚上的起步价格分别为 44 元和 55.2 元,计算油费均价为 0.5 元,最高时速 60 km/h,最低时速 30 km/h。根据问题一的结论,我们将搜集到的浦东机场的数据代入公式,得到如下结果。

$$\begin{cases} t = 2.857 \text{ 小时(白天)}, \\ t = 4.967 \text{ 小时(晚上)}。 \end{cases}$$

则白天出租车等待时间上限为 2.857 小时,晚上等待时间上限为 4.967 小时。

模型结果完全包含了图 6.6 中的数据,晚上等待最长时间不超过 4.967 小时,白天不超过 2.857 小时。这说明普通时间段内,机场出租车实际等待时长与模型I结果相吻合。

下面进一步在特殊情景、超负荷载客量等情况下对模型Ⅰ进行论证,检验其是否具有适用性和普遍性。

(3)模型的论证。

①假期客流量高峰。2019年春运于1月21日启动,3月1日结束,为期40天。浦东机场春运期间航班起降架次约5.60万架次,同比增长1.21%,旅客吞吐量约860万人次,同比增长4.7%(日均起降1400架次,日均进出港旅客21.5万人次)。根据客流分布,进出港客运峰值将达到22.5万人次。假设出租车转乘率(0.05)、出租车载客率(1.6)及乘客上车时间(45 s)等车参数不变,求得节假日高峰期的实际等待时间上限为:$t=7.324$ 小时。

假期时间内的平均最长等待时间和理论值求解中得到的白天2.857小时、晚上4.967小时有较大差异,由此可以看出,机场出租车的等待载客的时长受节假日客流量高峰期的影响。

②天气影响。季节和天气对航班有直接且较大的影响。根据相关文献,以浦东机场为例,对天气和气候进行量化计算得出其对航班的影响。考虑天气原因,计算离港飞机的独立延误时长,参数和公式如下。

表6.2 参数表

AJ	航班计划离港时间	TM	航班最小过站时间
BJ	航班计划到达时间	TJ	航班计划过站时间
AS	航班实际离港时间	W_0	航班离场总延误
BS	航班实际到达时间	W_1	独立离港延误
C	航班过站松弛时间	W_2	波及延误

$$\begin{cases} TJ = AJ - BJ \\ C = TI - TM \\ W_2 = BS - BJ - C \\ W_0 = AS - BS \\ W_1 = W_0 - W_2 \end{cases}$$

以秋冬季节为例,上海处于沿海城市,易受暴雨、冰雹、雷电等极端天气影响。文献[4]分析了不同天气的得分和飞机独立离港时长延误的关系,得到正相关的线性关系:在一定范围内,极端天气的长时间延续恶化,会导致飞机延误的时间增加。由于天气和气候对航班延误有较明显的影响,间接影响了站台候车客流量,因此在考虑出租车的载客选择时,需要考虑天气和气候的影响因素。

6.5.3 问题三的分析与求解

1. 对问题三的分析

本问题提出在机场陆侧交通实际情况下,经常会出现出租车排队载客和乘客排队乘车的情况。假定某机场乘车区现有两条并行车道,管理部门应如何设置"上车点",并合理安排出租车和乘客,在保证车辆和乘客安全的条件下,使得总的乘车效率最高。

现今国内机场基础设施建设较为完善,但有些方面远远达不到高水平旅客服务建设程度。比如正在研究的国内机场乘客搭乘出租车排队问题,乘客下机后乘车效率低下。机场大多建在远离市中心的偏远位置,尤其到了晚上,选择出租车离港的乘客很多,如果没有精准完备的出租车管理系统,就会导致大量旅客滞留机场,对机场运行管理造成负担。若能对需求精准预判,实现车辆提前调配,旅客与出租车双方的利益就能最大化。

出租车排队系统类似于物流管理中的自动化物料搬运流程,大致可以分为四个步骤:

①出租车在蓄车区域等待需求指令;

②在收到乘车需求指令后开往指定泊车位载客(依次进入 A 或 B 车道,接上各自对应上车点的乘客);

③载上乘客的出租车进入驶离区域出发;

④出租车将乘客送到指定位置后,选择返回或者不返回机场蓄车位,若返回,则等待下一次的需求指令。详细流程如图 6.7 所示。

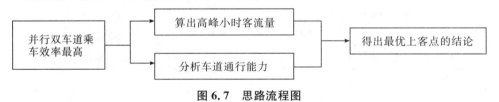

图 6.7 思路流程图

2. 对问题三的求解

模型Ⅱ——蒙特卡罗仿真模拟模型

(1)模型准备。

模型的原理。蒙特卡罗方法,又称随机抽样或统计模拟方法,是一种以概率统计理论为指导的数值计算方法。将蒙特卡罗方法用于仿真即为蒙特卡罗仿真。蒙特卡罗方法适用于两类问题:第一类是本身就具有随机性的问题,第二类是能够转化为某种随机分布的特征数,比如随机变量的期望值或随机事件出现的概率。

利用蒙特卡罗仿真模拟算法,模拟随机过程。设计 n 个上客点来模拟出租车和乘客的交互过程,综合分析乘客和出租车的排队等待时间,得出最优上客点数量。整个过程中,按照出租车的行驶状态可以分为:空车等待、空车行驶和载客行驶三种驾驶情况。

(2)模型的建立。

①研究车道边通行能力的约束条件。根据收集到的数据,可以观察到在正常情况下,每天的出租车乘车需求量都比较大,所以可以认为以下现象假设合理。

- 在客流量高峰期进行研究其最大通行能力,故假设在一定时间内有充足的乘客乘车,不会出现没人等车排队的现象。
- 在一定时间内,蓄车位有充足的出租车排队等待接客,不会出现车辆短缺的现象。
- 上客区有 n 个泊车位,且有 n 名乘客同时上车。
- 每个泊位上客时间服从负指数分布,排序靠后的泊位上客时间比前一个多 $T(h)$。

② 蒙特卡罗仿真模型。在上述约束条件的前提下,单车道单通道车道边通行能力问题可转化为:随机仿真模拟 x 次,每次生成 n 个(n 为上客点泊位数)随机变量,到达时间间隔服从参数为 $\frac{1}{t_0}$,$\frac{1}{t_0+T_0}$,$\frac{1}{t_0+2T_0}$,$\cdots$,$\frac{1}{t_0+(n-1)T_0}$ 的负指数分布,t_0 为第一个上客点的时间,记 R_i 为第 i 次生成的 n 个随机变量中的最大值,即:

$$R_i = \max(r_{i1}, r_{i2}, r_{i3}, \cdots, r_{in})$$

则单车道单通行能力可计算为:

$$x = \text{int}(x),$$
$$g_x = \sum_{i=1}^{n} R_i + xT_b = 3600,$$
$$C = \omega n x,$$

其中:x 取整数,C 为单车道单通道车道边通行能力(人/小时);β 为出租车平均每车的载客人数,根据文献,一般取 1.4。

重复上述计算过程,可得到 y 个独立的通行能力数据,其分别可计为 $C_{11} + C_{12} + C_{13} + \cdots + C_{1y}$,取其均值即得到最终的通行能力:

$$\bar{C} = \frac{C_{11} + C_{12} + C_{13} + \cdots + C_{1y}}{y}$$

在给定的出租车上客泊位设置规模(上客泊位数 N)的情况下,根据出租车在上客区的行驶规则和乘客的上车规则,模拟出租车和乘客的到达过程、出租车载客驶离过程、待客出租车重排过程和乘客上车过程,以计算乘客和出租车的平均排队时间[5]。

对于车道边出租车排队载客问题,简化成图 6.8 和图 6.9 所示。

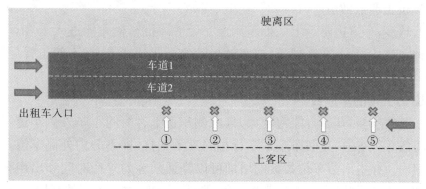

图 6.8 传统上客区单侧上车车道

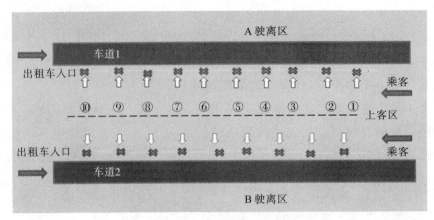

图6.9 上客区双侧上车车道

(3)模型的求解。

车道边通行能力

出租车先在蓄车位进行等待,得到指令后,开往乘车区的车道边,可以认为有连续的专用通道将蓄车区域与泊车位相连接。乘客和出租车的排队时间越小,车道边通行能力越大,即乘车效率越大。影响平行并发式车道的通行能力的主要因素有以下4个方面。

A. 延误驶出

平行双车道互相独立,不会相互干扰。但是每个车道边各设置几个上客点,求解最优上客点可以大大提高车道边的通行能力,在最短的时间内运送最大的客流,减少旅客和出租车排队等待时间,以提高乘车效率。

B. 车道边上旅客上车时间

上车时间是其直接影响因素,上车时间取决于一个车道中几个上客点用时最长的那个乘客,上车时间越长,泊车位车辆停靠时间越长,通行能力越小,乘车效率越小。乘客上车时间可以包含放置行李,上车开关门,说明目的地,车辆启动所需的时间。问题二已经说明乘客的平均上车时间为45 s。

C. 泊车位上客点数量

上客点越多,车道边要等上一批车辆走完后下一批车辆才能进入上客点,导致后排车辆等待时间变长,且上客区越长,乘客对车辆的干扰越严重,越不便于管理,且前期的建造费和中间的设备维修费也随之增多。因此,存在最优上客区问题。

D. 车道边上客的车道数

一般车道数越多,其通行能力越大,且内侧车道上车时间小于外侧车道。而本题我们只研究两条并行车道,且两条车道平行相隔互不干扰,即不分内外侧车道。不同车道上客时间相同。并指出泊车位数量的增加可以降低乘客的等待时间和出租车在蓄车区等待时间,但同时增加了出租车的闲置时间,降低了泊车位出租车的利用率。

最优上客点规律

用 MATLAB 进行编程,对上述流程进行仿真模拟,可以得到上客点为 n 时单车道的通行能力。设置 n 的取值范围是 [1 20],超过 20 个上车点,会导致上客区过长,不符合实际情况。对 $n=1,2,\cdots,20$,进行仿真模拟,可以得到有 n 个上客点时其对应的车道边通行能力。即可认为当 $\max C_i = \{\overline{C_1}, \overline{C_2}, \overline{C_3}, \cdots, \overline{C_{20}}\}$ 时,即对应有 i 个上客点时其车道边通行能力最大,故单车道的最优上客点个数为 i。

单车道单通道车道边是其他车道边组合的基本单元,计算得到其通行能力,即可得到其他交通组织形式的通行能力[5]。

模拟仿真结果分析。图 6.10 是两车道并行车道边在不同上客点数量时的通行能力。

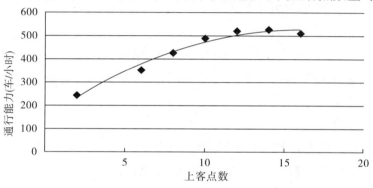

图 6.10 两车道并行通行能力分析

当泊车区域的上客点数超过 10 个以后,由于后排上客完毕的出租车受前车阻挡时间增加,通行能力图像出现拐点。

原因分析:上客点越多,意味着上客区越长,越不利于上客区的管理,行人对车辆的干扰越严重,也越容易引起上客区的混乱。同时上客点数量过多也会导致穿越的旅客对靠人行道的内侧车道造成较大干扰。

给管理部门的建议

(1) 上客点个数要适中。若上客点数设置得太少,在部分高峰时期会造成客流和车流的堵塞,乘车效率低下。适当地增加上客点数,可以有效避免交通枢纽堵塞。一般随着上客点数的增加,车道边通行能力提高。但是上客点的数量过多,会导致乘车区过长,不便于工作人员管理,而后排车辆要等待前方所有的车辆都驶出后才能进入上客点,等待时间较多。

根据上述仿真模拟结论,双车道并行车道边通行能力在上客点个数小于 10 之前线性增长,但上客点个数大于 10 以后,通行能力开始缓慢增长,最后甚至开始下降。从建造成本和维护成本来看,每增加一个上客点,成本费用就会持续增加,而通行能力的增长缓慢,总的乘车效率得不到提高,还会出现资源浪费。

(2) 对于两条并行车道,应根据高峰客流量与车道边的通行能力,在容许乘客等待一定的时间下,建议设置的上客点不宜超过 10 个,以使总的乘车效率最高。

(3) 双车道并行模式创新规划(模式见图 6.9)。双车道中间是乘客上车点,左右两边规划车道,此时两车道互不影响。而当双车道紧挨在一起,那么乘坐外侧车道内车辆的乘客需穿越内侧车道,易造成车道内管理混乱,延误上车时间,乘车效率没有前者的高。

6.5.4 问题四的分析与求解

1. 对问题四的分析

本问题要求我们给出机场短途载客出租车的可行的"优先"安排方案,使得出租车选择长途短途载客的收益尽量均衡。

本问题完成的主要步骤为:

① 以上海市出租车市价为例构造分段函数;
② 利用理想市场理论,构建出租车与乘客的供给函数;
③ 通过对短途出租车司机决策进行模拟,选择出最优选择方案;
④ 采用运筹学排队模型对排队情况进行模拟;
⑤ 得出优先安排方案。

定义 1 出租车保有量:指机场出租车蓄车池内停放的出租车总数,是数量指标。

定义 2 短途行驶司机:指在机场周围 25 公里以内并能够在 1 小时之内返回机场的出租车司机。

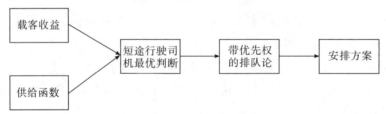

图 6.11 思路流程图

2. 对问题四的求解

模型Ⅲ——基于供求关系的短途出租车收益最大化

模型的准备

对于上海市出租车收费定价进行数据收集,在日间(5:00—23:00)及夜间(23:00—次日 5:00)两个时间段的基础上将行驶里程分为三段,具体数据见表 6.3。

表 6.3 上海市出租车收费标准

公里数	日间(5:00—23:00)	夜间(23:00—次日 5:00)
0—3 公里	14 元	18 元
3—15 公里	2.5 元/公里	3.1 元/公里
15 公里以上	3.6 元/公里	4.7 元/公里

得出关于行驶里程的分段函数 y。

日间出租车计价函数：

$$y = \begin{cases} 14, & 0 \leqslant s \leqslant 3, \\ 6.5 + 2.5s, & 3 \leqslant s \leqslant 15, \\ -10 + 3.6s, & 15 \leqslant s. \end{cases}$$

夜间出租车计价函数：

$$y = \begin{cases} 18, & 0 \leqslant s \leqslant 3, \\ 8.7 + 3.1s, & 3 \leqslant s \leqslant 15, \\ -15.3 + 4.7s, & 15 \leqslant s. \end{cases}$$

根据上海市浦东国际机场周围商圈划分情况，结合上海市出租车定价数据，对各个商圈价格进行估算，其中行驶里程≤25 km 为短途，行驶里程≥25 km 为长途，具体数据见表 6.4。

表 6.4 浦东机场出租车里程及参考价格

	目的地	里程(km)	日间参考车费(元)	夜间参考车费(元)
短途	川沙	23	73	93
	南汇	25	80	102
	迪士尼	24	76	98
长途	徐家汇	48	163	210
	陆家嘴	46	156	201
	浦东八佰伴	44	148	192
	外滩	47	159	206
	上海科技馆	40	134	173
	虹桥交通枢纽	60	206	267

(2) 模型的建立。

当可以实时获得航班信息以及客流量时，建立如下需求模型：

$$f(\alpha, t) = Q_t \alpha / \omega \text{ 且 } \omega \in (0, 4];$$

引入短途选择比 β，则短途出租车需求函数 $f(\alpha, t)$ 可改写为：

$$f(\alpha, \beta, t) = Q_t \alpha \beta / \omega, \beta \in (0, 1);$$

长途出租车选择比应为 $(1-\beta)$，有 $f(\alpha, \beta, t) = Q_t \alpha (1-\beta) / \omega, \beta \in (0, 1)$。

已知当 t_1 时刻出租车保有量如下：

$$M = M_{t+\Delta t} + \int_{t+\Delta t}^{t + \frac{f + \Delta f}{v_{0_{t_1}}}} v_{1 t_1} \mathrm{d}t$$

则在 t_1 时刻理论上出租车保有量与长短途乘客的总需求量的供求关系为：

① 供不应求。

$$M_{t+\Delta t} + \int_{t+\Delta t}^{t + \frac{f + \Delta f}{v_{0_{t_1}}}} v_{1 t_1} \mathrm{d}t + \Delta t * v_{0_{t_1}} < f(\alpha, t) + f(\alpha, t_1)$$

②供过于求。

$$M_{t+\Delta t} + \int_{t+\Delta t}^{t+\frac{f+\Delta f}{v_{1t_1}}} v_{1t_1} \mathrm{d}t + \Delta t * v_{0t_1} > f(\alpha,t) + f(\alpha,t_1)$$

③供求相等。

$$M_{t+\Delta t} + \int_{t+\Delta t}^{t+\frac{f+\Delta f}{v_{1t_1}}} v_{1t_1} \mathrm{d}t + \Delta t * v_{0t_1} = f(\alpha,t) + f(\alpha,t_1)$$

司机收益为:$G = G_车 - G_油$。

(3)模型的求解。

由题目要求收益相同进行场景模拟,通过问题二给出结论,选取等待时间和客流量的均值代入需求与供给两式后,得出需求为 165 辆/h,出租车供给为 200 辆/h,得出供大于求的结论。通过短途出租车选择式可以得出短途倾向旅客约为 25 辆/h,将数据代入长途出租车选择式可得长途倾向旅客约为 140 辆/h。

由问题二可知,当需求为 165 辆/h 而供给为 200 辆/h 时,所需等待时间应为 1.5 h,则可假设在等待后有 A、B 两辆车接到乘客,A 接到短途乘客,B 接到长途乘客,并假定油费固定在 0.4 元/公里,收益分析如下。

①A 车。以送至川沙为例:$G_1 = G_车 - G_油 = 73 - 23 \times 0.5 = 61.5$ 元。

以时速 60 km/h 速度计算,所需时间为 25 min。此时,不考虑其他方面,司机有空车回城或回机场两个选择。

以外滩为终点,此时:$G_2 = G_车 - G_油 = 0 - 20 \times 0.5 = -10$ 元,即回到外滩所消耗的时间为 2.4 h,总收益 G 为 51.5 元,即 21.5 元/h。

②B 车。假定外滩为终点,则:$G_3 = G_车 - G_油 = 159 - 47 \times 0.5 = 135.5$ 元。

所需时间为 57 min,此时,B 车所消耗时间为 2.45 h,即 55.3 元/h。

由上述计算可知若司机接短途客人则会产生较少利润,除去固定费用后可能会亏本,故某些短途载客的出租车司机会再次返回。

记返回至机场时总利润为 G_4,从机场载客后返回外滩,最终获得总收益为 G_5。

$$G_4 = G_1 - G_2 - 0.5 \times 23 = 40 \text{ 元},$$
$$G_5 = G_3 + G_4 = 40 + 135.5 = 175.5 \text{ 元}。$$

假设返回后再次排队的时间为 t_2,B 车此时在市区以 30 min 为一周期拉得起步价 14 元的乘客,由问题二可知若要收益相等则可得出如下公式:$175.5 \div (3.3 + t_2) = (148 + t_2/0.5 \times 12.5) \div (2.6 + t_2)$。

可以求出 t_2 均为负数,则当 t_2 作为理想状态无限趋近于 0 时,A 车的收益将由 21.5 元/h 转为 53.2 元/h。

即当短途司机返回机场后,等待时长与每小时收益成反比,短途司机回机场后接到长途旅客为理想状态,如若所接依旧为短途旅客则会降低边际收益。

模型Ⅳ——基于非强占型排队论的出租车模型

(1)模型的选择与假设。

通过附件一的数据可以看出,出租车的数量呈 Poisson 分布,通过优待措施可将出租车分为享有优先权和无优先权两个等级,由问题二可知每分钟可搭载的乘客数为 $\mu=1.33$ 人,出租车每分钟到达 $\xi=2.75$ 辆。当无优先权的出租车等待 $\theta=3$ 个享有优先权的出租车优先载客后,可进行插队,由模型 I 可知出租车属于享有优先权的概率为 0.15,无优先权的概率为 0.85。

(2)模型的建立。

对于整个系统而言,系统的平均队长为[6]:$L_s = \dfrac{\xi}{\mu - \xi}$。

系统的平均等待队长为

$$L_q = \dfrac{\xi}{\mu - \xi} \cdot \dfrac{\xi}{\mu} = \dfrac{\xi^2}{\mu(\mu - \xi)},$$

对于短途出租车 A 进站后可能情况进行如下讨论。

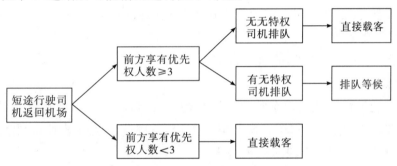

图 6.12 短途出租车 A 排队情形

A 前方有较多等待车辆,有足够无优先权司机按间隔为 3 插队[7]:$W_q = T_1 + T_2 = \rho_1 W_1 + \rho/\mu$。

A 前方有较多等待车辆,无足够无优先权司机按间隔为 3 插队:$W_q = T_1 + T_2 + T_3 = \rho_1 W_1 + \rho_2 W_1 + \rho/\mu$。

最终可知短途出租车 A 在机场的逗留时间(平均排队时间+平均服务时间)应为:$W = W_q + 1/\mu$。

所得出的等待队长应为:$L_1 = \lambda_1 W_1$。

(3)模型的求解。

对于上述模型进行 Lingo 模拟,得出模拟结论,见表 6.5。

表 6.5 输出结果示意表

上车效率	前方短途出租车车辆数(台)	平均队长(台)	平均逗留时间(min)
45 s/人	0—2	0.58	0.5
	3—5	3.15	2.40
	6—8	6.07	4.56
	9—11	9.06	6.80
	12—15	13.06	9.8

续表

上车效率	前方短途出租车车辆数(台)	平均队长(台)	平均逗留时间(min)
40 s/人	0—2	0.54	0.43
	3—5	2.99	2.05
	6—8	5.85	3.94
	9—11	8.82	5.91
	12—15	12.81	8.59

通过表6.5可以看出，当短途出租车司机返回时，由于享有优先权依旧可在10 min内接到下一位乘客，以模型Ⅳ中A出租车司机为例，等待后的时薪为51.6元，与直接载客回市区的出租车收益差距较小。

提高乘客的上车效率可在一定范围内减少出租车的排队时间，从表6.5可以看出，每人减少5 s上车时间，可使出租车等待时间减少13%。

通过输出结果可知当前方有较多返回的短途出租车时，出租车的有效到达率高于1，处于较好状态，推断能够在一定程度上达到收益尽量均衡。

3. 对问题四的解答

通过对于现有机场实施措施进行改进。

(1) 提出如下建议。

①GPS轨迹追踪。从出租车驶离机场开始计时，当出租车返回排队系统入口时，获取该车号码牌，并根据该出租车上次行驶的GPS轨迹信息及行驶时间判断是否满足短途行驶条件，若满足，则可走快速通道进入。

②分流引导。对于满足短途行驶条件的出租车可进行分流引导，具有非强制型的后到先进优先服务权，但应在考虑短途出租车的同时考虑排队等候的其他出租车，可在短途行驶出租车数量较多时实行每走3辆短途出租车后，其他无特权出租车可进行插排。

③大数据+互联网。通过后台大数据进行监控，对于旅客到站人数进行预测，并进行展示屏展示使得蓄车池内出租车司机以及送客至机场的出租车司机能够以此作为参考进行决策。

(2) 最终得出方案如下。

Step 1：司机驶出机场时获取其车牌号码。

Step 2：司机驶回机场后根据出租车行驶的GPS轨迹以及行驶时间判断是否满足短途行驶条件（来回50公里并在1小时以内）。

Step 3：若司机满足短途行驶条件，则对于出租车进行分流，进入免进蓄车池通道并开启相应的道闸栏杆。

若司机不满足短途行驶条件，则会提示该出租车应驶向蓄车池等待排队，并开启相应的道闸栏杆。

Step 4：若司机进入分流通道后，前方有较多满足在短途行驶条件的司机则需进行排队，实行三辆享有特权出租车一辆无特权出租车方式进行排队载客。

若司机进入分流通道后,如果前方仅有不超过 2 辆满足短途行驶条件的出租车则可根据提示进行载客。

Step 5:司机接到乘客并驶离机场。

(3)方案优点。

①利用 GPS 轨迹检测可在一定程度上减少人工工作量,降低成本,得到更为真实的短途出租车需求量数据。

②利用大数据和互联网使司机能够第一时间获得数据,并分析数据,做出适当决策。

③将短途司机与正常排队司机交叉排队,既满足短途司机的优先权又能够在一定程度上保证正常排队司机的权益。

④整体方案在现有的首都国际机场、浦东国际机场的现有方案上进行改进,具有可行性。

6.6 误差分析与灵敏度分析

6.6.1 误差分析

(1)在问题一和问题二中,由于航班每天数量不可控,旅客人数也较难精准获得,上海浦东国际机场可选择的换乘交通工具较多,由于各个交通工具开放时间有差距,故会使得在夜间乘坐出租等交通工具的乘客较多,为了方便计算,选取的出租车上座率为每日出租车上座率均值,后期会产生一定的误差。

(2)在问题三中,蒙特卡罗仿真实验是模拟随机过程,由于数据量太大,运行时间较长,故运行次数较少,而模拟次数不同结果会有所差异,故对最后的结论造成一定的误差。

(3)在问题四中,由于每日客流量不可控,每日到港旅客对于短途的需求量也较不可监测,为方便运算,文中选取短途旅客概率为日常均值,对于享有优先权的出租车等待时间以及平均队长造成一定的误差。

6.6.2 灵敏度分析

问题一根据数据测算,得到拟合曲线:$y=2.5x+16$。

对于基准行驶价格进行讨论,针对不同的系数 a,我们运用 MATLAB 软件进行灵敏度分析。

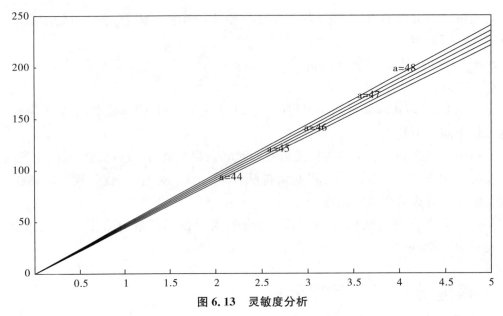

图 6.13 灵敏度分析

由灵敏度分析可知,当行驶路程超过 3 小于 15 时,变动系数 a 可发现,当变动单位量的系数时获得的单位价格差距较小,故此次灵敏度分析效果较好。

6.7 模型的评价与推广

6.7.1 模型的评价

1. 模型的优点

(1) 问题三蒙特卡罗模型的优点:由于出租车和乘客的到达都具有随机性和偶然性,蒙特卡罗仿真模拟对随机过程模拟程序结构简易;

(2) 蒙塔卡罗仿真模拟误差容易确定,收敛速度较快;

(3) 蒙特卡罗仿真模拟比算式解法更加准确,进行出租车与乘客的排队时交互式模拟,可以对不同模式下的车道边情况进行改进、对比分析。

2. 模型的缺点

(1) 对于一些数据,我们对其进行了一些必要的处理,这会带来一定的误差;

(2) 模型中为使计算简便,使所得结果更理想化,忽略了一些次要影响因素;

(3) 蒙特卡罗仿真时需要模拟次数较大,所以导致仿真时间过长,收敛速度慢;

(4) 问题三模型误差具有概率性。

6.7.2 模型的推广

1. Laplace 准则的推广

Laplace 准则应用广泛,在医学临床的科研中有较大的应用,可用其诠释一些病理现象。如呼吸内科、五官科、内科等均用 Laplace 准则解释生理现象并推测其发展可能性，在大血管重构的研究中，Laplace 准则对脉压对劲动脉的压迫有重要研究意义。

2. 排队论模型的推广

在模型中将拟合较好的数据进行模拟，可得到具有一定参考意义的平均队长、平均等待时间的结论，不论是运用在车辆排队、医院看病、超市服务、银行服务均具有可行性，具有一定的推广性。

3. 蒙特卡罗仿真模型的推广

随着电子计算机的发展，蒙特卡罗方法在解决科技问题时也越来越广泛。它不仅较好地解决了许多高难度和复杂的数学计算问题，而且在统计物理、核物理、真空技术、地质、医学等领域都得到成功应用。

6.8 模型的改进

由于问题四中非强占型排队论研究问题较为复杂，对于问题四只采用了一个上车点，故可对于上车点的数量进行改进，以 M/M/n/∞ 为例，进行该种情况下上车点数量多少的影响为进一步改进问题四中的非强占型排队论提供参考。

表 6.6 对于多上车点的 M/M/n/∞ 模拟

上车点	平均排队长（台）	平均等待时间（min）
3	3.31	1.2
4	2.34	0.8
5	2.17	0.79
6	2.13	0.77
7	2.11	0.77

由表 6.6 可看出，上车点并不是尽可能多就好，所以进一步改进非强占型排队论的上车点时，可将上车点控制在 7 个以内。

参考文献

[1] 胡运权. 运筹学基础及运用[M]. 北京：高等教育出版社，2014.

[2] 耿中波，宋国华，赵琦，高永，万涛，辛然. 基于 VISSIM 的首都机场出租车上客方案比选研究[J]. 中国民航大学学报，2013，31(06)：55—59.

[3] 李嘉靖，李江. 大型交通枢纽出租车预警系统分析[J]. 交通科技与经济，2009，11(01)：59—61.

[4] 刘彤丹. 基于天气影响的离场航班延误分析及预测[A]. 中国科学技术协会、中华人民共和国交通运输部、中国工程院. 2019 世界交通运输大会论文集(下)[C]. 中国科

学技术协会、中华人民共和国交通运输部、中国工程院:中国公路学会,2019:13.

[5] 柳伍生,谭倩.交通枢纽出租车车道边通行能力仿真研究[J].计算机仿真, 2012,29(04):357-361.

[6] 陆传赉.排队论[M].北京:北京邮电大学出版社,1993.

[7] 黄业文,邝神芬.非强占有限优先权 M/M/n/m 排队系统[J].应用概率统计, 2018,34(04):364-380.

◆ 建模特色点评

❋ 论文特色 ❋

◆标题定位:从赛题"机场的出租车问题"到"机场出租车返程载客综合研究",标题具有针对性,标题定位准确、合理、简洁。

◆方法鉴赏:文章基于出租车和乘客的到达都是具有随机性和偶然性,突出了蒙特卡罗模型仿真模拟的随机过程。蒙塔卡罗仿真模拟误差容易确定,收敛速度较快。蒙特卡罗仿真模拟比算式解法更加准确,进行出租车与乘客的排队时交互式模拟,可以对不同模式下的车道边情况进行改进、对比分析。

◆写作评析:论文写作格式规范,内容全面,能有针对性地按建模思路、模型的建立与求解、结果分析等步骤处理,行文有条不紊且图文并茂。

❋ 不足之处 ❋

在细节处理方面不够细致,存在如文献标注没有上标等一些小问题。

第 2 部分

美国大学生数学建模竞赛优秀论文评析

内 容 简 介

一、美国大学生数学建模竞赛简介

美国大学生数学建模竞赛(MCM/ICM)由美国数学及其应用联合会主办,是唯一的国际性数学建模竞赛,也是世界范围内最具影响力的数学建模竞赛。MCM/ICM 是 Mathematical Contest In Modeling 和 Interdisciplinary Contest In Modeling 的缩写。MCM 始于 1985 年,ICM 始于 2000 年,由 COMAP(the Consortium for Mathematics and Its Application,美国数学及其应用联合会)主办,得到了 SIAM,NSA,INFORMS 等多个组织的赞助。赛题内容涉及经济、管理、环境、资源、生态、医学、安全等众多领域。竞赛要求三人(本科生)为一组,在四天时间内,就指定的问题完成从建立模型、求解、验证到论文撰写的全部工作,体现了参赛选手分析问题、解决问题的能力及团队合作精神。它是现今各类数学建模竞赛之鼻祖。MCM/ICM 着重强调研究和解决方案的原创性、团队合作、交流及结果的合理性。2019 年,共有来自美国、中国、加拿大、英国、澳大利亚等 17 个国家和地区共 25370 支队伍参加,它们来自哈佛大学、普林斯顿大学、麻省理工学院、清华大学、北京大学、上海交通大学等国际知名高校。

二、安徽财经大学参赛获奖情况介绍

安徽财经大学从 2010 年起组织学生参加美国大学生数学建模竞赛,至今已累计获美国赛国际奖 225 项,其中 56 项为 Meritorious Winner 奖,169 项为 Honorable Mention 奖。由于篇幅有限,本书仅从 2010—2019 年所获得的 56 项 Meritorious Winner 奖论文中筛选出 5 篇。它们分别获得:2011 年国际一等奖;2015 年国际一等奖;2016 年国际一等奖;2017 年国际一等奖;2018 年国际一等奖。

第 7 篇
电动汽车对环境和经济的影响如何？其广泛使用是否可行？

◆ 竞赛原题再现

Problem C How environmentally and economically sound are electric vehicles? Is their widespread use feasible and practical?

Here are some issues to consider, but, of course, there are many more, and you will not be able to consider all the issues in your model(s):

• Would the widespread use of electric vehicles actually save fossil fuels or would we merely be trading one use of fossil fuel for another given that electricity is currently mostly produced by burning fossil fuels? What conditions would need to be put in place to maximize the savings through use of electric vehicles?

• Consider how much the amount of electricity generated by alternatives such as wind and solar would need to climb during the twenty-first century to make the widespread use of electric vehicles feasible and environmentally beneficial. Assess whether or not the needed growth of these alternate sources of electricity is likely and possible.

• Would charging batteries at off-peak times be beneficial and increase the feasibility of widespread use of electric vehicles? How quickly would batteries need to charge to maximize the efficiency and practicality of electric vehicles? How would progress in these areas change the equation regarding the environmental savings and practicality of widespread use of electric vehicles?

• What method of basic transportation is most efficient? Is the efficiency of different methods dependent of the nation or region in which it is used?

• Pollution caused directly by electric vehicles is low, but are there hidden sources of pollutants associated with electric vehicles? Gasoline and diesel fuel burned in internal combustion engines for transportation account for nitrites of oxygen, vehicle-born monoxide and carbon dioxide pollution but are these bi-products something we really should worry about? What are the short and long term effects of these substances on the climate and our health?

• How would the pollution caused by the increasing need to dispose of increasing numbers of large batteries effect the comparison between the environmental effects of electric vehicles versus the effects of fossil fuel-burning vehicles?

• You also should consider economic and human issues such as the convenience of electric vehicles. Can batteries be recharged or replaced fast enough to meet most transportation needs or would their ranges be limited? Would electric vehicles have only a limited role in transportation, good only for short hauls (commuters or light vehicles on short trips) or could they practically be used for heavier and longer-range transportation and shipping? Should governments give subsidies to developers of electric vehicle technologies and if so, why, how much, and in what form?

Requirements:

• Model the environmental, social, economic, and health impacts of the widespread use of electric vehicles and detail the key factors that governments and vehicle manufacturers should consider when determining if and how to support the development and use of electric vehicles. What data do you have to validate your model(s)?

• Use your model(s) to estimate how much oil (fossil fuels) the world would save by widely using electric vehicles.

• Provide a model of the amount and type of electricity generation that would be needed to support your recommendations regarding the amount and type of electric vehicle use that will produce the largest number of benefits to the environment, society, business, and individuals.

• Write a 20-page report (not including the summary sheet) to present your model and your analysis of the key issues associated with the electric vehicle and electricity generation. Be sure to include the roles that governments should play to insure safe, efficient, effective transportation. Discuss if the introduction of widespread use of electric vehicles is a worthwhile endeavor and an important part of an overall strategy to address global energy needs in the face of dwindling fossil fuel supplies.

获奖论文精选

A Comprehensive Analysis for Widespread Use of Electric Vehicles

Summary: This paper mainly investigates the environmental and economic impacts of worldwide use of electric vehicles. It can be concluded that governments have good reason to promote the use of electric vehicles.

By analyzing large amount of data, advantages of electric vehicles over traditional fuel automobiles are obvious in economic cost and carbon emission on the premise that the correlation techniques are mature and electronic generation is mainly from clean energy.

Further, the global vehicles population is evaluated with the help of grey forecasting model and the amount of oil saving is estimated through approximate calculation. After that, based on the game theory, the amount and types of electricity generation needed by electronic vehicles are established.

In addition, some government incentive policies are designed to promote the adoption of electric vehicles, including technical innovation, subsidy system, and technology standardization and so on.

The main advantage of this paper is that the model is very plain and comprehensive, and allows us to see the prospects evidently. Meanwhile, the disadvantage still exists due to the insufficiency of statistical data, which leads to that some results are not accurate enough.

7.1 Introduction

An electric vehicle (EV), also referred to as an electric drive vehicle, uses one or more electric motors for propulsion. Electric vehicles include electric cars, electric trains, electric Lorries, electric airplanes, electric boats, electric motorcycles/scooters and electric spacecraft. For simplicity, an EV refers in particular to an electric car in this paper.

- **Carbon dioxide emissions in the automotive industry**

During the last few decades, increased concerns over the environmental impact of the petroleum-based transportation infrastructure, along with the peak oil, has led to

---

① 本文获 2011 年国际一等奖 M 奖。队员：周钰，董朝阳，赵晓梅；指导教师：庄科俊。

renewed interest in an electric transportation infrastructure [1]. In fact, the amount of carbon dioxide emissions from motor vehicles increased dramatically and accelerated the trend of global warming. As estimated by International Energy Agency (IEA) in 2006, about 15.9% of global carbon dioxide was from automobile exhaust, and the total carbon dioxide emissions will add to six billion tons in 2020 from three billion tons in 1990 as shown in Fig. 7.1 [2].

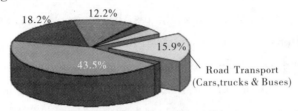

Fig. 7.1　Transport CO_2 Emissions Compared to Total Man-Made Emissions

- **Global energy situation and prominent contradiction**

Relevant data shows that global energy supply situation tends to become more tense day by day. The proved reserves of oil in 2009 only increased 0.5% than that in 2008, but the oil extraction rate and consumption growth rate reached 2.5% and 1.4% respectively. In addition, Fig. 7.2 shows that the global natural gas and petroleum reserves to production ratios were both below 50%. For coal, the reserves to production ratio was only just over 100%. In terms of energy consumption, the proportion in transportation is much greater than other industries. As Fig. 7.3 shows, in terms of transportation, the consumption proportion of oil was as high as 60.5% in 2006.

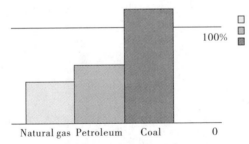

Fig. 7.2　Fossil Fuel Reserves to Production Ratios at the End of 2009

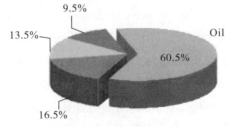

Fig. 7.3　Structure of Global Oil Consumption (2006)

All these crises show the necessity and urgency for the rapid development of new energy vehicles.

7.2　Problem Statement and Analysis

Faced with the increasing environmental and economical pressure, electric vehicles have come into view and are becoming more and more popular. However there are a lot of problems followed which one shouldn't turn a blind eye on. Now what we should consider are the issues as follows.

Firstly, the main reasons for developing EVs are related to environment and economy, so there is a dispute whether the EVs really play an evident role in saving the non-renewable resources and protecting environment. After all, there are many ways to generate electricity such as solar power, nuclear energy, wind power hydropower and fossil fuels.

Secondly, the widespread use of EVs requires adequate sophisticated infrastructures. To analyze the market attractiveness, we take the convenience and profits into account for consumers and manufactures respectively. Besides, the construction of charging stations is particularly important. How to arrange drivers to charge reasonably becomes the key point.

Thirdly, governments and manufactures will play a critical role in the popularization of EVs. What can they do to promote the use of EVs?

In view of the above descriptions, we need to solve the following problems:

• Search the data we need and model the environmental, social, economic, health impacts of the widespread use of EVs.

• Establish a model to estimate how much oil the world would save by widely using EVs. Then according to the amount and type of EVs widespread use, give the amount and type of electricity generation.

• Based on the models we have established and the kinds of impacts we have analyzed, explain whether the governments should support the promotion of EVs and what they have to do.

As is known to all, EVs promotion may lead to many complicated problems. From the above mentioned, the impacts mainly include environment, economic, society and health. Then we will evaluate the value of promoting EVs due to the results that how much oil can be saved by widely using EVs.

Although there are many types of EVs, hybrid vehicles and pure electric vehicles are the most popular. For hybrid vehicles, it can reduce oil dependency and carbon dioxide emissions but still not completely. On the other hand, equipping with double power systems not only adds the cars costs, makes handling difficult and complex, but also increases the weight of the car and reduces the overall energy efficiency. So this is only expediential in near future. Further, according to the US Department of Energy report, only 12.6% energy provided by steam and diesel are used to drive vehicle, internal combustion engine loss is up to 62%, idling, brake and transmission losses are about 20%. Without the external energy input, the hybrid can increase energy efficiency 20%, so the space for fuel-efficient is limited. In the long run, pure electric vehicle is the ultimate goal of new energy automobile industry.

So, in this paper, we shall focus on pure EVs, which only use electric motor for propulsion. We assume that the correlation techniques are mature and electronic generation is mainly from clean energy when the EVs are widely used.

Firstly, we can put economy and society factor together and establish models based on macro and micro. Our vantage point is to discuss whether EVs widespread use beneficial to energy saving and emission reduction or not.

For micro, we compare the costs of EVs and fuel vehicles. We will select some vehicle brands to study the emissions of CO_2 from EVs and fuel vehicles with the same specification. Then we illustrate the potential economic of EVs mainly about charging stations and arrangement of charging time. For macro, we use the knowledge of Econometrics to study the relationship between EVs and green economy.

Secondly, the problem whether the widespread use of EVs can reduce the consumption of oil and how much it can save should be taken into consideration. To solve it, what must be done at first is to gain EVs ownership. Here we establish GM(1, 1) model to estimate the global vehicles ownership based on gray prediction theory and then EVs ownership is worked out utilizing forecasted ratio after the widespread use of EVs. The next step is calculating oil savings if EVs have been promoted and analyzing EVs energy compared with fuel vehicles.

Thirdly, given the amount of EVs after its widespread use, we can use Fuzzy Math Comprehensive Evaluation System to calculate the type of EVs and market share of different types needed in the future.

Fourthly, we can divide the EVs marker into three subjects: the government, electric vehicle service providers, electric vehicles users. Then we can use the game theory to make the three parts reaching a balanced combination of maximum efficiency. Then according to the purchase probability of EVs users, we will find out the annual total number of EVs and the number to be used each year. As long as we can find the additional power to generate electricity, we will find the power form.

At last, based on the analysis above, we will give a recommendation about the role of governments to promote EVs.

7.3　Main Models

7.3.1　Economic Cost Model

(1) Economic Contrast between EVs and Fuel Vehicles

For the BYD E6 pure electric cars and BYD F3 traditional cars, we compare the costs of these two cars under the same circumstances. With the help of relevant conversion knowledge, we can get the model as follows:

$$F_1 = a * f_1 \quad F_2 = b * f_2$$

where F_1 and F_2 denote the traveling costs ($) for fuel vehicles and EVs per hundred kilometers respectively; f_1 is the diesel price ($/L), f_2 is the valley electricity price ($/kWh), a is the automobile diesel consumption (L) per hundred kilometer, b is automobile electric consumption (kWh) per hundred kilometers.

By BYD website[4], we get the technical parameters a and b for these two types of cars. Substitute a and b into the model, then we can get the final results (see Table 7.1).

Table 7.1　The Contrast Between Fuel Vehicles and Electric Vehicles

Automobile type	Technical parameters	Acquisition costs (US$)	Fee (US$)	The cost of driving 15,000km p.a. (US$)
Fuel vehicles	a=8.2(L/100 km)	15 168	8.487	1 273.05
EVs	b=18(Kwh/100 km)	45 505	2.0205	303.075
Saving costs	—	30 337	6.4565	969.975

Thus, we can conclude that the proportion of driving cost per hundred kilometers for electric vehicles to fuel vehicles is about 25%. Comparatively speaking, EVs can save energy as much as 76.08%. If the annual mileage is 15,000 kilometers, then EVs will save 969.975 dollars than fuel vehicles.

In addition, we can analyze the economy from the perspective of price changes. Based on the useful UP statistical review of energy[2], we can get prices of coal, oil, gas, and electricity in United States, Britain, Japan, and other countries.

Table 7.2 Global Average Price of Fossil Fuels

Year	Natural Gas		Coal		Oil		Electricity	
	Price	Year-on-year Growth Rate	Price	Year-on-year Growth Rate	Price	Year-on-year Growth Rate	Price	Year-on-year Growth Rate
1990	1.345	——	45.203	——	22.893	——	——	——
1991	1.190	−0.119	44.087	−0.025	19.390	−0.153	——	——
1992	1.375	0.019	41.627	−0.079	19.020	−0.169	——	——
1993	1.905	0.411	39.597	−0.124	16.783	−0.267	——	——
1994	1.685	0.248	40.223	−0.110	15.923	−0.304	——	——
1995	1.290	−0.044	41.993	−0.071	17.180	−0.250	——	——
1996	1.917	0.420	42.597	−0.058	20.450	−0.107	——	——
1997	1.950	0.444	41.397	−0.084	19.310	−0.157	——	——
1998	1.787	0.323	37.920	−0.161	13.107	−0.427	——	——
1999	1.950	0.444	34.303	−0.241	18.177	−0.206	——	——
2000	3.563	1.640	35.193	−0.221	28.357	0.239	——	——
2001	3.617	1.679	43.503	−0.038	24.393	0.066	0.124	——
2002	2.757	1.042	35.620	−0.212	24.973	0.091	0.121	−0.024
2003	4.597	2.405	41.230	−0.088	28.893	0.262	0.133	0.075
2004	5.113	2.788	65.980	0.460	37.800	0.651	0.143	0.151
2005	7.807	4.783	73.330	0.622	53.487	1.336	0.142	0.145
2006	6.820	4.052	73.510	0.626	64.220	1.805	0.142	0.145
2007	6.377	3.723	76.063	0.683	70.927	2.098	0.146	0.177
2008	9.210	5.822	148.497	2.285	97.220	3.247	0.163	0.312
2009	4.040	1.993	102.187	1.261	61.660	1.693	0.116	−0.065

Resource: US Energy Information Administration & BP Statistical Review of World Energy June 2010.

Using the data from Table 7.2, we draw up the line graph about energy prices year-on-year change rate with the help of Excel.

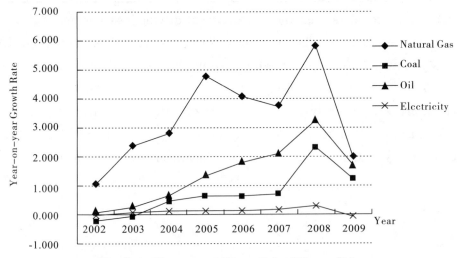

Fig. 7.4 Year-on-year Change Rate of Energy Price

From Fig. 7.4, it is shown that the prices of three major non-renewable energy are rising. However, the electricity price is mainly stable. It is not difficult to find that the three major fossil fuel prices grew relative to 1990, and the growth rates were as high as 199.3%, 126.1%, 169.3% respectively. Meanwhile, the electricity price fell by 6.5%. Obviously, electricity has great strength in price.

(2) Economics of Electric Vehicle Charging Infrastructure

- **Potential Economic Benefit of Charge-station**

The EVs not only have advantages in saving energy, but also bring huge benefits for related industry chain, such as the charging stations, special cables, chargers and other manufactures.

The types of charging for EVs include external and internal sources. What we talk is charging from an external source which leads to three levels in Table 7.3.

Table 7.3 EV Charging Equipment Nomenclature

Charging level	Specification	Typical use	Time to charge battery
Level 1 (Slow)	120V/13A	Charging at home/office	7—8 hours
Level 2 (Fast)	240V/32A	Charging supermarket, gym	3—4 hours
Level 3 (Rapid)	Up to 500V/200A	Like a normal gas station	30 minutes

Resource: Policy options for electric vehicle charging infrastructure in C40 cities.

In view of the above three kinds of charging forms, the requirement of prior investment for Level 1 and Level 2 may be low, but the number is large and it is not convenient enough. But relatively, the cost of construction for Level 3 is large, and the investment yield is high. For instance, the return on investment may reach 115% as shown in Fig. 7.5[6]. In the long run, it still has great attraction to the market.

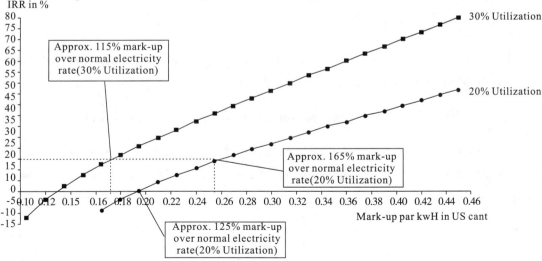

Fig. 7.5 Rate of Return Based on Varying Mark-up Prices-Level 3

- **Economic Benefits of Charging at Off-peak Time**

According to the CEC data of 2008, the national calibration electricity generation is 3.4334 trillion kWh, and the total electricity consumption is 3.4268 trillion kWh. The waste quantity of electricity, which was not fully used, was about 340 trillion kWh and enough to charge 680 million electric cars. At the same time the state grid also win new sales income.

(3) EVs on the Contribution of the Green Economy

Modern society is seeking the green living, people-oriented, which reflects environment, economy and society harmonious development.

We analyze the impact popularity of electric vehicles to green economic, from view of macroscopic. Then Human Development Index is introduced. HDI are based on life expectancy, education level, and quality of life. We combine the economic indicators and social indicators to reveal the imbalance of economic and social development.

Now take the example of China. The output value of EVs industry represents the development of EVs as an explanatory variable, and HDI represents the development of green economy as an explained variable. Using the Econometrics information, analyze their relationship.

Through related information, we establish Logarithmic Model based on the searched data:

$$HDI = \alpha \ln x + \beta$$

Here x is the output value of EVs industry, α and β are parameters. Through EVIEWS, we get the result:

$$HDI = 0.0184 \ln x - 0.7063$$

And the result gets through some tests.

The result shows when the output value of EVs industry increases 1%, HDI increases 0.0184. EVs are beneficial to the development of green economy.

7.3.2 The Emissions Comparison between EVs and Fuel Vehicles

Still for the BYD E6 pure electric cars and BYD F3 cars, we compare the emissions of these two kinds of cars under the same circumstances. We establish the following calculation model:

$$C_1 = a * \alpha$$
$$C_2 = b * \beta$$

where C_1 (kg) and C_2 (kg) respectively denote the amount of carbon dioxide produced by fuel vehicles and EVs per 100 km, α(kg) is the carbon dioxide emission caused by diesel per liter, β(kg) is the carbon dioxide emission caused by electricity generated by coal per degrees.

Table 7.4 Contrast on Emissions Caused by Different Types of Vehicles

Vehicles type		Technical parameters	CO_2 emissions per 100 km(kg)	CO_2 reduction(kg)
Fuel vehicle		a=8.2(L/100 km)	21.566	—
EVs	Thermal power	b=18(kwh/100 km)	17.946	3.62
	Clean energy power		0	21.566

It is shown in Table 7.4 that each EV, with electricity from thermal power generation and clean energy power generation, displaces a conventional car produces respectively savings of approximately 3.62 kg and 21.566 kg of CO_2 per hundred kilometers, compared to a petrol-powered car.

7.3.3 Prediction of the Number of EVs

(1) Forecast of Vehicle Population

As the data obtained is little, we will use the gray prediction theory to overcome it. So we first preprocess the data to get the series x_0 which meets the requirements. Then do a cumulative averaging to x_0:

$$y_0(k) = \lambda x_1(k) + (1-\lambda)x_1(k-1) \quad k=2,3,4,5$$

where $\lambda=0.5$. Establish grey differential equation:

$$x_0(k) + \lambda y_0(k) = \gamma \quad k=2,3,4,5$$

where $x_0(k)$ is known as the grey derivative, λ is the development system, $y_0(k)$ is the albino background value, and γ is grey action. Substituting $k=2,3,\cdots,6$ into the above equation:

$$\begin{cases} x_0(2) + \lambda y_0(2) = \gamma \\ x_0(3) + \lambda y_0(3) = \gamma \\ x_0(4) + \lambda y_0(4) = \gamma \\ x_0(5) + \lambda y_0(5) = \gamma \end{cases}$$

the corresponding albinism differential equation is in the form of:

$$\frac{dx_1}{dt} + \lambda x_1(t) = \gamma$$

$$\text{Set } Y_N = \begin{pmatrix} x_0(2) \\ x_0(3) \\ x_0(4) \\ x_0(5) \end{pmatrix}, u = \begin{pmatrix} \lambda \\ \gamma \end{pmatrix}, B = \begin{pmatrix} -y_0(2) & 1 \\ -y_0(3) & 1 \\ -y_0(4) & 1 \\ -y_0(5) & 1 \end{pmatrix}$$

The above matrix can be rewritten $Bu=Y_N$.

By the least squares method, we can make $J(\hat{u})=(Y_N-Bu)^T(Y_N-\hat{B}u)$ reach the minimum requirements.

$$\hat{u}=\begin{bmatrix}\hat{\lambda}\\\hat{\gamma}\end{bmatrix}=(B^TB)^{-1}B^TY_N.$$

Then we have the following predictive value:

$$x_1(k+1)=\left[x_0(1)-\frac{\hat{\gamma}}{\hat{\lambda}}\right]e^{-\hat{\lambda}k}+\frac{\hat{\gamma}}{\hat{\lambda}} \quad k=1,2,\cdots,5$$

Substituting the global vehicles ownership from 2004 to 2008[7] into it, we have

$$x_1(k+1) = 323819262477.9 * e^{0.02733k} - 31534485259.9$$

We obtain the forecast data (see Table 7.5).

Table 7.5 Predictive Value of Cars, Ownership Ratio of EVs

Year	The amount of cars	Ownership ratio of EVs	The amount of EVs	Year	The amount of cars	Ownership ratio of EVs	The amount of EVs
2011	1056859545	1%	10568595	2021	1388885765	12%	166666292
2012	1086130975	2%	21722620	2022	1427353196	14%	199829447
2013	1116213125	3%	33486394	2023	1466886046	16%	234701767
2014	1147128448	4%	45885138	2024	1507513821	18%	271352488
2015	1178900020	5%	58945001	2025	1549266848	20%	309853370
2016	1211551558	6%	72693093	2026	1592176292	22%	350278784
2017	1245107433	7%	87157520	2027	1636274183	24%	392705804
2018	1279592692	8%	102367415	2028	1681593436	26%	437214293
2019	1315033075	9%	118352977	2029	1728167879	29%	501168685
2020	1351455038	11%	148660054	2030	1776032276	30%	532809683

Calculating residuals series,

$$\varepsilon_1 = \left|\frac{X-x1}{x1}\right|.$$

Then $\varepsilon_1=(0.03\%, 0.009\%, 0.13\%, 0.07\%)$.

Fig. 7.6 is a scatter plot chart about the practical and estimated values. By doing this, we can verify that this model has certain accuracy. And results of this model have high reliability.

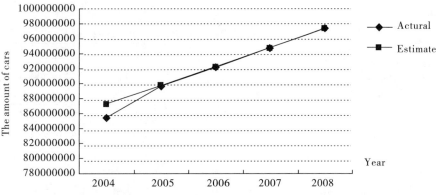

Fig. 7.6 The Graph of Actual and Estimated Value

(2) The Promotion of EVs in the Proportion and the Number Forecasts

According to Ref. [8], we can establish electric vehicle retain ratio prediction model as follows:

$$\frac{df}{dt} = c \frac{f(1-f)^2}{1-(1-d)f}$$

where f denotes EVs possession, c means the internal factors, d means retardation coefficient and denotes the relatively optimistic degree.

The promotion factors of a EVs market include environmental protection degree, onward travel distance, fuel supply the degree of difficulty, vehicle economic performance, driving conveniences, comfortableness, social effect, popularize strength and other indexes. Thus we can get the proportion and number of EVs (see Table 7.5).

7.3.4 Estimation of Oil Savings and Benefits of Energy Balance

Assume that the vehicles in the world are all fuel vehicles before the widespread use of EVs, we focus on the oil consumption every year and establish the model:

$$Q = q * m * v.$$

Here Q denotes the vehicle oil consumption per year, q means oil consumption per km, m means the distance a vehicle travels per year, v is the vehicle ownership.

When EVs are widely used, certain fuel vehicles will be replaced by EVs and the resource of electricity will be regarded to completely come from the clean energy. It produces fuel substitution effect and the model as follows:

$$Q_s = q * m * v_s,$$

where Q_s denotes oil savings per year of EVs, v_s means EVs ownership.

Fuel vehicles consume 8.2 L oil per 100km and travel 30,000 km per year. Based on the EVs and vehicles ownership that have been estimated, we obtain oil consumption of fuel vehicles and oil savings caused from the widespread use of EVs every year (see

Table 7.6).

Table 7.6 Oil Consumption of Fuel Vehicles and Oil Savings of EVs

Year	Before promotion		After promotion	
	Amount of vehicles (100 million)	Oil consumption per year(100 million L)	Amount of EVs (100 million)	Oil savings per year(100 million L)
2011	10.5686	25998.756	0.1057	260.022
2015	11.789	29000.94	0.5895	1450.17
2030	17.7603	43690.338	5.3281	13107.126

The problem that what extent the EVs' usage reaches is the widespread use is determined by the policy of governments. For example [1], President Barack Obama's goal is to put one million plug-in hybrid vehicles on the road by 2015 with an annual growth rate of 39.1% (excluding in this estimation electric hybrids); In the Republic of Ireland the government has set a target of 10% for all vehicles on Irish roads to be electric by 2020; German government wants to develop Germany into a leading market for electric mobility with 1 million electric vehicles on its streets by 2020. Hence the ownership ratio of EVs of 30% in 2030 is regarded as the widespread use of EVs. And the world may save fossil fuels of 1,310,712.6 million L annually with 532.81 million EVs after the widespread use of EVs.

Analyzing the results in 2011, 2015, 2030, the fuel consumption reduces 1%, 5%, 30% respectively and chain growth rate of fuel savings, which reaches 4.8 in 2015 and 8 in 2030, has becoming faster and faster. Therefore the promotion of electric vehicles contributes significantly to energy savings.

7.3.5 On the Types of EVs

Here, we only consider the market share of hybrid vehicles and pure EVs due to the battery technological deficiency in near future. To evaluate the strength and weakness of two kinds of vehicles, we use fuzzy comprehensive evaluation and choose four main indexes in Table 7.7.

Table 7.7 The Basic Evaluation Index

Index Type	Efficiency x1	Price x2	Range x3	Emission x4
BYD pure EVs	0.14	30	300	0.3
BYD hybrid vehicles	0.195	14.98	430	0.2

1. Ideal alternative

$$u = (u_1^0, u_2^0, u_3^0, u_4^0) = (0.195, 14.98, 430, 0.3),$$

where

$$u_i^0 = \begin{cases} \max\{a_{ij}\} & a_{ij} \text{ is beneficial} \\ \min\{a_{ij}\} & a_{ij} \text{ is cost} \end{cases}$$

2. Relative deviation fuzzy matrix

$$R = \begin{bmatrix} r_{11} & r_{12} & r_{13} & r_{14} \\ r_{21} & r_{22} & r_{23} & r_{24} \end{bmatrix}$$

where

$$r_{ij} = \frac{|a_{ij} - u_i^0|}{\max\{a_{ij}\} - \min\{a_{ij}\}}$$

3. Weight numbers w_i ($i=1,2,3,4$)

$$v_j = s_j/\bar{x}_j, \quad w_j = v_j / \sum_{j=1}^{4} v_j.$$

The four weight numbers are: 0.19, 0.38, 0.20, 0.23.

4. Comprehensive evaluation model

$$F_i = \sum_{j=1}^{4} w_j r_{ij}.$$

If $F_t < F_s$, then F_t is in the front. By direct calculation, we have: $F_1 = 0.7704$ and $F_2 = 0.2296$. Thus, hybrid vehicles are better and the proportion of pure EVs is: $l_1 = F_2/(F_1+F_2) = 0.23$, the proportion of hybrid EVs is: $l_2 = F_1/(F_1+F_2) = 0.77$. As a result, the market share of pure EVs and hybrid EVs are 23% and 77% respectively.

7.3.6 The Amount and Types of Electricity Generation

Parameters

S—The total cost that users using EVs pay

C—The total cost that charging station operators in the business carried out

Z—The total cost of government to promote the development of EVs

Y_1—The income of EVs users after promoting EVs

Y_2—The income of charging station operators after promoting EVs

Y_3—The income of government after promoting EVs

P_1—The probability of charging station operators carrying out business

P_2—The probability of EVs users purchasing EVs

P_3—The probability of government support

P_4—The probability of government support and success

Considering the following situations

• Users buy EVs and charging station operators provide the charging service but governments don't support them.

• Users buy EVs and charging station operators provide the charging service and governments support them but the business fails.

• Users buy EVs and charging station operators provide the charging service and governments support them and the business fails.

• Users don't buy EVs and charging station operators don't provide the charging service and governments support them.

• Users buy EVs and charging station operators provide the charging service but governments don't support them.

Game theory

(1) When the charging station to charge carriers with probability P_1, the expected return of the government supporting strategy is: $U_1 = P_1[(Y_3-Z)P_4 + (-Z)(1-P_4)] + (1-P_1)[P_4(-Z) + (1-P_4)(-Z)]$.

The expected return of the government not supporting strategy is $U_2 = 0$.

When there is no difference between the expected return of support and nonsupport from governments, namely $U_1 = U_2$, we gain the most probability of charging station operator carrying out business in governments' game equilibrium. The maximum probability is $P_1 = (Z/Y_3)P_4$.

(2) When the probability of buying EVs is P_2, the expected return of the government supporting strategy is $U_3 = P_2[(Y_3-Z)P_4 + (-Z)(1-P_4)] + (1-P_2)[P_4(-Z) + (1-P_4)(-Z)]$.

The expected return of the government not supporting strategy is $U_4 = 0$.

When there is no difference between the expected return of support and nonsupport from governments, namely $U_3 = U_4$, we gain the most probability of charging station operator carrying out business in governments' game equilibrium. the most probability is $P_2 = (Z/Y_3)P_4$.

(3) When the probability of government support is P_3, the respectively expected return for charging station operators to charge or not is: $V_1 = P_3[(Y_2+Z_2-C)P_4 + (Y_2+Z_2-C)(1-P_4)] + (1-P_3)(Y_2-C), V_2=0$.

When there is no difference between the expected return of charge and non charge from changing station operators, namely $V_3 = V_4$, we gain the optimal decision of governments supporting in changing station operators' game equilibrium.

We can get $P_3 = (C-Y_2)/Z_2$.

(4) When the probability of government support is P_3, the respectively expected return for users to buy electric vehicles or not are: $V_3 = P_3[(Y_1+Z_1-S)P_4 + (Y_1+Z_1-S)(1-P_4)] + (1-P_3)(Y_1-S), V_4=0$.

When there is no difference between the expected return of purchase and non purchase from EVs users, namely $V_3 = V_4$. We gain the optimal decision of governments supporting in changing station operators' game equilibrium. We can get: $P_3 = (S - Y_1)/Z_1$.

To sum up, mixed strategy Nash equilibrium of game model is:
$$P_1 = P_2 = (Z/Y_3)P_4, P_3 = (C - Y_2)/Z_2 \text{ or } P_3 = (S - Y_1)/Z_1$$

Result

Thus, our model can meet the environmental, social, business and personal benefit-maximizing requirements.

Let the probability of buying EVs be P_2, the vehicle population be N_i. About 1% of the EVs need charging and the power consumptions for pure and hybrid EVs are 20.8kWh and 6.93kWh. Then the amount of electricity generation can be estimated by the following formula:
$$W = (20.8 * 0.23 + 6.93 * 0.77) * N_i * P_2 * 0.01 = 36.94 * N_i * P_2$$

The types of electricity generation mainly include solar, nuclear, hydroelectric and wind.

Table 7.8 National Energy Structure Data in the World[9]

Power generation	Installed capacity (10 thousand kW)	Annual energy output (1000 billion Kwh)	Cost of electricity (RMB/kW)	Equipment utilization hours(h)
Solar power	1000	300	0.73	3000
Water power	27000	9250	0.31	2100
Nuclear power	4000	2610	0.906	6525
Wind power	3000	630	0.38	3425

To determine the annual electricity generation types, we adopt the econometric model as follows:
$$\ln^f = \rho_0 + \rho_1 \ln^{f_1} + \rho_2 \ln^{f_2} + \rho_3 \ln^{f_3}$$

7.4 Government

This paper comprehensively studies the environmental, social, economic, and health impacts of the widespread use of EVs through micro and macro perspectives. Hence governments and manufactures should support the development of EVs and governments play a critical role.

Make the development goals of coordinate with the development plan of national economy and society, make the development goals and planning and offer information instructing investment. Besides, make adjustments in a certain time interval.

Because the benefits caused by the widespread use of EVs are different according to the different types of electricity generation, governments should formulate the right development directions to guide promotion of EVs based on the primary type of electricity generation. For example, they can take their advantage products as development priorities; advocate those countries which use clean energy such as wind power, water power to generate electricity to use EVs and so on.

- **Provide the platform of technology innovation**

Recently, what restricts the development of EVs seriously is the battery technology innovation. To achieve a breakthrough, governments should invest adequate capital for study.

- **Establish the standard systems of EVs and related industry**

The prerequisite of EVs commercialization and industrialization is that security and unity is assured. Governments should establish safety regulation about EVs, battery and supporting facilities, improve charging network system and changing technology standards.

- **Use the tax industrial policy reasonably**

As a macro regulator of the market economy, that governments use tax industrial policy physically may well simulate development of EVs. For tax industrial policy, adopt the tendency to industry and regional and take the rational distribution of the industry dependent on area resources, humanities and geographical advantages. For consumer policy, if governments can subsidize the consumer purchase appropriately, or relieve consumption tax and others, it must greatly stimulate the desire for consumption, which is explained in the economic model that the buying price difference between EVs and fuel vehicles will be made up by energy saving through 30 years. What's more, this kind of support may not cause losses to the government financials for government governance in the atmosphere will substantially reduce the capital investments when EVs are widely used.

7.5 Conclusion

We get electric vehicles on the savings in running costs as high as 76.08 percent, according to compare EVs with fuel vehicles on the economic performance when EVs has been used widely. In recent years, the price of the three major fossil fuels continues to grow while prices are stable and the advantage in price is obvious. When the electric vehicle industry outputs for each additional 1%, HDI increases 0.0184. So EVs is conducive to the development of a green economy and the improvement of economic efficiency. Currently, facing the world's fossil fuel supply shortage, we estimate that fuel consumption will reduce 30% in 2030. It greatly increases social benefits.

From an environmental point of view, each EV with electricity from thermal power generation and clean energy power generation that displaces a conventional car produces respectively savings of approximately 3.62 kg and 21.566 kg of CO_2 per hundred kilometers, compared to a petrol-powered car. It greatly reduces carbon dioxide emissions and increases environmental benefits.

No matter from the environmental point, or the economy and the social point of energy conservation, the government will vigorously support the widespread use of electric vehicles.

Under the government's strong support for the widespread use of EVs, we just consider pure electric vehicles and hybrid vehicles. Thus the pure electric vehicles and hybrid vehicles respectively proportion of the total electric we get are 0.23 and 0.77. According to the electric vehicles proportion prediction model, we can calculate the electric vehicles ownership from 2011 to 2030, and the annual electricity generation is estimated. Comprehensive consideration of the game behavior of government, charging suppliers and trolley user to get Nash equilibrium for mixed strategy. Finally, we would achieve the public benefit maximization, the supplier economic maximization, and EVs user personal utility maximization.

At last, we consider water power generation, nuclear power generation, wind power generation, and solar power generation, to determine the power form according to the actual need and estimate the amount of electricity generation needed by EVs.

7.6 Evaluation

7.6.1 Advantage

- The game model was established in close contact with the actual, full account of the different stages of reality, so that the model is closer to reality and versatility.
- It is simple, intuitive and fast to use EXCEL software for data processing.
- In the discussion of EVs on the contribution of the green economy, a model is established using HDI to explain the influence the development of EVs make on green economy, which combines economy with environment and society.

7.6.2 Disadvantage

- In game model, with no specific data, we can only give scheme for generating forms, but we can decide which form to use.

Reference

[1] http://en.wikipedia.org/wiki/Electric_vehicle

[2] BP Statistical Review of World Energy June 2010[Z]. http://www.bp.com/liveassets/bp_internet/globalbp/globalbp_uk_english/reports_and_publications/statistical_energy_review_2008/STAGING/local_assets/2010_downloads/statistical_review_of_world_energy_full_report_2010.pdf

[3] http://www.oica.net/wp-content/uploads/geneva-presentation.ppt

[4] Liu Bin, Yang Ping. *The New Energy Automotive Battery Industry Report*[R]. Gou Lian Securities, 2009.

[5] http://www.byd.com

[6] Policy Options for Electric Vehicle Charging Infrastructure in C40 Cities[Z].

[7] State Council Development Research Center[Z]. http://www.drcnet.com.cn/DRCnet.common.web/DocView.aspx?DocID=2430146&LeafID=17133&ChnID=4484

[8] Min Haitao, Cheng Meng. Environment Impacts and Energy Efficiency Analysis For New Energy Vehicles[J]. *Tractor & Farm Transporter*, 2007, 34(4): 105—108. (in Chinese)

[9] Huang Xiang, Sun Yuwen. Different Forms of Energy Generation Impact on the Environment[J]. *Electric Power*, 2008, 41(2): 48—50. (in Chinese)

◆ 建模特色点评

❋ 论文特色 ❋

◆标题定位:"A Comprehensive Analysis for Widespread Use of Electric Vehicles"为"电动汽车广泛使用综合分析",定位准确,融问题和方法于一体,贴切、恰当、简洁。

◆方法鉴赏:电车相关分类较细,共有六个模型:①电动汽车与燃料汽车的经济性对比;②电动汽车充电设施的经济性;③电动汽车对绿色经济的贡献;④能源平衡的节油效益估算;⑤论电动汽车的类型;⑥发电量和类型。基本方法较一般。

◆写作评析:文章条理性好,内容简洁,有一定的技术面,整体效果较好。

◆后继方面:该文后继研究在国外EI期刊上发表,作者之一的周钰已保送中科大,博士毕业留校。

❋ 不足之处 ❋

方法较简单,技术面较一般,缺少误差分析、灵敏度分析等。

第 8 篇
它是可持续的吗？

◆ 竞赛原题再现

Problem D Is it sustainable?

Problem background

One of the largest challenges of our time is how to manage increasing population and consumption with the earth's finite resources. How can we do this while at the same time increasing equity and eradicating poverty? Since the beginning of the modern environmental movement in the 1960's, balancing human needs with the earth's health has been a topic of considerable debate. Are economic development and ecosystem health at odds? In order to reconcile this difficult balance, the concept of sustainable development was introduced in the 1980s.

Sustainable development is defined by the 1987 Brundtland Report as "development that meets the needs of the present without compromising the ability of future generations to meet their own needs". Since its conception, sustainable development has become a goal for international aid agencies, planners, governments, and non-profit organizations. Despite this, striving towards a sustainable future has never been more imperative. The United Nations (UN) predicts the world's population will level at 9 billion people by 2050. This, coupled with increased consumption, places a significant strain on the earth's finite resources. Understanding that the earth is a system that connects both time and space is critical to sustainable development. Development must focus on needs (e.g., reducing the vulnerability of the world's poor) and limitations (e.g., the environment's ability to detoxify wastes). In 2012, the UN conference on sustainable development recognized that: "poverty eradication, changing unsustainable and promoting sustainable patterns of consumption and production and protecting and managing the natural resource base of economic and social development are the overarching objectives of and essential requirements for sustainable development." Decreasing personal poverty and vulnerability, encouraging economic development, and maintaining ecosystem health are the pillars of sustainable development.

Problem statement

The International Conglomerate of Money (ICM) has hired you to help them use their extensive financial resources and influence to create a more sustainable world. They are particularly interested in developing countries, where they believe they can see the greatest results of their investments.

Task 1: Develop a model for the sustainability of a country. This model should provide a measure to distinguish more sustainable countries and policies from less sustainable ones. It can also serve to inform the ICM on those countries that need the most support and intervention. Some factors may include human health, food security, access to clean water, local environmental quality, energy access, livelihoods, community vulnerability, and equitable sustainable development. Your model should clearly define when and how a county is sustainable or unsustainable.

Task 2: Select a country from the United Nations list of the 48 Least Developed Countries (LDC) list (http://unctad.org/en/pages/aldc/Least%20Developed%20Countries/UN-list-ofLeast-Developed-Countries.aspx). Using your model and research from Task 1, create a 20-year sustainable development plan for your selected LDC country to move towards a more sustainable future. This plan should consist of programs, policies, and aid that can be provided by the ICM within a country based on their demographic, natural resources, economic, social and political conditions.

Task 3: Evaluate the effect your 20-year sustainability plan has on your country's sustainability measure created in Task 1. Predict the change that will occur over the 20 years in the future by implementing your plan in your evaluation. Based on the selected country, you may need to consider additional environmental factors such as climate change, development aid, foreign investment, natural disasters, and government instability. The ICM would like to get the "most bang for their buck", so determine which programs or policies produce the greatest effect on the sustainability measure for your country. Identifying highly effective strategies to be implemented is the ultimate goal of the ICM to create a more sustainable world.

Task 4: Write a 20-page report (summary sheet does not count in the 20 pages) that explains your model, your sustainability measure, your sustainability development plan, and the effect of your plan based on your model and the country's environment. Be sure to detail the strengths and weaknesses of the model. The ICM will use your report to invest in sustainability development intervention strategies for specific LDC countries. Good luck in your modeling work!

获奖论文精选

The Evaluation and Prediction of Sustainable Development

Summary: Nowadays, sustainable development is playing a significant role in managing increasing population and consumption with the earth's finite resources. Therefore, the method of sustainability evaluation should be improved to identify highly effective strategies, which serves the sustainable development.

In this paper, after analyzing the influencing factor of sustainable development, we adopt genetic algorithm (GA), the general contrast algorithm, and analytic hierarchy process(APH) to calculate the universal evaluation formula. And then, we use this formula to evaluate sustainability of Cambodia. Further, combining the actual situation of Cambodia, a 20-year sustainable development plan is created and $0-1$ linear programming model was established to assess the effectiveness of sustainable strategies. Finally, we established system dynamics (SDy) model to optimize our solution and identify highly effective strategies more scientifically and reasonably.

For Problem 1, firstly, based on the characteristics of the indicator systems of the sustainable development, we get the exponential universal formulae of classified indicators by using GA. Secondly, combining the general contrast algorithm with AHP, the weight of different subsystems is obtained and the sustainability evaluation model was established. Finally, utilizing MATLAB to calculate the sustainability of Cambodia, which indicates that Cambodia' sustainable condition is poor, and it's urgent to improve its sustainability.

For Problem 2, based on SDy, three subsystems are established. Combining the internal relations of each subsystem with Cambodia general situation, we put forward a 20-year sustainable development plan, including strategies in population, economy, natural resource and society. Then, we predicted the effect our strategies have on the sustainability of Cambodia, and comparing with the original model, the rationality of our plan was proved.

For Problem 3, we chose indicators, including population growth, energy productivity, GINI coefficient and GDP, to build efficiency matrix. Then $0-1$ linear programming model is established and using Hungarian method we find that in the first stage of

---

① 本文获 2015 年国际一等奖。队员：侍冰雪，周立敏，王晨睿；指导教师：方国斌。

sustainability improvement, more attention is supposed to be paid upon economic strategy.

Finally, we use SDy to optimize our method of identifying highly effective strategies. By implement our plan, the multiple objectives of Cambodia' development can be pursued sustainably to achieve a better quality of life for every citizen, now and for generation to come.

Keywords: sustainable development; effective strategies; GA optimum; system dynamics; Stella

8.1 Introduction

8.1.1 Background

With the development of national economy, as well as the advancement of global trade, people has gradually realized that the earth resources limit human development, and declare that the earth resources need to be more configured reasonably. What's more, one of the largest challenges of our era is how to coordinate the increasing population consumption with the earth's finite resources. In order to reconcile this difficult balance between increasing population and finite resources, the concept of sustainable development was introduced in the 1980's. Sustainable development is defined by The Brundtland Commission as the "ability to make development sustainable—to ensure that it meets the needs of the present without compromising the ability of future generations to meet their own needs".[1] The concept of sustainable development is now enshrined on the masthead of *Environment* magazine in the years following.

The United Nations (UN) predicts the world's population will level at 9 billion by 2050. This, coupled with increased consumption, places a significant strain on the earth's finite resources. Governments have to realize the sustainable development of the social economy and the environment is the inevitable choice of human survival and development.[2] Many countries not only accepted the theory of sustainable development, but also took sustainable development as the goal of political, economic and cultural development.

8.1.2 Introduction of System Dynamics

The System Dynamics has played an increasing important role in the past few years since there is a growing awareness of environmental and resources problems. And the surge of interest towards SD, many scholars have written series of works, form which we could reasonably receive the conclusion that SDy is an ideal methodology to analyze the complexity which is normally present in sustainability issues. When it comes to the measure distinguishing more sustainable countries and policies from less sustainable ones, we can reference the Urban Dynamics.

In the late 1960s, Forrester put forward Urban Dynamics theory, which mainly research in the results of the theory and application of the study concerning American cities' blossom-decline[3]. Yu-feng Ho has established the SDy model for the sustainable development of science city, and points out that more attention should be paid to the harmonious and coordinated development among the economic, social and environmental[4]. Jorge a. Duran took AURA in Puebla Mexico as object, established a complete framework of the SDy model [5]. And in the case of the Puerto Aura project, SDy has proved to be an excellent method and tool for helping key urban actors, from diverse disciplines, to relate actions from the different dimensions involved in a situation as complex as this project.

After reading the works about the urban sustainable development, we concluded that sustainable development is mainly divided into there parts: social, economic and environment, and these subsystems have interaction and mutual influence to each other. The subsystem is shown as follows.

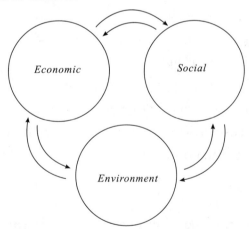

Fig. 8.1　Three Subsystems of Sustainable Development

8.1.3　Introduction of Least Developed Country

A least developed country (LDC) is a country that, according to the United Nations, exhibits the lowest indicators of socioeconomic development, with the lowest Human Development Index ratings of all countries in the world. The concept of LDCs originated in the late 1960s and the first group of LDCs was listed by the UN in its resolution 2768 (XXVI) of 18 November 1971. LDC criteria are reviewed every three years by the Committee for Development Policy of the UN Economic and Social Council (ECOSOC). The classification (as of 24 January 2014) applies to 48 countries [6], which can be shown as follows.

8.2 Problem Statement and Analysis

8.2.1 Problem Statement

The International Conglomerate of Money (ICM) expects us to evaluate the sustainability of countries, and then give a 20 year sustainable development plan for the less sustainable country we selected. By carrying out this plan, the selected country should becoming more sustainable form the aspects of demographic, natural resources, economic, social and political conditions, and the world will be a more sustainable one. In order to fulfill the 4 tasks scientifically and reasonably, we need to solve the following problems.

• Develop a sustainability evaluation model to distinguish more sustainable countries and policies from less sustainable ones. What's more, the model should clearly define when and how a county is sustainable or unsustainable. Finally Using the model and research from Task 1 to evaluate the sustainability of the less sustainable country which is selected form the 48 LDC list.

• Put forward a 20-year sustainable development plan based on the evaluation result and valuate the effect our plan has.

• Evaluate the effect our 20-year sustainability plan has through the model established above and analyze which programs or policies in the plan produce the greatest effect on the sustainability measure and finally identify highly effective strategies to be implemented to create a more sustainable world.

8.2.2 Overall Analysis

Firstly, three subsystems should be divided: environment, economy, and society. Considering the influence factor in each subsystem, and then based on the factors chosen, we can establish the sustainability evaluation model to distinguish the more sustainable countries and policies from less sustainable ones. Secondly, after evaluating the selected country, we can give the improvement plan form the aspects of demographic, natural resources, economic, social and political conditions. Thirdly, valuate the effect of each strategy, and try to figure out the highly effective strategies.

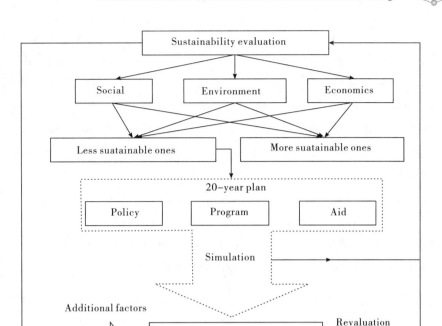

Fig. 8.2　Modeling Flow Chart

8.3　Basic Assumption

Assuming the indicators chosen can fully reflect the countries' sustainable development level and won't influence the following simulation.

• Assuming the total investment for each strategy supported by ICM is fixed.

• Assuming the total cost of the strategy implementation is fixed.

• In order to simplify 0－1 linear programming model, our article assumes that the interaction of each strategy can be neglected.

• Assuming there is no deviation among the inner relationship of indicators in system dynamics analysis.

8.4　Glossary & Symbols

8.4.1　Glossary

• Renewable resources: resources that be consumed faster than the rate at which they are renewed.

• Non-renewable resources: resources that must not be consumed at a rate faster than that which they can be substituted for by a renewable resource.

- GINI coefficient: a ratio of the areas on the Lorenz curve diagram that theoretically range from 0 (complete equality) to 1 (complete inequality), which consider it to be a measure of social inequality.

- Genetic algorithm: A search heuristic that mimics the process of natural selection. This metaheuristicis routinely used to generate useful solution to optimization and search problems.

- Energy production: refers to forms of primary energy—petroleum (crude oil, natural gas liquids, and oil from nonconventional sources), natural gas, solid fuels (coal, lignite, and other derived fuels), and combustible renewable and waste—and primary electricity, all converted into oil equivalents.

8.4.2 Symbols

Tab. 8.1 Variables and Their Meanings

Number	Sign	Significance
1	x_i	The single index we selected to reflect the corresponding indicators
2	ω_i	The weight of indicators after using general contrast algorithm
3	n	The number of indicators in each subsystem
4	y_{ij}	The normalized value of indicators calculated through Genetic Algorithm
5	γ	The synthesis scores of sustainability evaluation
6	C	Efficiency matrix in 0—1 linear programming model

8.5 Problem I Sustainability Evaluation Model

8.5.1 Analysis Approach

The interpretation of the trend in each block of time may be quite different (*Simon Bell and Stephen Morse*, 2008)[7]. Therefore, we should put forward a graded standard to judge which sustainability degree countries belong as well as shed light how a county is sustainable or unsustainable. What more, the time scale over is a further dimension, which helps to explain when a county is sustainable or unsustainable and reveals the changes of countries' sustainability? The problem analysis is shown below, x_i represent indictors of each subsystem.

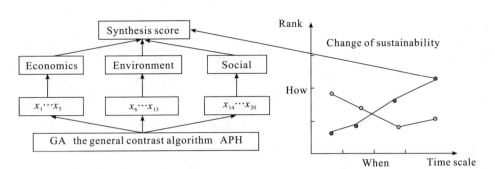

Fig. 8.3 Thinking Flowchart

8.5.2 Model Preparation

1. Indicator Systems of the Sustainable Development

When it comes to the evaluation of sustainability, the evaluation indictor system can be divided into their subsystem, social, economic and environment. In order to select specific indictors of each subsystem, we should follow the principles of scientificity, conciseness, harmony and integrity. We referenced international articles and establish an optimistic, scientific, predictable and comparable variable set. They are shown in Tab. 8.2.

Tab. 8.2 Indicator System

Sector	Short	Indicator	Significance
Economy	ET	x_1	Ratio of Export tax to total tax (%)
	IOD	x_2	Import of goods(BoP,Present price $)
	GDP	x_3	Gross Domestic Product
	ROF	x_4	Ratio of Foreign direct investment outflows to GDP (%)
	ROTG	x_5	Ratio of Tax to GDP (%)
Environment	PG	x_6	Poverty gap, measured at $1.25 a day(PPP)(%)
	PCCL	x_7	Per capita cultivated land(hectare)
	PCRF	x_8	Per capita renewable inland freshwater resources(steres)
	EY	x_9	Energy yield(thousand tons of oil equivalent)
	ECOD	x_{10}	Energy consumption of per Unit GDP(2005 constant purchasing power parity dollars/kg of oil equivalent)
	PCGY	x_{11}	Per capita grain yield (tonnes)
	PCCD	x_{12}	Per capita carbon dioxide emissions(tonnes)
	ROFA	x_{13}	Ratio of forestry area to total land area)
Society	GINI	x_{14}	The GINI coefficient
	ROYT	x_{15}	Ratio of young people to total unemployment (the annual percentage)
	ROPG	x_{16}	Rate of population growth (%)
	PD	x_{17}	Population density(population per kilometer land area)
	TU	x_{18}	Total unemployment(the percentage of labour total amount)(simulate labor organization to estimate)
	LE	x_{19}	Life expectancy at birth(year)
	TIOM	x_{20}	The incidence of malnutrition(the percentage of total population)

2. Steps of Genetic Algorithm

Generally, there of three stages during the process of social and economic development, slow, fast, and saturated stages. Therefore, when describing the development degree of each subsystem by using these indicators, the s-shaped growth curve formula is applied universally and UA is used to optimize by transforming different parameter into the shared value.

The steps of genetic algorithm are: (1) set up the objective function for the given optimization problem; (2) set the change interval of estimated parameters; (3) encode the estimated parameters to binary code; (4) generate initial population randomly; (5) calculate and evaluate the fitness of individuals in each population; (6) the crossover of parent; (7) the mutation of filial generation; (8) repeated iteration, until meet specific indicators or stop at a certain genetic algebra.

3. The General Contrast Algorithm

When synthesize the single indicators into comprehensive subsystem ones, we should impair the function of indicators whose value of y_i is smaller, and strength the function of indicators whose value of y_i is bigger. Therefore, the general contrast algorithm should be applied to calculate the weight of three subsystems.

8.5.3 Model Establishing—Sustainability Evaluation Model Based on Genetic Algorithm

Step 1: universal formulae of classified single indicators

Firstly, uses-shaped growth curve formula to describe the development degree

$$y = 1/(1 + a_j e^{-b_j x_j}) \tag{8-1}$$

Secondly, optimize the parameter by GA

$$y = 1/(1 + a e^{-b x_i}) \tag{8-2}$$

Step 2: calculate the synthesis scores of three subsystems

We should calculate the weight as

$$\omega'_i = \begin{cases} (u_i/2)^{0.5} & 0 \leqslant u_i < 0.5 \\ 1 - [(1-u_i)/2]^{0.5} & 0.5 \leqslant u_i \leqslant 1 \end{cases} \tag{8-3}$$

Secondly, normalize ω'_i to obtain the weight ω_i and calculate synthesis scores as

$$y_i = \sum_{i=1}^{n} \omega_i y_i \tag{8-4}$$

where y_i is calculated by formula.

Step 3: calculate the synthesis scores of countries

Firstly, build array R according to AHP and then calculate ω_{ci}.

$$R = \begin{Bmatrix} C_{11} & C_{12} & C_{13} \\ C_{21} & C_{22} & C_{23} \\ C_{31} & C_{32} & C_{33} \end{Bmatrix} \qquad (8\text{-}5)$$

Finally, calculating

$$Y = \sum_{k=1}^{3} \omega_G y_i \qquad (8\text{-}6)$$

8.5.4 Model Solving-evaluating Cambodia' Sustainability

The graded standard and normalized value of 20 single indicators for each subsystem in Cambodia (2002~2013) are shown in Table 8.3.

Tab. 8.3 Graded Standard and Normalized Value of Indicators

Sector	Indicator	Evaluation Criterion			Cambodia's Indicator Values		
		Grade 1	Grade 2	Grade 3	2002	2008	2013
Economy	ET	3	5	10	1.35	3.85	7.24
	IOD	7	10	20	6.27	5.34	11.46
	GDP	5	7	10	3.26	4.53	6.45
	ROF	1	1.5	3	0.61	0.87	1.47
	ROTG	3	5	10	1.43	1.61	3.24
	PG	8	12	15	7	9	11
	PCCL	10	16.67	50	13.95	25.71	23.61
	PCRF	6.67	10	20	6.28	7.92	11.09
Environment	EY	20	33.3	83.3	19.21	51.02	35.28
	ECOD	1.56	10	62.5	7.16	23.96	53.76
	PCGY	9.09	43.48	200	7.01	29.14	50.17
	PCCD	8.3	33.3	100	6.91	27.78	25.38
	ROFA	1.43	5.6	20	3.46	10.67	12.33
	GINI	8.33	10	12.5	6.19	10.97	9.24
	ROYT	9.8	9.95	10	5.13	8.62	9.96
	ROPG	10	15	30	9.81	10.95	13.64
Society	PD	15	20	30	13.6	17.2	21.44
	TU	10	15	25	5.09	10.4	17.34
	LE	15	25	50	6	9	18
	TIOM	15	30	50	13	25	37

Note: In order to keep consistent among the sort of evaluation standard, we exchange the rank 1 with rank 3 as regard to the indicator of Per capita carbon dioxide emissions (tonnes).

①Establish the objective function of three subsystems:

$$\min f(x) = \min \sum_{j=1}^{3} \sum_{i=1}^{n} |y_{ij} - y_i| \tag{8-7}$$

$$y_j = I_0 (I_9/I_0)^{j/9}$$

where $I_0 = 0.01$, $I_9 = 0.99$, when $j = 1, 5, 9$, $y_1 = 0.0106$, $y_2 = 0.594$, $y_3 = 0.99$

②Calculating the universal formula by GA:

$$Y_{1j} = \frac{1}{1 + 167.45 e^{-0.1955 x_i}} \tag{8-8}$$

$$Y_{2j} = \frac{1}{1 + 200 e^{-0.3542 x_i}} \tag{8-9}$$

$$Y_{3j} = \frac{1}{1 + 200 e^{-0.0827 x_i}} \tag{8-10}$$

③Calculate the weight of each indictor

There are three ranks, when y_i belong to rank 1, rank 2, rank 3, the corresponding value of u_i is 0.3, 0.5, 0.7, and when y_i is the beyond the ranks, u_i is 0.9. Therefore, we figure out ω'_i is 0.3873, 0.5, 0.6128 and 0.7764.

④Calculating weight according to AHP:

$$R = \begin{matrix} & C_1 & C_2 & C_3 & \\ & \begin{pmatrix} 1 & 4/3 & 3/4 \\ 4/3 & 1 & 1 \\ 3/4 & 1 & 1 \end{pmatrix} & \begin{matrix} C_1 \\ C_2 \\ C_3 \end{matrix} \end{matrix}$$

⑤Through the calculation of MATLAB: $\omega_{C1} = 0.3019$, $\omega_{C2} = 0.3323$, $\omega_{C3} = 0.3658$

Thus, the sustainability evaluation formula is

$$Y = 0.3019 Y_{i1} + 0.3323 Y_{2i} + 0.3685 Y_{3i} \tag{8-11}$$

The graded standard and the synthesis scores of three subsystems in Cambodia is shown in the Table 8.4.

Tab. 8.4　Graded Standard and the Synthesis Scores of Each Subsystem

Subsystem	Graded standards			Cambodia's scores		
	I	II	III	2002	2008	2013
Economic	0.0347	0.3015	0.5643	0.0521	0.0934	0.1055
Environment	0.2365	0.3478	0.6217	0.1954	0.2891	0.2548
Social	0.1023	0.3285	0.5931	0.0964	0.3047	0.2372
Y	0.1268	0.3277	0.5955	0.1162	0.2366	0.2039

By using similar, dissimilar and inverse connection numbers and their state sorting analysis[8], we figure out the corresponding relationship between the degrees of sustainable development and the synthesis scores of sustainability evaluation (shown in Table 8.5).

Tab. 8.5 The Corresponding Relationship

Ranks	Weaker	Weak	Sustainable	Strong
Score range	(0, 0.1268]	(0.1268, 0.3276]	(0.3276, 0.5955]	(0.5955, 1]

8.5.5 Result Analysis

According to the results shown in Table 8.4 & Table 8.5, the sustainability of Cambodia couldn't reach the sustainable level, even worse there is a downward trend. Combining with the real-world, owing to the boom of Cambodia construction industry, agriculture and tourist industry, its economic had growth about 10 percent, which contributing to the rise in 2008. And because of globe recession and the economic growth was slowing down in the following years, the rate of Cambodia's economic growth decrease to 7 percent. What's the more, the increase of the net population rate limit Cambodia's sustainable development. As for the society and policy, Cambodian social political stability has been enhanced since the third-term government was established in July 2004, and the foreign investors are full of confidence to the future economic development[9].

In addition, reviewing the general situation of this country, Cambodian is a traditional agriculture country, weak in industry. Its energy utilization technology is backward. And due to the imbalance of the ecological environment and overfishing, the aquatic resources decreased in the recent years. The result of our model is accordance with the real world, which proves the scientificity and accuracy of the sustainability evaluation measure.

8.5.6 Model Test

Similarly, we use the Sustainability evaluation model to assess the sustainability of France, Guinea and Japan. If the result is in accordance with the real-world, the scientificity and rationality of the model established above can be proved. The synthesis scores of France, Guinea and Japan are shown below.

Tab. 8.6 Synthesis Scores of France, Guinea and Japan

	France	Guinea	Japan
Economy	0.4526	0.2156	0.6425
Environment	0.2105	0.1252	0.3163
Society	0.3151	0.3425	0.5426
Y	0.3227	0.2027	0.6023

Compared with the graded standard in Table 8.6, we know that the sustainability of Guinea, France and Japan is weak, sustainable and strong respectively. As we know, France is industrial power, and its agricultural technology is very advanced. What's more, its GDP index is in the first class of the world, which makes its economic index high. But the lack of mining of mineral resources and the resources that are difficult to exploit make its environment low. Combining these two kinds of the situations, the fact is in accord with the result of our sustainability evaluation model. Guinea and Japan can be analyzed in the same way, and their situations are also in accord with the model, so the rationality and applicability of the model can be proved.

8.6 Problem II The 20-year Sustainability Plan

8.6.1 Problem Analysis

In order to put forward a scientific and reasonable plan, we should take the internal relations of each subsystem into consideration, combining with Cambodia general situation. What's more, considering that Cambodia is agricultural country, and the development of manufacturing industry and service industries is poor, what's worse, its energy development is insufficient. Considering the complexity of indicator system, we analyze the effect of the strategies by selecting the GDP index of economic aspect, the energy productivity of environment aspect, the GINI of social aspect and growth rate of population.

8.6.2 Model preparation

Take the subsystem of population for example(see in Figure 8.4), as for the arrow, indicator in the head of arrow has impact on the other side one.

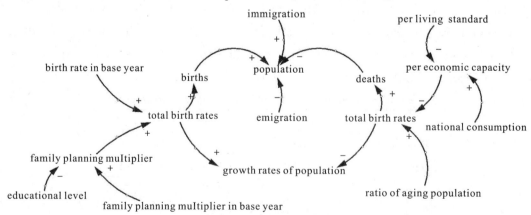

Fig. 8.4 Causal Loop Diagram

8.6.3 Model Establishing and Solving

For sensitivity analysis, we mainly leverage Stella to perform. Take population and natural resource for example, firstly we establish population and natural resources sector respectively, by analyze the relation between the two part we can finally establish the population and natural resource model, which is shown in Figure 8.5.

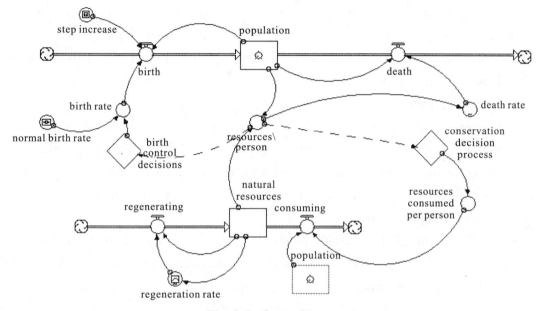

Fig. 8.5 System Diagram

Input the data of energy production, the output of Stella is shown in the Figure 6, form which we can see that the fast growth of population lead to the excessive consumption of the energy production, and with the time goes by, the shortage of energy may cause the increase of death rate and finally result in the decrease of population.

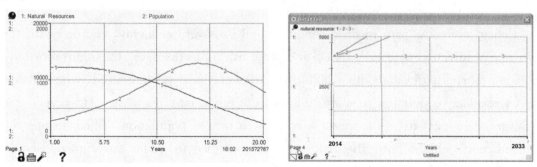

Fig. 8.6 Output of Stella   **Fig. 8.7 Sensitivty Analysis Result**

The analysis result can be seen in the Figure 8.7, series 1 & 2 represent the population size of 9 and 11 million reprehensively. Under these circumstances, the resources are on the rise. And when the size of Cambodia's population is 13 million, equilibrium state is arrived. Namely, when the population size is smaller than 13 million, resource would increase then, on the contrary, when it's bigger than 13 million, the resources would decrease.

These subsystems are mutual linked and influenced. Similarly, cultivated area and water shortages would reduce grain yield, and then affect the national health, population growth and economic development, even have influence on the development and use of energy. We use the same method to analyze the relationship between each subsystem. Based on the analysis above, combining with the actual situation, we put forward the strategies form 4 aspects:

- **Economic strategy**

Accelerate the development of industrial manufacturing industry (Policy)

Cambodia is an agricultural country, and its industrial manufacturing industry is relatively backward, so industrial manufacturing should be developed to coordinate the development of agriculture and finally support the development of agriculture.

- **Natural resource strategy**

Upgrade the energy structure: Building hydropower and gas station (Program)

Cambodia energy production is especially rich in hydropower resources. But as for the shortage of development, as well as the backwardness of supporting infrastructure, the cost of water, electricity and gas is high. Therefore, it's essential to focus on the development of available energy means a great deal to Cambodia. The government can make medium-term plan to develop all potential hydropower and gas station. In the long term, the country will realize diversified energy supply and reduce the dependence on oil.

- **Demographic strategy**

Improve healthcare (Policy)

Policy: as for the poor medical condition and backward economy, coupled with the high rate of infection with HIV/AIDS and other infectious diseases, the infant mortality and adult mortality of Cambodia is high. And life expectancy also reduced, which makes the Cambodian population stay in the risk of abnormal decrease. However, the backward economy causes the increase of growth rate of population, which brings the Cambodian population into the risk of excessive growth. The government should integrate the two factors, and improve the health care policies, to sustain Cambodian population growth rate in the normal level.

• **Social and political strategy**

Eliminate poverty (Policy & Aid)

The GINI coefficient is extent index to judging whether income distribution is fair. The measures are as follows. a. Make reasonable tax policy. b. Intensify the poverty alleviation, and constantly enhance the knowledge and technological level of the vulnerable group, and broaden their income channels.

Considering the complexity of indicator system, we analyze the effect of the strategies by selecting the GDP index, energy productivity, GINI and growth rate of population to assess. The result is shown below.

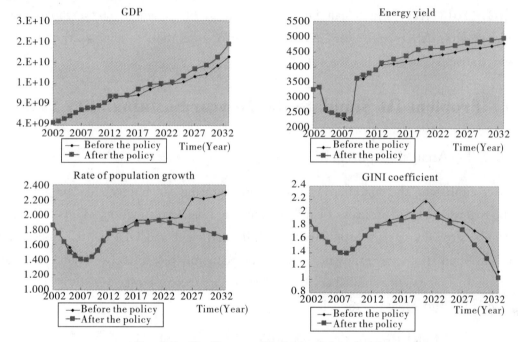

Fig. 8.8 The Changes of Indicator Before & After

8.6.4 Result Analysis

Model simulation period is 2002—2033. Comparing with the historical data in 2002—2013 is a kind of calibration of the model. After 2013, we use the model's scenario simulation to analyze the effects of different strategic choices for the future.

①GDP: According to the latest data of the World Bank in 2013, in the case of implementation of policy, Cambodia's GDP is obviously higher than the expected level that without policy implementation.

②Energy productivity: According to the latest data of the World Bank in 2013, the energy yield of Cambodia is 4085.267(One thousand tons of oil equivalent). After the implementation of policy, the energy yield of Cambodia will reach 4964.387(One thousand tons of oil equivalent).

③Population growth rate: Shown in Figure 8.8, the Cambodian population growth rate tended to rise before the implementation of policy. However, after the implementation of the policy, he population growth rate will be tardily adjust and tend to go down, that will fell to 1.72 in 2033 year.

④GINI coefficient

Factors include the level of economic development, traditional social cultural, political system, etc. In addition, improving the GINI coefficient can narrow the gap between poverty. Model simulation period is 2002—2033. And there are historical data in 2002—2013 year, which can be the calibration of the model by compared with simulated data. After 2007 year, use the scenario simulation to analyze the effects of different strategic choices for the future.

8.7 Problem III Simulate the 20-year Sustainability Plan

8.7.1 Analysis Approach

Accomplishing the maximum efficiency and figuring out highly effective strategies can be regarded as the progress of maximizing the objective function. Thus, we chose population growth, energy productivity, GINI coefficient and GDP as the indicator to represent population, environment, society and economic respectively. Then 0—1 linear programming model is established and efficiency matrix is built. Finally, Hungarian method is used to solve the model.

8.7.2 Model Preparation—Building Efficiency Matrix

Set the effect of population growth on population is 1, similarly, the effect of Energy productivity on natural resources is 1, the effect of GINI coefficient on society is 1, and the influence of GDP on the economy is 1. Refer to the World Bank Data, build the efficiency matrix about the population, natural resources, economic, social and political conditions.

$$C = (c_{ij}) = \begin{bmatrix} 1 & 1/2 & 1/4 & 1/3 \\ 1/2 & 1 & 1/4 & 1/3 \\ 1/4 & 1/3 & 1 & 1/3 \\ 1/3 & 1/2 & 1/6 & 1 \end{bmatrix}$$

8.7.3 Model Establishing and Solving

Step 1: set $b_{ij} = \begin{cases} 1 & select\ the\ j\ program\ or\ policy \\ 0 & not\ select\ the\ j\ program\ or\ policy \end{cases}$ $(i=1,2,3,4; j=1,2,3,4)$

Step 2: Calculating the objective function using Hungarian method

$$\max f = \sum_{i=1}^{4} \sum_{j=1}^{4} a_{ij} b_{ij}$$

$$s.t. \begin{cases} \sum_{j=1}^{4} b_{ij} \leqslant 4 \\ \sum_{i=1}^{4} b_{ij} = 4 \quad (i=1,\cdots,4; i=1,\cdots,4;) \\ b_{ij} \leqslant b_{jj} \\ b_{ij} = 0\ or\ 1 \end{cases}$$

Obtaining the optimum solution

$$(b_{ij}) = \begin{pmatrix} 0 & 0 & 0 & 1 \\ 0 & 0 & 1 & 0 \\ 0 & 1 & 0 & 0 \\ 1 & 0 & 0 & 0 \end{pmatrix} = \begin{pmatrix} economic \\ society \\ resourse \\ population \end{pmatrix}$$

8.7.4 Result Analysis

Assuming invests time is divided into four stages, according to the results, the investment strategy is as follows.

Stage Ⅰ: In this stage, the programs or policies in the economic subsystem have the greatest effect on the sustainability improvement. In other word, the first stage of investment will focus on economic aspects. According to the calculation formula of GDP based on expenditure method, GDP=consumption+investment+(exports−imports). So that ICM can create a more sustainable world by intervening consumption, investment and export, import aspects of interested countries.

Stage Ⅱ: The programs or policies in the social and political subsystem have the greatest effect on the sustainability improvement. Based on GINI coefficient's influence on the social aspect, attention is supposed to be focused on how to narrow the gap between rich and poor. As the growth of the national economic, the gap between rich and poor will increase simultaneously. In order to have a more sustainable future, government should take the initiative to make policies to reduce the income gap, such as Change the way of distribution.

Stage Ⅲ: In this stage, natural resources should be focused on. The related programs or policies have larger impact on sustainable development in this period. Some developing countries are rich in hydropower resources, but because of the lack of mining technology in relevant field, which leads to low energy production. The ICM should tender an olive branch to the developing country, which helps to create a more sustainable world.

Stage Ⅳ: Programs or policies in the population subsystem are more effective for sustainable development in this stage. In terms of population, government can control the birth rate and mortality rate to make growth rate of population more reasonable, such as carrying out family planning policy. On the other hand, infection of disease are more serious in some developing countries, government can implement related programs or policies to improve medical and healthcare system.

8.8 Optimization Solution for Problem III

8.8.1 Model Preparation

The sustainable development is closely related to the countries' demographic, natural resources, economic, social and political conditions. The problem of evaluate the effect of our development strategies requires a systematic and flexible methodology, including an indicators system and most importantly, a simulation model to relate the social, ecological and economic aspects of urban planning together as a whole[10]. System dynamics can assist in strategy assessment and provides insights into possible changes in the system during policy implementation (Sterman, 2000).

8.8.2 Model Solving

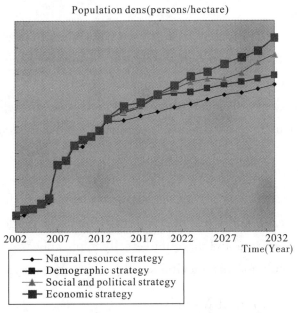

Fig. 8.9 The Effect of Strategy on Population Dense

In order to make the country more sustainable development, formulate the corresponding policy from population, natural resources, economic, social and political conditions. Then we select population density, and gain the effect of the country's population density on policies. As shown in Figure 8.9, you can see the population policy follows, and the influence of economic policy is the minimum.

8.9 Evaluation and Spread of the Model

8.9.1 Evaluation of the Model

1. Strengths

Problem 1: Sustainability evaluation model based on Genetic Algorithm

• The model is simple in computation without being limited by the number of indicators. Additionally, the evaluation method is intuitive, practical with strong comparability.

• The weights are calculated by the general algorithm and AHP respectively for different levels. The former method is well-defined and calculated strictly; based on the importance principle, the latter one can make appropriate adjustment according the actual situation. Therefore, the weight calculation is normative, flexible and scientific.

Problem 2: The 20-year sustainability plan

• Combing the fundamental realities of the Cambodia, the strategies would be more practical and scientific.

• Using a variety of mathematical software (like Matlab, Stella), learn from each other in mutual emulation, to make the calculation results more accurate.

Problem 3: Simulate the 20-year sustainability plan

• During the process of identifying highly effective strategies, control variate method is used to calculate the influence of each single indicator, and greatly simplified the solution.

2. Weaknesses

• If the graded standard is highly disparate from the normalized value in table 3, the optimized parameters a and b will be different. Thus, the graded standard should be improved to be more accurate and more universally applicable.

• The weight calculation has inevitable subjectivity and may cause error.

8.9.2 The Extension of Models

SDy emphasizes the connection, development and movement of system, considering the behavior patterns and characteristics of the system mainly rooted in its internal dynamic structure and feedback mechanism. In addition to the application for the analysis of factors influencing the sustainable development of inner link, it can also promote regional planning policy to energy management enterprise management marketing strategy transportation and other fields.

Reference

[1] World Commission on Environment and Development (WCED). *Our Common Future*[M]. New York: Oxford University Press, 1987.

[2] Lan Guoliang. Study on Construction and Application of Indexes System of Sustainable Development[D]. Tianjin University, 2004.

[3] Forrester, J. W. *Urban Dynamics*[M]. Cambridge MA: Productivity Press, 1969.

[4] Ho Yufeng, Shusone Wang. System Dynamics Model for the Sustainable Development of Science City[C]. The 23rd International Conference of the System Dynamics Society, Boston, US, 2005.

[5] Duran-Encalada J., Paucar Caceres A. System Dynamics Urban Sustainability Model for Puerto Aura in Puebla Mexico [J]. *System Practice and Action Research*,

2009,22(2):77—99.

[6] http://en.wikipedia.org/wiki/Least_developed_country

[7] Bell Simon, Stephen Morse. *Sustainability Indicators*: *Measuring the Immeasurable*[M]. Earthscan, London, 2008.

[8] Li Zuoyong, Gan Gang, Shen Shi-Iun. Appraisal Index Systems and Assessment Model of Social, Economical and Environmental Coordinative Development [J]. Chengdu University of Information Technology. 2001,16(3). (in Chinese)

[9] http://globserver.cn

[10] Ho Yufeng, Wang Shusone. System Dynamics Model for the Sustainable Development of Science City [C]. The 23rd International Conference of System Dynamics Society, Sloan School of Management, MIT, USA. 2005.

[11] Legasto, Augusto A., Jay Wright Forrester, and James M. Lyneis, eds. *System dynamics*[M]. New York: North-Holland publishing company, 1980.

[12] Wang Qifan. *System Dynamics*[M]. Beijing: Tsinghua University Press, 1994. (in Chinese)

◆ 建模特色点评

❀论文特色❀

◆标题定位：The Evaluation and Prediction of Sustainable Development 即"可持续发展的评价与预测"，定位较好，融主题与方法为一体，简洁、贴切、完美。

◆方法鉴赏：方法有针对性，比如遗传算法的可持续性评价，计算简单，不受指标数限制，评价方法直观、实用、可比性强。权重计算具有规范性、灵活性和科学性。结合柬埔寨的基本现实，检验20年可持续发展，使计划战略更加实际和科学。数学软件使用熟练（如Matlab、Stella），可相互学习、相互仿真，使计算结果更加准确。给出了应用分析和误差分析，并对模型给出了合理适当的评价。

◆写作评析：整体思路清晰，内容结构安排合理，有一定的技术面，语言顺畅。

❀不足之处❀

分级标准与标准值相差很大，应改进分级标准，使其更准确、更具普遍适用性。权重的计算带有主观性，会造成一定的误差。

第 9 篇
我们的星球正在走向干涸吗？

◆ 竞赛原题再现

Problem E Are we heading towards a thirsty planet?

Will the world run out of clean water? According to the United Nations, 1.6 billion people (one quarter of the world's population) experience water scarcity. Water use has been growing at twice the rate of population over the last century. Humans require water resources for industrial, agricultural, and residential purposes. There are two primary causes for water scarcity: physical scarcity and economic scarcity. Physical scarcity is where there is inadequate water in a region to meet demand. Economic scarcity is where water exists but poor management and lack of infrastructure limits the availability of clean water. Many scientists see this water scarcity problem becoming exacerbated with climate change and population increase. The fact that water use is increasing at twice the rate of population suggests that there is another cause of scarcity- is it increasing rates of personal consumption, or increasing rates of industrial consumption, or increasing pollution which depletes the supply of fresh water, or what?

Is it possible to provide clean fresh water to all? The supply of water must take into account the physical availability of water (e.g., natural water source, technological advances such as desalination plants or rainwater harvesting techniques). Understanding water availability is an inherently interdisciplinary problem. One must not only understand the environmental constraints on water supply, but also how social factors influence availability and distribution of clean water. For example, lack of adequate sanitation can cause a decrease in water quality. Human population increase also places increased burden on the water supply within a region. When analyzing issues of water scarcity, the following types of questions must be considered. How have humans historically exacerbated or alleviated water scarcity? What are the geological, topographical, and ecological reasons for water scarcity, and how can we accurately predict future water availability? What is the potential for new or alternate sources of water (for example, desalinization plants, water harvesting techniques or undiscovered aquifers)? What are the demographic and health related problems tied to water scarcity?

Problem Statement

The International Clean water Movement (ICM) wants your team to help them solve the world's water problems. Can you help improve access to clean, fresh water?

Task 1: Develop a model that provides a measure of the ability of a region to provide clean water to meet the needs of its population. You may need to consider the dynamic nature of the factors that affect both supply and demand in your modeling process.

Task 2: Using the UN water scarcity map, pick one country or region where water is either heavily or moderately overloaded. Explain why and how water is scarce in that region. Make sure to explain both the social and environmental drivers by addressing physical and/or economic scarcity.

Task 3: In your chosen region from Task 2, use your model from Task 1 to show what the water situation will be in 15 years. How does this situation impact the lives of citizens of this region? Be sure to incorporate the environmental drivers' effects on the model components.

Task 4: For your chosen region, design an intervention plan taking all the drivers of water scarcity into account. Any intervention plan will inevitably impact the surrounding areas, as well as the entire water ecosystem. Discuss this impact and the overall strengths and weaknesses of the plan in this larger context. How does your plan mitigate water scarcity?

Task 5: Use the intervention you designed in Task 4 and your model to project water availability into the future. Can your chosen region become less susceptible to water scarcity? Will water become a critical issue in the future? If so, when will this scarcity occur?

Task 6: Write a 20-page report (the one-page summary sheet does not count in the 20 pages) that explains your model, water scarcity in your region with no intervention, your intervention, and the effect of your intervention on your region's and the surrounding area's water availability. Be sure to detail the strengths and weaknesses of your model. The ICM will use your report to help with its mission to produce plans to provide access to clean water for all citizens of the world.

获奖论文精选

Are we heading towards a thirsty planet?[①]

Summary: Global water scarcity is becoming more and more serious. How to alleviate the scarcity of water resources has been widely concerned in the world.

For Task 1, we first pick 8 indexes that can represent water supply and demand volume and then collect 15 years' data in 9 cities. Then, by using principal component analysis, we determine the weight of each index with the help of MATLAB. Moreover, we build the comprehensive evaluation model and obtain the evaluation index. Comparing the water supply level, it is easy to judge the water supplying ability of the region.

For Task 2, we focus on the water scarcity in Hohhot in China. It can be concluded that the physical reasons for lack of water in Hohhot are short of supply, climate drought, bad water quality and huge demand. The economic reasons are short of basic equipment, poor management and serious pollution.

For Task 3, we predict the 8 indexes in the next 15 years in Hohhot by virtue of the BP neural network prediction model. Based on the results in Task 1, we can find that the lack of usable water in Hohhot will be more and more severe in the future 15 years. Furthermore, water scarcity interferes with the life of the residents through the daily water, economy, environment and other aspects.

For Task 4, we design Plans A and B, two intervention plans from the supply and demand aspects due to all elements of water ecology system in Hohhot. For Plan A, we describe the result, its advantages and disadvantages qualitatively. For Plan B, we build the system dynamics model and use the VENSIM to simulate the result of the intervention plan. The intervention plan can distinctly ease the water shortage and the corresponding advantages and disadvantages are given. We also analyze the influence of all aspects in Plan B by introducing the index of water resource development ability.

For Task 5, we simulate the future usability of water in Hohhot under both conditions with and without the intervention plan. It reaches to the conclusion that Hohhot is easily affected by water shortage. Without the plan, water shortage will appear in 2020 and it will be delayed for 27 years in 2047 with the plan.

---

① 本文获 2016 年国际一等奖。队员:林根,何玲,黄晓东;指导教师:庄科俊。

Keywords: Finally, we comprehensively evaluate the advantages and disadvantages of every model, make the sensitivity analysis and make the improvement and popularization.

9.1 Introduction

9.1.1 The Status of Water Resources

Currently, the state of water resources has become the focus of world attention. The distribution of water resources in the world is uneven, due to geographical location, abundant water resources, Europe has enough water. However, other continents, to some extent, have serious water shortage areas, especially in sub-Saharan Africa's landlocked countries.

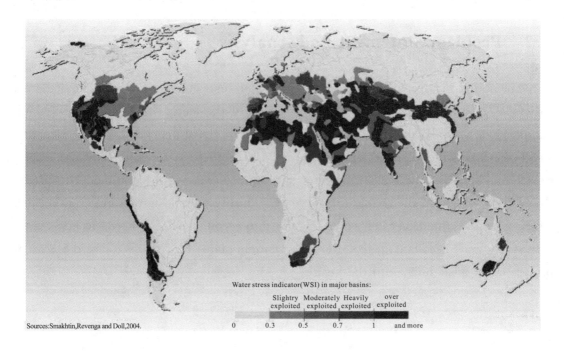

Fig. 9.1 Water Resource

9.1.2 The Reasons of Water Scarcity

Water scarcity for many reasons, including a shortage of physical and economic scarcity. Physical scarcity is where there is inadequate water in a region to meet demand; economic scarcity is where water exists but poor management and lack of infrastructure limit the availability of clean water. However, UNESCO believes that the current Earth's freshwater resources is generally adequate, because of uneven distribution, poor management, environmental change, and inadequate investment in infrastructure and other reasons, about 20% of the world's people lack access to safe drinking water.

9.1.3 The Call for Protection of Water Resources

Water is the necessities of life, the lack of water resources will seriously affect the living conditions of the residents, the lack of water resources may lead to a variety of problems, therefore, the global appeal to protecting water resources and conserve water. Based on the research of the method of water saving, such as seawater desalination, development of groundwater, etc., the purpose is to obtain more available water resources. In fact, water pollution is an important reason for the lack of water resources. And therefore, human should reduce pollution and cherish water resource.

9.2 Problem Statement and Analysis

The ability of a region to provide clean water restricts the development of human society. Firstly, we consider the dynamic nature of the factors that affect both supply and demand. And develop a model to measure the ability of providing water. Secondly, we choose area which is short of water heavily. And then we measure the water supply capacity and analyze the reason of water shortage. Next, we use the model to predict the water situation in the region for 15 years. Thirdly, we design an intervention plan which taking all the drivers of water scarcity into account, then it needs to analyze the advantages and disadvantages of the intervention plan. Finally, we use the plan to project water availability into the future.

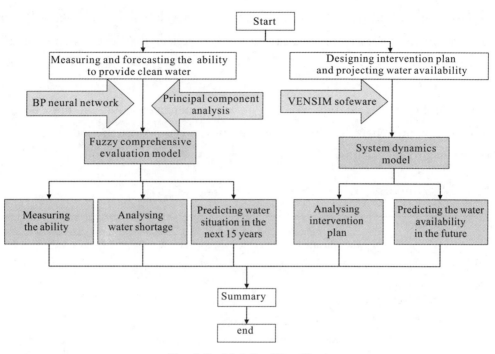

Fig. 9.2 Modeling Flow Chart

9.3 Basic Assumptions

• Clean water can be used in the areas about production, life and ecology.

• In the process of utilization of water resources, the water resources will not be reduced.

• The weather will not change in the selected research time, nor will it be affected by global warming.

• Environment only affects the surface water and groundwater resources in a region in finally.

• There is no transformation between surface water and ground water.

9.4 Glossary & Symbols

9.4.1 Glossary

• System dynamics: System dynamics simply interprets as a system simulation method, which is used to study the information feedback of the system. The understanding of system dynamics problem is based on the close relationship between the system behavior and the internal mechanism.

• The index of water resources available for development: The index of water resources available for development can be used to reflect the development capacity of the water resources in a region. Its essence is the average number of the fixed base development index.

• Comprehensive evaluation index: The comprehensive evaluation index is obtained by the fuzzy comprehensive evaluation model, which can reflect the index of the ability of a region to provide clean water.

9.4.2 Symbols

Number	Symbols	Symbol Description
1	w_0	i-th indicators
2	α	Eigenvalues
3	λ	Eigenvectors
4	r_{ij}	Correlation coefficient of index i and j
5	w_i	The weight of i-th indicators
6	F	Comprehensive score of water supply capacity
7	η	The value of i-th indicators at time T
8	$C_i(T)$	Water resources carrying capacity of i-th indicators at time T

9.5 The Evaluation and Forecast of Water Resources

9.5.1 Analysis Approach

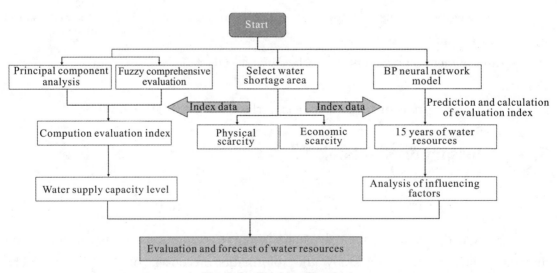

Fig. 9.3 Thinking Flowchart

9.5.2 Data and Preparation

We first collect the data about surface water, groundwater, irrigation water, urban public water, domestic water and ecological water data of Hohhot in China from 2001 to 2015. These data can be used to measure the region's ability to provide clean water to their population and predict water conditions in the region in the next 15 years. Then, we search the relevant information on this region's water resources and explain why and how water is scare in Hohhot. Finally, we find those data in different countries and regions. Based on the evaluation results, the ability to provide the clean water can be classified.

The relevant data in 2001 to 2015 is listed in the following table. Because of the limit of the page, we just list the data about Hohhot in Table 9.1.

Tab. 9.1 15 Years of Relevant Data in Hohhot

	surface water	groundwater	agriculture	FAF	industry	urban public area	life	ecology
2001	2.5	10.1	6.63	0.44	0.98	0.08	0.42	0.01
2002	4.3	11.5	6.43	0.51	0.65	0.09	0.49	0.01
2003	4.1	13.62	6.18	0.87	1.2	0.11	0.52	0.01
2004	3.53	10.2	6.34	0.51	0.91	0.1	0.58	0.01
2005	3.26	10.71	6.26	0.61	1.45	0.13	0.62	0.02
2006	2.75	9.76	6.37	0.5	1.7	0.13	0.65	0.02
2007	3.8	13.02	6.1	0.78	1.59	0.16	0.69	0.02
2008	3.7	12.08	6.05	0.56	1.85	0.17	0.72	0.02
2009	1.92	10.09	6.23	0.42	1.73	0.28	0.79	0.03
2010	2.84	10.65	6.15	0.44	0.44	0.29	0.81	0.03
2011	2.55	9.03	6.52	0.47	1.82	0.34	0.88	0.04
2012	3.68	12.6	6.26	0.82	1.73	0.41	0.87	0.11
2013	2.85	11.19	6.1	0.48	2.02	0.4	0.89	0.17
2014	1.53	10.3	6.27	0.48	1.51	0.57	0.92	0.11
2015	2.71	10.03	6.62	0.68	1.99	0.62	0.92	0.14

Unit: one hundred million cubic meters

FAF: Forestry, animal husbandry and fishery

9.5.3 The Construction of the Model

1. Relative Deviation Fuzzy Matrix Evaluation

$U = \{u_1^0, u_2^0, \cdots, u_m^0\}$ is assumed as a collection of m objects which need to be evaluated. $V=\{v_1, v_2, \cdots, v_n\}$ is a collection of factors that will be evaluated. We measure each factor in U. And we obtain an observation matrix A, where a_{ij} represents the value of the j-th program indicators on the i-th evaluation factors.

(1) Establishing an ideal solution: $u=(u_1^0, u_2^0, \cdots, u_m^0)$

where $u_i^0 = \begin{cases} \max\{a_{ij}\} & \text{when } a_{ij} \text{ is cost index} \\ \min\{a_{ij}\} & \text{when } a_{ij} \text{ is benefit index} \end{cases} \quad i=1,2,\cdots,m$

(2) Establishing the relative deviation fuzzy matrix, $\underset{\sim}{R} = \begin{pmatrix} r_{11} & r_{12} & \cdots & r_{1n} \\ r_{21} & r_{22} & \cdots & r_{2n} \\ \vdots & \vdots & \cdots & \vdots \\ r_{m1} & r_{m2} & \cdots & r_{mn} \end{pmatrix}$

where $r_{ij} = \dfrac{|a_{ij} - u_i^0|}{\max\limits_{1 \leq j \leq n}\{a_{ij}\} - \min\limits_{1 \leq j \leq n}\{a_{ij}\}}$ $(i=1,2,\cdots,m; j=1,2,\cdots,n)$

(3) The weight of each index is obtained by principal component analysis

$$w_j (i=1,2,\cdots,m)$$

(4) Establishing a comprehensive evaluation model[4]

$$F_j = \sum_{i=1}^{m} w_i r_{ij} \ (j=1,2,\cdots,n)$$

And if $F_t > F_s$, then t-th program was better than the s-th.

2. Principal Component Analysis

The capacity of n sample observations is $x_i = (x_{i1}, x_{i2}, \cdots, x_{ip})^T (i=1,2,\cdots,n)$. The correlation matrix are S and R, and they are regarded as the estimated $\sum$ and ρ, respectively.

About the main component in sample, we have the following conclusions.

Assuming that $S_{p \times p}$ is sample covariance matrix, so it's eigenvalues are $\hat{\lambda}_1 \geq \hat{\lambda}_2 \geq \cdots \geq \hat{\lambda}_p \geq 0$. Corresponding orthogonal unit eigenvectors is $\hat{e}_1, \hat{e}_2, \cdots, \geq \hat{e}_p$. the first main component about k-th sample is:

$$y_k = \hat{e}_k^T x = \hat{e}_{k1} x_1 + \hat{e}_{k2} x_2 + \cdots + \hat{e}_{kp} x_p.$$

And eigenvalues are $\alpha = (\alpha_1, \alpha_2, \cdots, \alpha_p)$. Therefore, the contribution of each sample's main components is:

$$b_i = \lambda_k / \sum_{i=1}^{p} \hat{\lambda}_i.$$

Generally, we select first m main ingredients, when their cumulative contribution rate is more than 85%.

Therefore, original indicators can be replaced by the first m principal components. In other words, index coefficients can be expressed by these i principal components variance contribution rates. By taking a weighted average of the coefficients in the linear combination of these i main components, we obtain

$$a_i = \sum_{j=1}^{p} \hat{e}_{ij} b_j / \sum_{j=1}^{p} b_j.$$

Then the comprehensive score model is

$$Y = a_1 x_1 + a_2 x_2 + \cdots + a_p x_p.$$

Normalizing the index weights, we get the following each index weight

$$w_i = a_i / \sum a_i.$$

3. BP Neural Network

The learning process of BP neural network[4] consists of two processes.

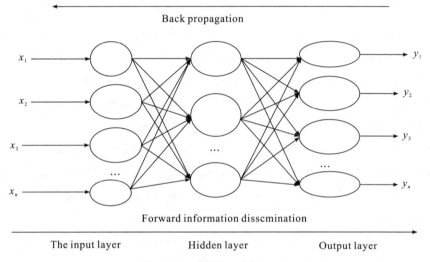

Fig. 9.4 BP Neural Network

BP algorithm is with gradient. For the neural network, the BP algorithm consists of two processes, which are the forward calculation of the data stream (forward) and the error signal. In the forward direction, the propagation direction is the input layer, the hidden layer and the output layer, and the state of each layer of the layer only affects the next layer of neurons. If the output layer is less than the desired output, the error signal is the reverse transmission process.

Specific algorithm:

(1) Network initialization and parameter learning is hidden layer and the output layer each node of the connection weights and threshold neurons give $[-1,1]$ interval of a random function.

(2) Provide training mode, that is, a training mode is selected from the set of training mode, and the input mode and the expected output is sent to the network.

(3) Forward propagation process, that is, given the input mode, from the first

layer of the hidden layer, the output mode of the network, and the obtained input model and the expectation model, If there is an error, the execution of a third step (4), otherwise, returns to step (2).

(4) The reverse propagation process, that is, from the output layer back to the first hidden layer, the connection a weight of each element is corrected by the layer by layer.

(5) Returns to the step (2), repeat every training mode to the training mode to repeat the steps (2) to (3), until each training pattern of the training mode is satisfied with the desired output.

9.5.4 The Solutions of the Problems

(1) TASK 1: measure the ability of providing clean water

In the 8 indicators, the total amount of surface water and the total amount of groundwater resources are benefit indexes. The greater its value, the more abundant water resources in the region and the stronger the ability to provide clean water. The agricultural irrigation, forestry, animal husbandry and fisheries, industry, urban public area, life and ecological water consumption are cost indexes. The greater its value, the greater water demand in the region and the weaker the ability to provide clean water.

The maximum value of the correlation coefficient was 0.89 for all indicators. So there is no exact linear relationship, without rounding variables. Each indicator is made dimensionless, and cost index is made into efficiency index.

In the fuzzy comprehensive evaluation model, the weight of each index on the final result plays a decisive role. So it is extremely important by using the appropriate method for solving. Here we construct principal component analysis model to calculate weights.

Using the data which was made dimensionless and MATLAB software, we can obtain orthogonal unit eigenvectors and eigenvalues.

Tab. 9.2 Eigenvalue and Eigenvector

Index	Eigenvalues	Eigenvectors				
		y_1	y_2	y_3	y_4	...
I_1	0.3247	0.384	0.2897	0.1092	−0.0298	...
I_2	0.2423	0.1887	0.5195	−0.2024	0.0975	...
I_3	0.0633	−0.0984	0.4282	−0.3721	−0.7094	...
I_4	0.0504	−0.2074	0.5739	−0.1388	−0.4118	...
I_5	0.0318	0.3049	0.2426	−0.8518	0.2561	...
I_6	0.0235	0.4542	−0.0897	0.2108	−0.291	...
I_7	0.0049	0.5225	−0.1558	0.1123	0.1202	...
I_8	0.0013	0.4381	−0.2056	0.0825	−0.3899	...

The ratio of eigenvalues and the total sum of the eigenvalues is the contribution rate of each sample's main ingredient. Samples' contribution rates are 0.4374, 0.3264, 0.0853, 0.0679, 0.0428, 0.0317, 0.0066, 0.0018. The cumulative value of the previous four principal components has reached 91.71%. Therefore, we only need to select the first four principal components.

As for comprehensive score model, index coefficients can be regarded as the weights of these four principal components variance contribution rates. By taking a weighted average of the coefficients, we get the comprehensive score model with each index as follows

$$Y = 0.294x_1 + 0.263x_2 + 0.018x_3 + 0.347x_4 + 0.001x_5 + 0.182x_6 + 0.213x_7 + 0.115x_8.$$

By normalizing the index weights, we can obtain each index weights as shown in Table 9.3.

Tab. 9.3 The Weight of Each Index

surface water	groundwater	agriculture	FAF	industry	urban public area	life	ecology
0.2051	0.1836	0.0128	0.2417	0.0008	0.1274	0.1486	0.0799

Substituting the weight into fuzzy comprehensive evaluation model, we can get the evaluation index in 2001 to 2015.

Tab. 9.4 The Evaluation Index in 2001 to 2015

Year	2001	2002	2003	2004	2005	2006	2007	2008
Evaluation Index	2.63	3.28	3.72	2.90	2.97	2.88	3.56	3.32
Year	2009	2010	2011	2012	2013	2014	2015	
Evaluation Index	2.58	2.88	2.56	3.54	3.03	2.62	2.88	

According to the evaluation index, we draw a line chart as shown in Figure 9.5.

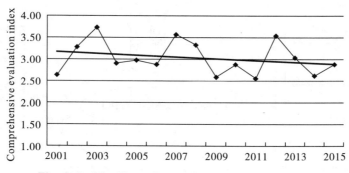

Fig. 9.5 The Chart of Evaluation Index Change Trend

From Table 9.4 and Figure 9.5, Hohhot's evaluation index changes with changing years, so its supply and demand for clean water is also changing. Further, we also find that the ability to provide clean water in Hohhot gradually reduces. According to the data of other selected areas, the average evaluation index in Tokyo, Shanghai, New

Delhi, Paris, Chicago, Toronto, Taipei, Jakarta and other cities in 2001 to 2015 years are calculated.

Tab. 9.5 The Evaluation Index in Different Areas

Tokyo	Shanghai	New Delhi	Paris	Chicago	Toronto	Taipei	Jakarta
0.78	0.56	5.32	6.74	9.75	10.62	13.1	14.59

Tab. 9.6 Water Supply Capacity Classification Table

Water shortage situation	Grade	Index
Extreme	★★★★★	0~2
Heavy	★★★★	2~4
Moderate	★★★	4~6
Slight	★★	6~8
Adequate	★	>8

From Table 9.6, Hohhot is a heavy water shortage area.

In summary, it is not difficult to measure a country's ability to provide clean water by using the main component analysis and fuzzy comprehensive evaluation method. Absolutely, you will use the ability to supply clean water classification table to help you measure.

(2) TASK 2: analysis of regional water shortage reasons

According to the preparatory phase, we choose Hohhot City in Inner Mongolia to analyze the water scarcity. Hohhot City is heavily short of water.

There are reasons from two aspects leading to this situation. One is physical scarcity, the other one is economic scarcity. It's easy to explain why this city is short of water from these two aspects.

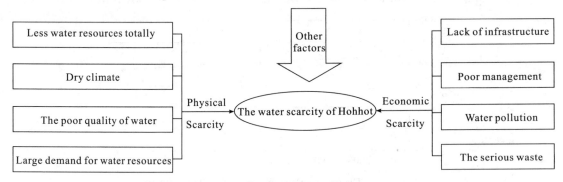

Fig. 9.6 Causes of Water Shortage

◆ **Physical scarcity**

a. The shortage of total water resources: the total amount of water resources in Hohhot is 339.972 million cubic meters. While its average water resource amount per capita is only 450 cubic meters, which is 1/24 of the world average and is one of the

city. Hohhot is severely short of water. The average annual precipitation in Hohhot is lower than that in the whole country. The water resources are less.

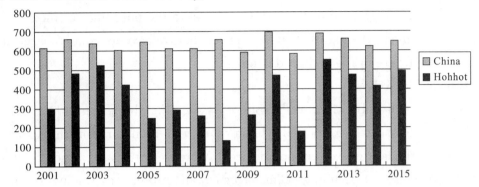

Fig. 9.7   Annual Average Precipitation in China and Hohhot

b. The weather effect: Hohhot is in the interiors of the continent, far away from the sea. It's a typical continental monsoon climate characteristic in the semi-arid mid-temperature zone. Climate features: winter is cold and long, summer is hot and short, spring is dry and windy, and autumn is cool and sunny. The changeable temperature often leads to some agricultural disasters. The rainfall can be influenced by the climate and causes drought.

c. The water quality problem: Because of the long-term pollution, the quality of the Hohhot water is very bad. The main pollution parameters are dissolved oxygen, potassium permanganate index, ammonia nitrogen, nitrite nitrogen, mercury, petroleum, etc., among which ammonia nitrogen and petroleum pollution are rather severe. Some indexes can be as high as hundreds of times. A large portion of the surface water can't be used because of the ovenproof of mercury.

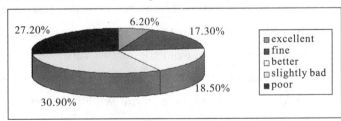

Fig. 9.8   The Proportion of Different Water Quality

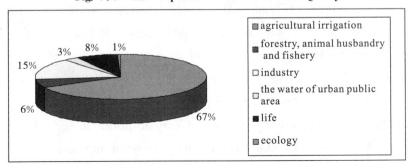

Fig. 9.9   The Use Direction of Water Resources in Hohhot

d. The great demand of water: the natural water resources in the Hohhot is short, especially the surface water resources according to the data. However, the large population and the large need of the water may lead to the unbalance of the supply and demand. According to the Figure 9.9, we can see that we have to exploit a large amount of groundwater to use in life and production.

◈ **Economic scarcity**

a. The lack of infrastructure: Hohhot is rich in the groundwater resource. However, people can't use it well, because of the immature of the exploitation technology and the lack of exploitation devices. Moreover, the infrastructure was built long time ago and hasn't been in good management and protection, leading to the loss of a portion of the water.

b. Poor management: the poor management and the ignorance of the recycle cause a series of problems like water waste and water pollution. The water is cheap, so many companies waste water and don't pay attention to the recycle of water. Although the industrial sewage has been dealt with, it's still not up to the standard of agricultural irrigation. The high price of domestic water restricts water saving severely.

c. The severe situation of water pollution and waste: in recent years, with the shortage of surface water, the long time over-exploitation of groundwater, the discharge of the water sewage, the diffusion of the pesticide residue, the water on the surface and underground has been polluted to varying degrees. The water sewage, due to the lack of dealing devices, has been discharged into river directly, leading to the severe pollution of water on the surface and underground. The crises of water resources are not only in the deficient of water quantity, but also in the water pollution, the deterioration of the water quality and the function reduction even the lost of function.

(3) TASK 3: the prediction of water condition

The environmental drivers' effect is a term which reflects the environmental changes' influence on water situation. It mainly includes the change of precipitation, landform and so on, which can finally affect surface water resources amount and ground water resources amount of an area. Accordingly, we can simply regard the environmental drivers' effects as the impact of the change of surface water and ground water resources amount on water situation.

Meanwhile, the increasing demand of public for water resources also affects water situation, which mainly embodied with water consumption of agricultural irrigation, forestry, livestock and fishing, industry, town, livelihood and ecology.

Thus we decide to construct BP neural network prediction model to predict 8 indexes affecting water situation of Hohhot City in the next 15 years. Then we implement fuzzy comprehensive evaluation with predictive data, and compare the evaluation results with situation 15 years ago and grade partition table.

We take surface water resources amount of Hohhot City in 2001—2012 as training sample and take the data in 2013, 2014 and 2015 as inspection items.

With the help of MATLAB software, error analysis of model prediction results is shown in Table 9.7.

Tab. 9.7 Error Analysis Table for Prediction Results

Particular year	Actual data	Forecast data	Relative error
2013	2.85	2.71	0.0491
2014	1.53	1.68	0.0980
2015	2.71	2.91	0.0738

Since the result of inspection is fine, we can use all data in prediction. Because of the different iterations, the results of prediction are different. In order to reduce error of prediction, we predict numerical value of 3 years every time and get a multiple forecast average. We predict surface water resources amount of Hohhot City in 2016, 2017 and 2018 will be 2.8679, 2.1678, 3.0512. Then we add data in these 3 years and data 15 years ago to predict data in 2009, 2010 and 2011. Finally, we get the average of surface water resources in the next 15 years.

In this way, we can get predictive values of 8 indexes in the next 15 years. Using these data, we make comprehensive estimation for Hohhot City's capacity of providing clean water and obtain integrated assessment value of water situation data over the next 15 years as is shown by the Tab. 9.8.

Tab. 9.8 The Comprehensive Evaluation Value of the Next 15 Years in Prediction

Year	2016	2017	2018	2019	2020	2021	2022	2023
Evaluation Index	2.33	2.51	3.42	2.21	2.67	2.58	2.16	3.02
Year	2024	2025	2026	2027	2028	2029	2030	
Evaluation Index	2.99	2.31	2.32	2.08	2.3	2.52	2.07	

Then we can draw a line graph as shown in Figure 9.10:

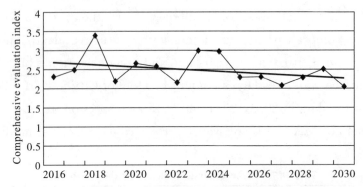

Fig. 9.10　The Comprehensive Evaluation Index of Water Resources in the Next 15 Years

From Figure 9.10, we can know that the comprehensive evaluation index of Hohhot City will decrease year by year in the next 15 years, which indicates that the water resources in the area are getting more and more intense. One of the reasons is the impact of environmental factors, along with the urbanization, there is more and more destruction of vegetation in the region, surface storage capacity of water resources degradation result that surface water resources quantity decrease year by year. Over exploitation of groundwater and water content of a sharp increase have caused the decrease of water resources supply and the increase of demand, the water resources problem of Hohhot has become increasingly serious.

We think that the negative water resources situation in the future will affect our lives in the following ways:

a. People and livestock's drinking water and domestic water will be affected. Because of water shortage and worse pollution, available clean drinking water will become less and less in 15 years.

b. Agriculture, forestry, livestock and fishery will be affected. The water shortage in the next 15 years will restrict the production of the four industries and reduce the food supply.

c. The industry development will be affected. The water shortage will restrict the development of industry so that the Hohhot's economy will be severely affected.

d. The ecological environment will be affected. Water shortage will make the surrounding environment become worse and worse so that people's lives will be uncomfortable.

9.6 Intervention and Prediction of Water Resources

9.6.1 Analysis Approach

First, we consider two aspects of supply and demand of water resources to make Plans A and Plan B. For Plan A, only the qualitative analysis of its impact is brought. In this plan, we just analyze its impact qualitatively. For Plan B, we set up system dynamics model(1) combined with the factors that influence the shortage of water resources, we simulate the intervention measures after the implementation of intervention program. Evaluation of the Plan B is indeed able to ease the shortage of water resource. PI index(3) introduced and used and we analyze the advantages and disadvantages of Plan B to bring about the surrounding area and water resource system. Finally, analyses of the advantages and disadvantages of Plan A and Plan B are made, and some useful suggestions are given.

9.6.2 TASK 4: The Design and Analysis of Intervention Plan

(1) The establishment of model

- **The establishment of SD model**[5]

This article takes Hohhot as an example and builds up a system dynamics model (SD model)

The advantage of this model is that it could reflect the stream flow regulation and the balanced situation between water resources demand and supply in each period. So, if we use this model to analyze water resources, economy and environment, we could make full use of water resources, promote economic development and protect environment at the same time.

①Hohhot's water resources systematic analysis

a. Population subsystem; b. Economy subsystem; c. Cultivated land subsystem; d. Livestock subsystem; e. Water resources system.

②SD model flow diagram

a. The diagram about Hohhot's water resource is based on the VENSIM.

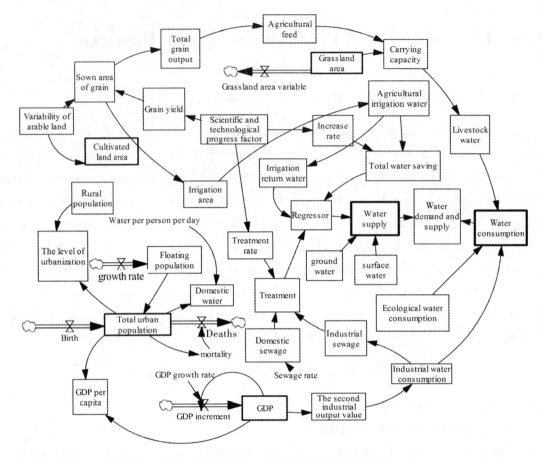

Fig. 9.11 The Diagram about Hohhot's Water Resource

In the module of water resources supply and demand, the two main elements are the quantity of water supply and water consumption. The water supply can be categorized into underground water, surface water and return water. Water consumption can be categorized into agricultural water, industrial water, domestic water and ecological water.

Establishment of the equation:

a. SD equation about available water resources is as follows:

Available water resources = surface water + underground water − the recycle of surface and underground water + unconventional water

b. SD equation about production water demand is as follows:

Agricultural water = agricultural irrigation water + forestry and animal husbandry water

Industrial water = industrial added value × unit of industrial added value water quantity

Tertiary industry water = tertiary industry added value × unit of tertiary industry value water quantity

c. SD equation about living water demand is as follows:

Total population=newly-born population+mechanical growth population−deaths+present total population

Urban population=urbanization rate×total population

Domestic water consumption= urban population×urban per capita domestic water consumption+rural population×rural per capita domestic water consumption

d. Ecological water: Ecological water demand is related to the corresponding ecological protection, restoration, ecological system's own demand, policy and planning.

- **Intervention plan design**

We design Plans A and B from water supply and demand based on these two aspects.

Plan A

We design Plans A and B from water supply and demand based on these two aspects.

Plan A is about intervening water supply. There's no explicit data, so we don't use model to do analogue simulation. We only use it to analyze the influences to the surroundings and the water resource environment.

a. Develop water transfer project. For example: China's South-North Water Transfer Project. Transfer the extra water resources from south to north through artificial channel.

b. Install the sea water desalting plant. Transfer the sea water into fresh water.

c. Bring in water collecting devices to increase water supply.

Plan B

Taking into account the demand of water resources, through consulting data, we make sure the various parameters of the data, the intervention program and water resources carrying capacity index system content.

Tab. 9.9 Water Resources Carrying Capacity Index System Content

Plan	Variables and Factors that Need to be Adjusted	Impact Point	Adjustment Method
Water conservation	Water quota for urban residents	−	Down 2%
	Water quota for rural residents	−	Down 2%
	Irrigation water use efficiency	+	Down 5%
	Water for irrigation per unit area	−	Down 5%
	Reuse ratio of industrial water	+	Down 2%
	Unit water of FAF	−	Down 2%
Bringing pollution under control	COD emissions per capita	−	Down 5%
	CDO emissions per unit of industry	−	Down 10%
	Centralized sewage treatment rate of domestic sewage	+	Rise 2%
	Industrial wastewater discharge coefficient	−	Down 5%

续表

Plan	Variables and Factors that Need to be Adjusted	Impact Point	Adjustment Method
Adjustment of industrial structure	First industrial added value growth rate	—	Down 1%
	Industrial added value growth rate	—	Down 100.2%
	Third industrial added value growth rate	+	Rise 1.5%
Adjustment of agricultural structure	Growth rate of crop planting area	—	Down 2%
	Forestry gross output value growth rate	+	Rise 1.5%
Comprehensive scheme	All of the above programs are adopted		

"+" expresses that the factors of sustainable development, improve the efficiency of water, environmental improvement has a role in promoting. "—" expresses inhibitory action.

2. The Solutions of the Model

- **Analysis of the impact of the plan**

Plan A:

a. The Hohhot's water transfer project can not only bring a large quantity of water resources, but also increase the surface and underground water resources of surrounding area. It could alleviate the water shortage in time, improve the ecosystem in a grand scale. However, changing the water recycle system randomly would also cause some negative impacts. For example, it would increase the water supply pressure of the water output area.

b. The rainwater harvesting technology can improve the water supply, but it makes it hard to supply the surface water in time.

c. Hohhot is in the interiors of the land and far away from the sea, so this scheme is not suitable for this city. Besides, this scheme may cost a great deal of money.

Plan B:

The impact of the Plan B needs to be analyzed by system dynamics model combined with the simulation results.

a. Data verification

The data is chosen from the industrial water consumption during 2008—2012. Relative error is less than 0.05.

Tab. 9.10 Error Analysis Table for Prediction Results

Year	Actual data of industrial water	Forecast data of industrial water	Relative error
2008	1.85	1.78	0.0378
2009	1.73	1.80	0.0405
2010	0.44	0.43	0.0227
2011	1.82	1.79	0.0165
2012	1.73	1.71	0.0231

When the relative error is less than 0.05, the test is passed. It is indicated that the system dynamics model is available and accurate.

b. Simulation

Suppose we carry out part of the intervention plan of Plan B and simulate the plan into the water resources supply and demand situation in 2025. By comparing the situation to the original status, we can get the following table.

Tab. 9.11 Simulation Results for 2025(Unit: Billion Per Cubic Meter)

Year	Available water resources	Agricultural water	FAF	Industry	Urban	Life	Other	Total	Difference
No intervention	11.5	6.34	0.41	2.52	0.98	1.55	0.23	12.03	−0.53
Intervention	13.86	6.20	0.38	2.45	0.80	1.47	0.36	11.66	1.60

c. According to the simulation results after the intervention plan, we found:

In the case of non intervention, the water resources of Hohhot city in 2030 can not meet the demand of water resources. However, the contradiction between supply and demand is prominent, and the water resources are overloaded.

In the case of the implementation of the intervention plan, the water supply in Hohhot city in 2030 increases significantly, and agriculture, forestry, animal husbandry and fishery water significantly decreases. That is, the water resources carrying capacity increase, supply and demand ease. So, water resource still can fully meet the needs of water resources with intervention.

d. System dynamics model shows that the comprehensive intervention plan B is conducive to ease the shortage of water resources. By introducing water resources for the development of the index[3], evaluate the pros and cons of the effect about the specific implementation details of the Plan B to the surrounding areas.

First, we select the water resources available for development index (PI). This indicator can indicate the development of water resources in a region.

The selected indicators $(Y_1(T), Y_2(T), \cdots, Y_i(T), \cdots, Y_n(T))$ are treated as follows:

$$H_i(T) = (Y_i(T) - Y_i(0))/Y_i(0), (1 \leqslant i \leqslant n)$$

$$P_i(T) = \begin{cases} H_i(T), & \text{Indicators point to positive} \\ -H_i(T), & \text{Indicators point to an inverse} \end{cases}$$

$$PI(T) = \frac{1}{n}\sum_{i=1}^{n} P_i(T),$$

where Y_i is the initial time of item i index, $P_i(T)$ is the sustainable development index of i-th index at T time.

Based on this model, we bring the data into the forecast of 2016 to 2030. Then we operate the results, and get a variety of intervention plan for the next 15 years of PI.

Tab. 9.12 Different Intervention Plans Within 15 Years of PI

Year	No intervention	water conservation	bring pollution under control	Adjustment of industrial structure	Adjustment of agricultural structure	Comprehensive plan
2016	0.077	0.133	0.075	0.112	0.12	0.181
2017	0.089	0.172	0.091	0.15	0.157	0.162
2018	0.101	0.211	0.103	0.191	0.186	0.21
2019	0.145	0.245	0.144	0.228	0.231	0.252
2020	0.216	0.272	0.218	0.265	0.268	0.312
2021	0.241	0.312	0.243	0.302	0.303	0.383
2022	0.275	0.355	0.277	0.339	0.344	0.427
2023	0.372	0.398	0.37	0.383	0.384	0.478
2024	0.403	0.412	0.408	0.423	0.424	0.52
⋮	⋮	⋮	⋮	⋮	⋮	⋮
2030	0.689	0.716	0.691	0.734	0.727	0.821

We draw a figure.

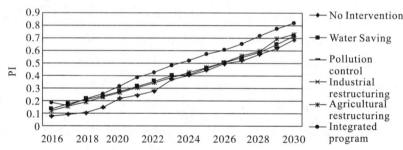

Fig. 9.12 PI Values under Different Scenarios

From the results of calculation, we can obtain:

ⅰ. Except for the intervention plan for pollution control, the PIs of other plans are all higher than the one without any measure. It means other plans are all able to significantly improve the water carrying capacity of Hohhot. Among them the comprehensive approach has the most negative impacts on economic development and water resource environment.

ⅱ. The PI of pollution control is basically the same as the one of regular mode. The plan of pollution control will increase the investment in economy and restrain the economic development to some extent, but will improve the environment. This plan will bring negative impact on economic development but will be good to ecological environment.

ⅲ. We can know through the data that the values of the water saving project before 2023 are all higher than the PI value of industrial and structural adjustment of agriculture but lower than the latter two projects after 2024, which means that water saving projects have obvious effects in a short term, but still need proper adjustment of industrial and agricultural structure in a long run. This also means structure adjustments are the best intervention measure to improve the lack of water in a long run.

- **The advantages and disadvantages of the intervention plan**

Tab. 9.13 The Advantages and Disadvantages of the Intervention Plan

Plan	Advantages	Disadvantages
Water transfer project	Alleviate the water shortage, improve the water environment	Large capital investment, may bring the negative impact of ecology
Rainwater harvesting technology	Increase the supply of fresh water resources	Need universal implementation
Sea water desalination	Measures to solve the water shortage in coastal areas	Inland areas can not be used, and large investment funds
Water saving	The implementation of the scheme is small, and the effect of carrying capacity of water resources is improved	Early effect is obvious, but long-term and can not fundamentally ease the lack of water resources.
Pollution control	Obviously improve the surrounding environment and water quality	Affect the regional economic development, the implementation of the cost of the program
Industrial restructuring, Agricultural restructuring	Conducive to long-term improvement of water ecological environment	Implementation cycle is long, the effect is not obvious, the implementation of cost
Integrated program	The effect is obvious, involving all aspects	Implementation cycle is long, the cost is big

In summery, according to the simulation results, the following implementation plans are suggested:

In terms of water supply, we could use water transfer project and rainfall collection project to increase water supply.

In terms of water demand, we should aim at some water saving measures in the earlier stage, for example, restricting the domestic water consumption, raising the water price, promote the irrigation rate. At the same time, it's important to bring in the advanced technology to tackle the pollution emission. In the long run, we should focus on the adjustment of the agriculture structure and the industrial structure.

9.6.3 TASK 5: Projecting Water Availability

Using the system dynamic model and the intervention plan, we simulate the water supply and demand of water resources in the coming decades, and analyze the results.

As shown in the following table 9.14:

Tab. 9.14 Water Resources Supply and Demand Difference (100 million /m^3) Intervention

Years	Implement of intervention plans	No implement of intervention plans
2016	4.79	1.97
2017	4.52	1.53
2018	4.29	1.20
2019	4.09	0.97
2020	3.82	0.72
2021	3.68	0.45
2022	3.56	0.23
2023	3.12	0.01
2024	2.70	−0.14
⋮	⋮	⋮
2050	−0.01	−3.1

The analysis:

• After carrying out intervention plan, Hohhot city will become less vulnerable to water shortages. Under the intervention plan, water supply and demand difference keeps positive throughout the year. It suggests that the situation of water shortage in Hohhot city has been greatly mitigated. Under normal circumstances there will be a shortage of water resources in 2020. But after the intervention plan implementation, there won't be a shortage in Hohhot city until 2050. So, the shortage situation will be delayed about 27 years.

• In the future there will still be a water shortage. Water shortage is still an important issue. According to the simulation results, there will be a water scarcity in Hohhot city in 2047. This situation shows that even if taking a lot of water resources intervention plans. The shortage of water resources will still restrict the development of the city in the future. So, we can't treat it lightly. On the contrary, we should continue to improve scientific and technological interventions to ease the shortage of water resources.

9.7 Sensitivity Analysis

Sensitivity analysis is used to study the influence of parameters on the behavior of the system. Through correcting parameters constantly, we predict the sensitivity of the parameters of the model.

Testing the effects of 3 variables (GDP, industrial added value, total water demand) on population growth rate, industrial wastewater discharge rate, urban water consumption, water consumption of rural population, water consumption per unit area of cultivated land and water consumption per unit of industrial added value to reflect the effects of these 3 variables on the whole system.

Test method: From 2001 to 2015, each parameter changes year by year 10%, examines its influence on 3 variables.

$$S_L = \left| \frac{\Delta L_t}{L_t} \times \frac{X_t}{\Delta X_t} \right|$$

where t expresses time S_L expresses the sensitivity of state variable L to parameter X, L_t express the value of the state variable L at the time t, X_t expresses the value of parameter X at time t, ΔL_t expresses the change of state variable at t time, ΔX_t expresses the change of parameter X at time t.

When the parameter X_j is changed, the sensitivity of the 1 to N variable $(L_1, L_2, \cdots, L_N)$ to X_j is expressed as $(S_{L1}, S_{L2}, \cdots, S_{LN})$. $(L_1, L_2, \cdots, L_N)$. So we can use these sensitivity means to represent the sensitivity S_{X_j} of the model with respect to the parameter X_j.

$$S_{X_j} = \frac{1}{N} \sum_{i=1}^{N} S_{Li}$$

Through calculation, we obtain 6 values of parameters. The result is shown in Figure 9.13.

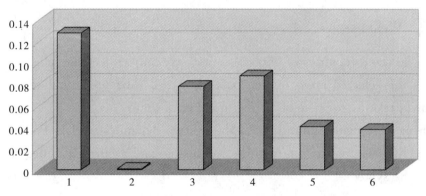

Fig. 9.13 Sensitivity Analysis

According to Diagram 4-9, the varies of parameter 1(rate of population growth) is higher than 10%. Others are lower than 10%, which suggests that the system is not sensitive to most of parameter changes. So, the model is good in effectiveness and can be used in the simulation of practical system.

9.8　Strengths and Weaknesses

9.8.1　Strengths

(1) In the fuzzy comprehensive evaluation model, there are many methods to determine the weight of each index. In this paper, the principal component analysis method is used to solve the weight. In this way, the evaluation indexes of the decisive function can be analyzed directly, and the subjective factors and the uncertain factors in the evaluation process can be reduced.

(2) In the prediction by BP neural network, the initialization of the weights and the threshold is random. The predictive results are different. The approach we take is that we only predict 3 years of data. It takes several training and the average of value of simulation finally. In this way, the prediction error can be effectively reduced.

(3) System dynamics is a closed system. It only requires data with low accuracy. System dynamics pay attention to general trends and do not care about the exact number of system variables in particular year. It just studies the causality diagram and system flow chart and basic data. The model not only considers the large number of systems, nonlinear relationships, but also pays attention to the role of the human.

(4) Task 4 introduce the PI index, which can reflect the favorable and adverse effects on the surrounding environment with the implementation about plan B.

9.8.2　Weaknesses

(1) Due to the difficulties of data collection, we only collect the indicators of water resources in the selected areas on the past 15 years. The predicting results are in a larger error in task 3, because of a small number of data.

(2) There exist many limitations in the system dynamics model: Parameters must be accurate. Model construction must be reasonable and accurately reflect the internal relations of the system.

(3) Models need to be close to the actual. Otherwise, the larger deviation of simulation results appears.

9.9 The Improvement and Development of the Model

9.9.1 The Improvement of the Model

(1) BP neural network can be combined with the gray system. Then, the introduction of grey numbers can improve the convergence rate of the neural network, so that the neural network training became more reasonable.

(2) The improvement of the System Dynamics Model is optimization the program. We use the method of Trial and Error which is to design the program first, simulate the progress and then choose the best program. It depends on the analyst's experience which means it is subjective. In order to be more objective, we can introduce the Genetic Algorithm (GA) to search for the best method. Therefore, we combine the System Dynamics Model and GA to find the optimization in mathematics.

9.9.2 The Development of the Model

(1) Fuzzy Comprehensive Evaluation Model is widely used in Fuzzy Mathematics. It is often used to make a comprehensive comment considering all the related factors. Therefore, it is popular in evaluating the operating conditions, the economic development of an area and the population carrying capacity, etc.

(2) BP neural network model is widely used in many areas such as speech analysis, image recognition, watermarking and computer vision due to its strength in dealing massive data parallelly and processing separately. It is also popular because of its self-organizing and self-learning strength. Therefore, it becomes a powerful tool for pattern recognition.

System dynamics model is widely used to reflect feedback system which can be used to study enterprise production management, project management and ecological environment which combines the natural science and social science together.

Reference

[1] Wang Weirong. Second Time Supply and Demand Balance Analysis of Regional Water Resources Based on System Dynamics[J]. *South-to-North Water Transfers and Water Science and Technology*, 2014, 12(1): 48—49.

[2] Chui Haishen. Prediction Research on Harbin Water Resources Carrying Capacity by System Dynamic Model[D]. Harbin: Harbin Institute of Technology, 2014, 34—38.

［3］ Wang Yinping. Study of Water Resources System Dynamics Model in Tianjin City［D］. Tianjin：Tianjin University，2007，27－30.

［4］ Li Bainian. *Matlab Data Analysis Methods*［M］. China Machine Press，Beijing，2011.

［5］ Hu Xueyuan. The Analysis on the Balance of Supply and Demand of Water Resources in Huhhot［D］. Huhhot：Inner Mongolia Normal University，2007，23－28.

［6］ UN Water Scarcity Map［Z］. http：//www.unep.org/dewa/vitalwater/jpg/0222-waterstress-overuse-EN.jpg.

［7］ Yan Guangle. Research on System Dynamics Model of Water Resources Sustainable Development in China［D］. Shanghai：University of Shanghai for Science and Technology，2010.

［8］ Tan Leyan. A Research on Demand-supply Analysis and Guarantee Countermeasures of Water Resources in Shandong Peninsula［D］. Nanjing：Hohai University，2006.

［9］ Yang Guiyuan. *Mathematical Modeling*［M］. Shanghai：Shanghai University of Finance and Economics Press，2015.

［10］ World Water Demand and Supply［Z］. http：//www.iwmi.cgiar.org/Publications/IWMI_Research_Reports/PDF/PUB019/REPORT19.PDF.

◆ 建模特色点评

❋论文特色❋

◆标题定位：Are we heading towards a thirsty planet？即"我们的星球正在走向干涸吗"，利用原始赛题，以疑问句命题，能够起到特殊的效果。

◆方法鉴赏：方法有针对性，用主成分分析法求解权重，直接分析评价指标的决定作用，减少模糊综合评价过程中的主观因素和不确定因素。对BP神经网络仅预测3年的数据中，采用多次仿真取平均值，来减小预测误差，可克服权值和阈值初始化中的随机性。对封闭系统的动力学，关注总体趋势，不关心特定年份系统变量的确切数量。研究因果关系、系统流程和基础数据等。不仅考虑了大量的系统、非线性关系，而且注重人的作用。PI指标反映了B方案实施对周围环境的有利和不利影响。

◆写作评析：整体思路清晰，内容结构安排合理，有一定的技术面，语言顺畅，图表清楚。

❋不足之处❋

只收集了过去15年选定地区的水资源指标。由于数据量较小，预测结果的误差较大。在系统动力学模型中存在许多局限性：参数必须是精确的。模型需要接近实际。否则，仿真结果会出现较大的偏差。

第 10 篇
收费后的车流合并问题

◆ 竞赛原题再现

Problem B Merge After Toll

Multi-lane divided limited-access toll highways use "ramp tolls" and "barrier tolls" to collect tolls from motorists. A ramp toll is a collection mechanism at an entrance or exit ramp to the highway and these do not concern us here. A barrier toll is a row of tollbooths placed across the highway, perpendicular to the direction of traffic flow. There are usually (always) more tollbooths than there are incoming lanes of traffic (see former 2005 MCM Problem B). So when exiting the tollbooths in a barrier toll, vehicles must "fan in" from the larger number of tollbooth egress lanes to the smaller number of regular travel lanes. A toll plaza is the area of the highway needed to facilitate the barrier toll, consisting of the fan-out area before the barrier toll, the toll barrier itself, and the fan-in area after the toll barrier. For example, a three-lane highway (one direction) may use 8 tollbooths in a barrier toll. After paying toll, the vehicles continue on their journey on a highway having the same number of lanes as had entered the toll plaza (three, in this example).

Consider a toll highway having L lanes of travel in each direction and a barrier toll containing B tollbooths (B > L) in each direction. Determine the shape, size, and merging pattern of the area following the toll barrier in which vehicles fan in from B tollbooth egress lanes down to L lanes of traffic. Important considerations to incorporate in your model include accident prevention, throughput (number of vehicles per hour passing the point where the end of the plaza joins the L outgoing traffic lanes), and cost (land and road construction are expensive). In particular, this problem does not ask for merely a performance analysis of any particular toll plaza design that may already be implemented. The point is to determine if there are better solutions (shape, size, and merging pattern) than any in common use.

Determine the performance of your solution in light and heavy traffic. How does your solution change as more autonomous (self-driving) vehicles are added to the traffic mix? How is your solution affected by the proportions of conventional (human-staffed) tollbooths, exact-change (automated) tollbooths, and electronic toll collection booths (such as electronic toll collection via a transponder in the vehicle)?

Your MCM submission should consist of a 1 page Summary Sheet, a 1—2 page letter to the New Jersey Turnpike Authority, and your solution (not to exceed 20 pages) for a maximum of 23 pages. Note: The appendix and references do not count toward the 23 page limit.

◆ 获奖论文精选

Design of Toll Plaza Based on Simulation[①]

Summary: The highway toll station plays a vital role as a bottleneck of the freeway traffic flow. Its capacity directly restricts the highway traffic capacity, thus the toll station directly affects the operation of the highway traffic. Therefore, solving the highway toll station congestion problem is the basis for the toll station design.

In this paper, the problem requires us to design the size, shape, and cost of the toll plaza. First, by considering that the honeycomb model which can make the material very economical, we got the toll plaza in the shape of two symmetrical "Y" type. Then take into account that toll plaza width on both sides and width of the middle road is different and the gradient of side slope. Using the slope angle's tangent of the toll plaza and the width of the toll plaza to simulate. Thus we can know that when the slope angle 's tangent of the side slope is 1/6, the toll square width is 65.2 m. The area of the toll plaza is 10 540 m^2. Every part of the toll plaza's unit charge is equal, so measure the level of the charging plaza's cost is to measure the area of the toll plaza.

In view of toll station throughput problem, we establish queuing theory model. And we have surveyed a standard bus service time for a certain day at a toll plaza on Interstate 5 in New Jersey. By using the toll station's basic capacity formula and delay time and VISSSIM software, we got the traffic flow in different charging way. Then we have a weighted average of each charging method, and finally according to the proportion of different charging methods, we got that the toll station throughput was 3

---

① 本文获 2017 年国际一等奖。队员：肖正，楼康，夏明明；指导教师：朱家明。

067 veh/h. For the above-mentioned solution in the different size of the car flow's performance, we use VISSIM to simulate the proportion of three types of charges and other data. Then we got the influence on delay time with different traffic flow and ETC charging mode. Next, we use EXCEL software mapping and find that in the case of different lane settings, with the ETC usage increasing, the delay time will go down. And when the ETC usage is large enough, delays in different lane settings are low and do not vary much. As more self-driving cars are added to increase Small vehicles and to increase the percentage of ETC, the proportion of manual toll stations will be reduced.

Keywords: toll plaza; queuing theory; simulation; VISSIM

10.1 Introduction

10.1.1 Background

The highway toll station plays a vital role as a bottleneck of the freeway traffic flow. Its capacity directly restricts the highway traffic capacity, thus the toll station directly affects the operation of the highway traffic. Therefore, solving the highway toll station congestion problem is the basis for planning the construction of toll stations, the basis for the toll station design and economic evaluation, and also the basis of research a variety of charging system. Therefore, reducing the overall delay of the toll station, the queue of vehicles and the cost of the toll station operation is the basic goal of the operation of the toll station[1].

Since the United States promulgated the federal aid Highway Act and road tax bill in 1956, the US highway construction has been developed rapidly. Until 1999, the US had 88,727 kilometers highway, and became the country which has the largest number of highways. Most of the United States toll highways use the way of open charging, where different rates are used in different periods[2]. Now, most of the toll highways are built by themselves, and charge by each of them. Each highway toll system follows the similar specifications and techniques.

10.1.2 Introduction of Queuing Theory

Queuing theory, or stochastic service system theory, derives the statistical rules of the number of indicators (wait time, queue length, busy period, etc.) from the statistical study of the arrival of the service object and the service time. These rules can improve the structure of the service system or re-organize the objects which should be served, so the service system can meet the needs of the clients and the cost of the organization can be the most economical. Using Queuing theory to study queuing service system, we must classify the various queuing system description first[3]. Every queue service system can be described as the following four aspects, shown in Figure 10.1.

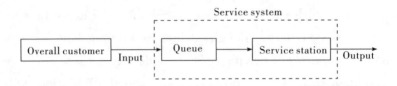

Fig. 10.1 Queuing Theory

(1) Input—refers to situation of the customer arrives at the service station. According to the arrival time interval: a certain time interval, a random time interval; from the number of customers arrive to see the situation: a single arrival, there are batches arrive; from the customer source overall: There are two sources of unlimited and limited customers, But as long as the total number of customers is large enough, the total number of customers can be limited to the total number of customers as a source of infinite approximation of the situation.

(2) Output—refers to the situation of customers from get the service to leave the service station, which have fixed or random service hours [4].

(3) Service principles of queuing theory—there are two systems including loss system and waiting system. Loss system means that when the customer arrives, the customer would automatically leave if all the service facilities are occupied. Waiting system refers to if the service facilities have been occupied, customers would stay to wait for the service, until the service is completed before leaving[5].

(4) Services station—refers to the number of service facilities, arrangement and mode of service. According to the number of service facilities, there are one or more of the points (often referred to as single-station service system and multi-station service system). According to the arrangement, Multi-station service system is divided into series and parallel. The parallel system of S service stations can serve S customers at the same time. In the case of tandem, each customer has to go through the service stations in turn, just like an object is processed through S-way processes. There are single and multiple services in the service mode[6].

10.1.3 Introduction of Basic Composition and Types of Expressway Toll Station

1. The basic structure of the toll station

Highway toll station is mainly composed of four parts: (1) Charge card door. Including toll island, toll booths, lanes, greenhouses and other parts of the charges. (2) The toll plaza. (3) Toll station room. (4) Power supply lighting.

2. The type of toll station

The toll station can be divided into the main toll station and the ramp toll station, as shown in Figure 10.2. The toll station can be divided into the main toll station and the ramp toll station, as shown in Figure 10.2.

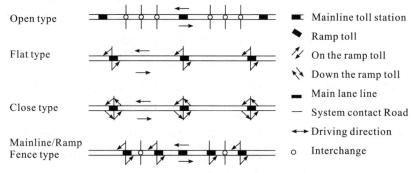

Fig. 10.2 The Main Toll Station and Ramp Toll Station

According to the different forms of parking fess, the toll station can also be divided into human-staffed tollbooths, automated tollbooths and electronic toll collection booths, such as electronic toll collection via a transponder in the vehicle [7].

10.2 Problem Statement and Analysis

10.2.1 Problem Statement

Multi-lane divided limited-access toll highways use "ramp tolls" and "barrier tolls" to collect tolls from motorists. A ramp toll is a collection mechanism at an entrance or exit ramp to the highway and these do not concern us here. A barrier toll is a row of tollbooths placed across the highway, perpendicular to the direction of traffic flow. There are usually (always) more tollbooths than there are incoming lanes of traffic. So when exiting the tollbooths in a barrier toll, vehicles must "fan in" from the larger number of tollbooth egress lanes to the smaller number of regular travel lanes. A toll plaza is the area of the highway needed to facilitate the barrier toll, consisting of the fan-out area before the barrier toll, the toll barrier itself, and the fan-in area after the toll barrier. For example, a three-lane highway (one direction) may use 8 tollbooths in a barrier toll. After paying toll, the vehicles continue on their journey on a highway having the same number of lanes as had entered the toll plaza (three, in this example).

• Consider a toll highway having L lanes of travel in each direction and a barrier toll containing B tollbooths (B > L) in each direction. Determine the shape, size, and merging pattern of the area following the toll barrier in which vehicles fan in from B tollbooth egress lanes down to L lanes of traffic. Important considerations to incorporate in the model include accident prevention, throughput (number of vehicles per hour passing the point where the end of the plaza joins the L outgoing traffic lanes), and cost (land and road construction are expensive). In particular, this problem does not ask for merely a performance analysis of any particular toll plaza design that may already be implemented. The point is to determine if there are better solutions (shape, size, and merging pattern) than any in common use[8].

• Determine the performance of the solution in light and heavy traffic. Evaluate how does the solution change as more autonomous (self-driving) vehicles are added to the traffic mix and how is the solution affected by the proportions of conventional (human-staffed) tollbooths, exact-change (automated) tollbooths, and electronic toll collection booths (such as electronic toll collection via a transponder in the vehicle).

10.2.2 Overall Analysis

The problem requires us to design the shape, size, and merging pattern of the area following the toll barrier in which vehicles fan in from B tollbooth egress lanes down to L lanes of traffic. Importantly, by the way, we should take considerations of accident prevention, throughput (number of vehicles per hour passing the point where the end of the plaza joins the L outgoing traffic lanes), and cost (land and road construction are expensive) to model; Then we need determine the performance of our solution in light and heavy traffic, what's more, we should analyze the change of our solution as more autonomous (self-driving) vehicles are added to the traffic mix and solution how affected by the proportions of conventional (human-staffed) tollbooths, exact-change (automated) tollbooths, and electronic toll collection booths (such as electronic toll collection via a transponder in the vehicle).

Considering the honeycomb structure materials are saved, the diamond of the two sides of the intersection of three adjacent hexagon is the toll plaza where the shape is foregone. According to the cost considerations, minimum area leads minimum cost, taking into account the different lane width size, giving a given for inner and outer lane width, the slope angle and the width of toll plaza is simulated to get optimal results, so that we can know the size of toll plaza and gain a map of toll plaza which the way of lane merging is reflected in. The width of the toll plaza can be given from the size of the toll plaza, we can investigate the waiting time of the toll plaza, and know the traffic flow of three kinds of charges with the he traffic flow theory, and then the ratio of the three charging methods can obtain the throughput.

Based on the solutions performance in car flow and wagon flow, the data which come from the toll plaza such as the ratio of three kinds of charges and the vehicle proportion can simulated by VISSIM, then we can gain the performance in car flow and wagon flow. According to analysis of simulation results, we can know the changes of charging methods when the road adds more self-driving cars. The modeling flow chart is shown in Figure 10.3.

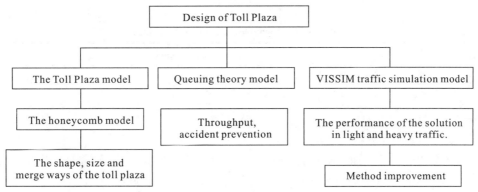

Fig. 10.3 Modeling Flow Chart

10.3 Basic Assumption

- Regardless of the toll station equipment installation and maintenance costs;
- Considering the speed limit for vehicles entering the toll station, so assuming all the vehicles go through the toll station's running speed are the same;
- Assuming that conventional (human-staffed) tollbooths, exact-change (automated) tollbooths, and electronic toll collection booths (such as electronic toll collection via a transponder in the vehicle) have the same length;
- Assuming that the cost of land in different sections of the toll plaza is equal;
- We assume that the selected address' plane and vertical line of the toll station has been taken into consider, and shall not be located on a long downhill or small radius curve to prevent traffic accidents happened due to a brake failure or poor sight;
- Regardless of the influence of the ordinary vehicle into ETC lanes unexpectedly.

10.4 Glossary & Symbols

10.4.1 Glossary

• Head distance: the distance between two cars when the traffic accident occurred.

• Throughput: number of vehicles per hour passing the point where the end of the plaza joins the L outgoing traffic lane.

• ETC System: using automatic identification technology to complete the wireless data communication between the vehicle and the toll station to achieve non-parking automatic electronic toll collection system.

10.4.2 Symbols

Tab. 10.1 Variables and Their Meanings

Number	Sign	Significance
1	μ	Time interval between two car stops
2	λ	Average service rate
3	L_s	Average queue length
4	L_q	Average vehicle length
5	W_s	Average residence time
6	W_q	Average queuing time
7	ρ_s	System load intensity coefficient
8	C_B	The basic traffic capacity of toll lanes
9	T_S	Standard car service time
10	T_G	Standard departure time
11	L_0	The length of the pavement in the toll plaza
12	L	The transition length of the toll plaza
13	d	The width of the multi-lane area before the tollgate in the toll plaza
14	d_0	Toll Square Width
15	S	The length of the multi-lane area before the tollgate in the toll plaza

10.5 Problem I Design of Toll Station

10.5.1 Analysis Approach

First, we consider the honeycomb structure can save material, so we regard the enclosed area of the two adjacent sides of the three adjacent hexagons serves as a toll plaza, whose shape and size can be learned. According to the cost considerations, minimum area leads minimum cost, taking into account the different width of the lane width, the width of the inner and outer lane is given, and the slope angle and the width of the toll plaza are simulated, and the optimal results are obtained. So we can know the size of toll plaza and gain a map of toll plaza which the way of lane merging is reflected in.

Then, the width of the toll plaza can be given from the size of the toll plaza, we can investigate the waiting time of the toll plaza, and know the traffic flow of three kinds of charges with the traffic flow theory, then combined with the proportion of the three charges, we can get throughput[9].

At the last, we based on the solutions performance in car flow and wagon flow, the data which come from the toll plaza such as the ratio of three kinds of charges and the vehicle proportion can simulated by VISSIM, then we can gain the performance in car flow and wagon flow. According to analysis of simulation results, we can know the changes of charging methods when the road adds more private cars[10].

10.5.2 Model Ⅰ:The Toll Plaza Model

1. Model preparation

Based on the honeycomb model, we consider the structure of the toll plaza. The honeycomb model is shown in Figure 10.4.

Fig. 10.4 The Honeycomb Model

In the honeycomb model, the toll stations squares are at the borderlines of these bordering hexagons. As we all known, the material used in the honeycomb structure is the least. Therefore, the square charge is the lowest. Based on the previous assumptions, the land unit cost of different toll plots is equal. Toll square is the smallest with minimum cost. Based on this situation, we have established the toll plaza model[11].

2. Model establishing and model solving

According to the honeycomb model, we got the toll plaza was in the shape of two symmetrical "Y" type. Considering the factors like the drainage conditions around the toll plaza, we set the side face of the toll plaza with a certain slope angle. The gradient is specified to be less than 1/3 generally. The longer length of the toll square transition section, the smaller of the gradient and the vehicles driving from the toll plaza to the standard sections are more naturally. Linear transition should be smooth and round. What's more, shall not make the vehicle drive too reluctantly.

In the design of the toll plaza, we should take account of the width of the different lanes. In general, the lane width usually be 3.5～3.75 m. And each side of the right lateral passage is the large vehicles and the maintenance of construction vehicles' passage. The width of this passage usually be 4.50 m. In this paper, the internal lane width is set at 3.6 m. The length of the multi-lane area and the fan-shaped area are 100 m. The structure of the toll plaza is shown in Figure 10.5.

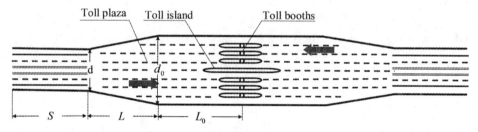

Fig. 10.5 The Structure of the Toll Plaza

Where,

L_0 is the length of the toll plaza;

L is the transition length of the toll plaza;

d is the width of the multi-lane area which is in front of the tollgate in the toll plaza;

d_0 is the width of the toll plaza;

S is the length of the multi-lane area which is in front of the tollgate in the toll plaza;

The length of the toll plaza pavement is 70 m and the toll plaza transition length is set at 60 m. The results of simulation is shown in Figure 10.6.

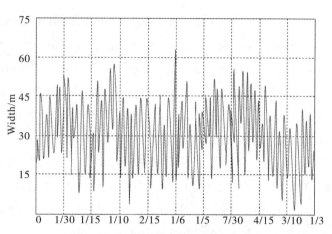

Fig. 10.6 The Results of Simulation

From the simulation results, we can know that when the slope angle's tangent of the toll plaza value is one-sixth, the width of the transition of the toll plaza is 10 m, the toll plaza width of the largest, it is 65.2 m. By calculation simple, total lane width is 45.2 m, after the removal of the right lateral channel width on both sides of the total width of the lane, it can be seen that the total width of the internal lane is 36.2 m, and internal lane width is set at 3.6 m, that is, the number of internal lanes is ten. With both sides of the right side of the outermost channel that there are 12 lanes, two directions on both sides of the highway have 6 lanes.

Above all, we can know that the area of the toll plaza is:
$$100 \times 45.2 \times 2 + 60 \times 10 \times 2 + 60 \times 70 \times 2 = 10\,540 \text{ m}^2$$

In addition, with the continuous economic developed, the traffic flow increased quickly, thus increasing the load of toll stations. The problems of toll stations operating capacity often arise. Moreover, some of the toll station equipments are aging and it's malfunctioning more and more frequently. Once the charging system is paralyzed, not only will affect upload of the charging data, resulting in economic and data loss, but also directly affect the operation of the entire highway. For a long time there will be congestion and queuing serious phenomenon. Therefore, in order to ensure the safety and smooth traffic on the expressway, it is necessary to make emergency charging system. When the traffic accident or the charging system paralyzed, we should use two kind of emergency charge system—the portable charging machine and the mobile power station toll car. The practical application of portable charging machine mainly has the following aspects:

Firstly, when the charging lane of the toll system paralyzed, the location of the emergency temporary charging facility is set to the following:

①Add the temporary toll stations at toll plaza. When the charging system is paralyzed, the charging work can not be carried out and the data can not be uploaded. At this point we can set two sets of portable charging machine in the toll plaza toll lane direction. The toll lanes are expanded into lanes that can pass both cars at once. And we can charge two vehicles tolls at the same time. Once the charge system back to normal, we can immediately remove the emergency portable charging machine[12].

②Add temporary toll points on the toll island. When the system is paralyzed, building tables and chairs in the charge island in order to the toll collectors work more convenient. Using a portable charging machine instead of the original toll lane charging system for the passing of the vehicle charges. Emergency charging machine can be evacuated after the charge system troubleshooting.

Secondly, when traffic is heavy and traffic accident occurs. In order to alleviate congestion at the entrance and exit of the toll station, the emergency temporary charging facility should be set up. Ambulances should be given priority to go through the toll booths and we should dispatch some traffic police to ease. There are two types of location for temporary charging facilities:

①Add portable charging machine in the toll plaza on the basis of the existing toll collection system. In the holiday or a specific period of peak hours, traffic volume in the toll station entrance increased dramatically, vehicle congestion is serious. At this point we can add one or two portable charging machine in the charging square. Since portable charging machine and toll booths working together, we can increase pass rate rapidly and solve the traffic congestion situation effectively. Once the traffic flow back to normal, we can immediately remove the emergency portable charging machine.

②Add temporary toll points on the toll island. The toll lanes are temporarily expanded into toll lanes with two toll sites. In order to improve the pass rate, the charging machines and the toll booths are working at the same time. Two cars can be released at the same time after charge.

Mobile power station charging car, which is simplified small lanes charging system, not only can guarantee the cost of collection, but also will upload the charge data to the center by use of computer network function.

It is inevitable that the problems of using highway toll station charging facilities may arise. when various factors lead to the power supply system and the charging system cannot run normally, the toll station will be paralyzed, mobile power station charging car can be used as emergency charging station for temporary paralysis of the toll station to complete charging and data upload task. When charging system turn to normal, mobile power station charges can be evacuated within a short time. The disadvantage of this approach is that the mobile power station charging car costs a little bit high[13].

In general, using portable charging machine is more convenient.

3. Result analysis

From the foregoing we can see the toll plaza was in the shape of two symmetrical "Y" type. When the slope angle of the tangent of the toll plaza value is one-sixth, the width of the transition of the toll plaza is 10 m, the toll plaza width of the largest, it is 65.2 m. By calculation simple, total lane width is 45.2 m, after the removal of the right lateral channel width on both sides of the total width of the lane, it can be seen that the total width of the internal lane is 36.2 m, and internal lane width is set at 3.6 m, that is, the number of internal lanes is ten. With both sides of the right side of the outermost channel that there are 12 lanes, two directions on both sides of the highway have 6 lanes. When the slope angle of the tangent of the toll plaza value is one-sixth, the width of the transition of the toll plaza is 10 m, and when the car out of the toll station will be symmetrical way, that is, the vehicle out of the toll station, also showed a fan-shaped and to the L-lane, and after the lane into the previous lane, there has no effect. It is worth mentioning that we are charging the direction of the car booth alignment, so that either in the direction of the entrance or the direction of the export, the charges are on the island of equipment in the same location, which looks neat and beautiful. What's more, this arrangement can make the pipeline of the export was arranged in a straight line, when the toll plaza need to set the sidewalk, the channel does not need twists and turns, the length is short, the project is small.

10.5.3 Model Ⅱ: Queuing Theory Model

1. Model preparation

①The vehicle arrival probability

According to the theory of traffic flow, the arrival time of vehicles obeys the Poisson distribution. The time headway obeys the negative exponential distribution. The service time and departure time follow the normal distribution.

Set a road with s lanes. For convenience, let's look at the M/M/1 model: The M/M/1 model is defined as when there is only one lane in the two directions of the road. Assuming that the number of vehicles on the road is infinite. And the arrivals of vehicles follow the principle of first-come-first-serve. The service times of the vehicles are independent with each other and obey the negative exponential distribution with the parameter m.

The probability of the arrival of the n vehicle in time $[t_1, t_2)$ is $P_n(t_1, t_2)$. In this case, the Poisson probability is

$$P_n(t) = \frac{(\lambda t)^n}{n!} e^{-\lambda t}$$

When the road traffic is busy, the time interval of two vehicles stay at the toll station is μ. According to traffic flow theory, m obey the negative exponential distribution $f(t) = \mu \cdot e^{-\mu \cdot t}$, and the traffic intensity is $\rho = \lambda/\mu$, where λ is the average service rate.

②The classification of toll station service level

The service level of the toll station is a standard to measure the service quality which is provided by the internal traffic flow of the toll station. In general, the indicators to evaluate the toll station service level are charging time, queue length and the vehicle delays in the toll station time. In different types of toll stations, the number of the toll lane is the same. The number of vehicles in the queue can lead to different toll stations. With the increase of the number of queuing vehicles, the toll stations can handle the number of vehicles is also increasing. This paper uses the toll station average delay as an indicator to evaluate the service level of the toll station. The service level is shown in the table 10.2.

Tab. 10.2   The Service Level

The service level	The average queuing vehicles	The average elapsed time
A	$L \leqslant 1$	$T \leqslant 15$
B	$1 < L \leqslant 2$	$15 < T \leqslant 30$
C	$2 < L \leqslant 3$	$30 < T \leqslant 45$
D	$3 < L \leqslant 6$	$45 < T \leqslant 60$
E	$6 < L \leqslant 10$	$60 < T \leqslant 80$
F	$L > 10$	$T > 80$

In view of the changing regularity of the toll station traffic capacity, the service level of the toll station can be divided into four levels, described as follows:

The first service level is level A or B: toll booths almost have no queue formation, most of the vehicles enter the charge to accept service without queuing channel, part of the vehicles need to wait for receiving a charge cycle. Drivers and passengers almost have no feeling through a toll booth not waiting for a long time, so they feel more comfortable and convenient;

The second service level is level C: toll booth have been forming line, but the queue length is shorter, most of the vehicles, need to wait for two or three charge cycles before through the toll station, part of the vehicle may be waiting a long time to through the toll station, only a few vehicles accept service through a toll booth directly. Although it takes the driver and passengers' some time, time is short enough so that drivers and passengers can understand.

The third service level is D: toll station has longer queue length, almost every vehicles need to wait for a long time to pass the toll station. Drivers and passengers feel they have waiting for a long time. Some drivers and passengers began to complain.

The forth service level is D: Toll station formed a long queue. Every vehicles have to wait a long time before they can pass through toll stations. The queue length of the situation even continued to grow. Drivers and passengers feel inconvenience obviously. Most of the drivers and passengers can not stand such a long wait.

2. Model establishing and model solving

For the model M/M/1, the model with only one lane in each direction. When the system is in a steady state, $P_n(t)$ is independent with time t. At this point it can be written as P_n, And the derivative of $P_n(t)$ to time is zero. Establish the difference equation for P_n. That is, each state is balanced, the state transition is shown in Figure 10.7.

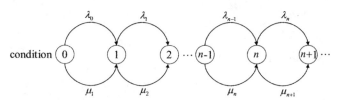

Fig. 10.7 The State Transition

The state transition equation is:

$$\begin{cases} \lambda P_{n-1} + \mu P_{n+1} - (\lambda + \mu) P_n = 0 \\ -\lambda P_0 + \mu P_1 = 0 \end{cases} \quad (10\text{-}1)$$

So $P_1 = \lambda/\mu \cdot P_0 = \rho \cdot P_0$.

According to Eq. (10-1), we can know that

$$P_n = \left(\frac{\lambda}{\mu}\right)^n \cdot P_0 \quad (n \geqslant 1)$$

According to $\sum_{n=0}^{\infty} P_n = 1$, we can get P_0、P_n, but we only discuss $\lambda/\mu < 1$ here.

When $\frac{\lambda}{\mu} < 1, \rho < 1$, we can know that $\begin{cases} P_0 = 1 - \lambda/\mu \\ P_n = (1 - \lambda/\mu)(\lambda/\mu)^n \; (n \geqslant 1) \end{cases}$

where $P_n = (1-\rho)\rho^n$.

We use L_s as the average captain, L_q as the average vehicle length, W_s as the average charge time, W_q as average queuing time, so

$$L_s = \sum_{n=0}^{\infty} nP_n = \sum_{n=0}^{\infty} n(1-\rho)P^n = \frac{\rho}{1-\rho} = \frac{\lambda}{\mu-\lambda}$$

$$L_q = \sum_{n=0}^{\infty} n(1-\rho)^n - \sum_{n=0}^{\infty} (1-\rho)^n = L_s - \rho = \frac{\rho\lambda}{\mu-\lambda}$$

$$W_s = 1/(\mu-\lambda)$$

$$W_q = \lambda/[\mu(\mu-\lambda)]$$

We call the relation of L_s, L_q, W_s, W_q the little formula.

Next, we establish and solve the model M/M/S.

If $\mu_n = \begin{cases} n\mu & (n \leq s) \\ s\mu & (n > s) \end{cases}$, we can know that $L_s = \lambda W_s$, $L_q = \lambda W_q$ by little formula.

As we can see from the above state transition equation:

$$\begin{cases} \mu P_1 = \lambda P_0 \\ (n+1)\mu P_{n+1} + \lambda P_{n-1} = (\lambda + n\mu)P_n & (n \leq s) \\ \mu P_{n+1} + \lambda P_{n+1} = (\lambda + s\mu)P_n & (n > s) \end{cases}$$

make $A_n = \begin{cases} \dfrac{(\lambda/\mu)^n}{n!} & (n \leq s) \\ \dfrac{(\lambda/\mu)^s}{s!}\left(\dfrac{\lambda}{s\mu}\right)^{n-s} & (n > s) \end{cases}$, $\rho_s = \dfrac{\rho}{s} = \dfrac{\lambda}{\mu s}$ (ρ_s represents the system load intensity factor)

When $\rho_s < 1$, we can know that

$$P_n = A_n \cdot P_0 = \begin{cases} \dfrac{\rho^n}{n!}P_0 & n \leq s \\ \dfrac{\rho^n}{s!s^{n-s}}P_0 & n > s \end{cases}$$

Above all, $P_0 = \left[\sum_{n=0}^{s-1}\dfrac{\rho^n}{n!} + \dfrac{\rho^s}{s!(1-\rho_s)}\right]^{-1}$.

When $n \geq s$, we can know that

$$\sum_{n=s}^{\infty} P_n = \sum_{n=s}^{\infty} \frac{\rho^n}{s!s^{n-s}}P_0 = \frac{1}{s!}\rho^s \sum_{k=0}^{\infty} \left(\frac{\lambda}{s\mu}\right)^k P_0 = \frac{\rho^s}{s!(1-\rho_s)}P_0 \qquad (10\text{-}2)$$

Eq. (10-2) is called Erlang wait.

Thus, we can know that

$$L_q = \sum_{n=s}^{\infty}(n-s)P_n = \sum_{k=0}^{\infty} kP_{s+k} = \sum_{k=0}^{\infty} k\frac{\rho^s}{s!}\rho_s^k P_0 = P_0 \frac{\rho^s}{s!}\rho_s \sum_{k=0}^{\infty} \frac{d\rho_s^k}{d\rho_s} =$$

$$P_0 \frac{\rho^s}{s!}\rho_s \frac{d\left(\dfrac{1}{1-\rho_s}\right)}{d\rho_s} = \frac{P_0 \rho^s \rho_s}{s!(1-\rho_s)^2}$$

So the average number of vehicles which are parking is

$$\sum_{n=0}^{s-1} nP_n + s\sum_{n=s}^{\infty} P_n = \sum_{n=0}^{s-1} \frac{n\rho^n}{n!}P_0 + s\frac{\rho^s}{s!(1-\rho_s)}P_0 = \rho P_0 \left[\sum_{n=1}^{s-1} \frac{\rho^{n-1}}{(n-1)!} + \frac{\rho^{s-1}}{(s-1)!(1-\rho_s)}\right]^{-1} = \rho$$

3. The model of further expand

The basic traffic capacity of toll lanes refers to the maximum traffic volume where each toll road can pass in a unit time under the ideal conditions of road and traffic. As the ideal road conditions, including the toll lane width which cannot less than 3 meters, the toll island width which cannot less than 2.2 meters, the toll island length which cannot less than 30 meters, toll plaza which has an open field of vision, good linear and road surface conditions. Ideal traffic conditions, is the standard flow of traffic for a single standard car, we will standard car as a small car[14].

The basic traffic capacity of a toll channel can be calculated by the following formula:

$$C_B = 3600/(T_S + T_G)$$

Above all, C_B is the basic toll channel capacity.

T_S is the standard service hours.

T_G is the standard car departure time.

From the formula, we can see that the basic traffic capacity of the toll lane is inversely proportional to the toll time. Based on this, we investigated the standard service hours of a certain day at a toll plaza in Interstate 5, New Jersey. The service hours are showed in Table 10.3.

Tab. 10.3 The Service Hours in Different Charge Types

Serial number	Charge types		
	Conventional tollbooths	Exact-change tollbooths	ETC
1	35.2	6.4	2.1
2	23.7	4.8	2
3	31.7	6.7	1
4	62.4	9.6	0.8
5	21.6	4.5	0.9
6	45.2	8.7	2
7	12.5	7.6	1.2
8	45.2	7.4	1.9
9	34.1	9.5	2.4
10	16.2	10.1	1
11	43.2	8.6	2
12	26.5	7.4	3.4
13	27.8	8.5	2.1
14	26.4	6.1	1.3
15	31.5	7.3	1.8
16	29.5	9.5	1.9

From the Table 10.3, we can know that the use of ETC channel is significantly better than the conventional tollbooths and exact-change tollbooths in the road capacity. For the throughput which is roadway capacity of the channel, the data lack standard car departure time, for which we introduce the correction factor g, it is a number less than 1, the revised charge channel basic capacity[15].

$$C_B = 3600g/T_S$$

In the toll plaza of Model I where there are six lanes in each direction, we get traffic flow when there are six lanes on each side simulated with VISSIM. The traffic flow is shown in Table 10.4.

Tab. 10.4 The Traffic Flow

Serial number	Charge types		
	Conventional tollbooths	Exact-change tollbooths	ETC
1	102	563	1714
2	152	750	1800
3	114	537	3600
4	58	375	4500
5	167	800	4000
6	80	414	1800
7	288	474	3000
8	80	486	1895
9	106	379	1500
10	222	356	3600
11	83	419	1800
12	136	486	1059
13	129	424	1714
14	136	590	2769
15	114	493	2000
16	122	379	1895

By the data weighted average, we obtain that the traffic flow of the conventional tollbooths is 131 veh/h, the exact-change tollbooths traffic flow is 495 veh/h and the ETC toll station traffic flow is 2415 veh/h.

Further analysis shows: When the conventional tollbooths, exact-change tollbooths, and ETC rates are known, we can know the throughput. Taking into account the actual situation, we take the ratio of manual charges, semi-automatic tolls, and ETC to 1 : 2 : 2. Thus we can calculate the throughput of six lanes in each direction, which is

$$131 \times 1.2 + 495 \times 2.4 + 2415 \times 2.4 = 3067.2 \text{ veh/h}$$

After rounding, we can know that throughput of six lanes in each direction is 306 7 veh/h.

10.5.4　Model Ⅲ: VISSIM Traffic Simulation Model

1. Model preparation

The speed limit of the general mainline toll station is 60km/h, which is very important for us to consider the rationality of traffic flow.

We assumed that:

• 400 meters in front of the toll station has been set up signs to remind the front of the toll lanes need to travel at a speed limit;

• ETC vehicles and MTC vehicles are traveling in accordance with their own way, ETC vehicles through the ETC or mixed toll lane without stopping, and all the ETC electronic tags can be correctly read by the system;

• MTC vehicles through MTC lane normal parking fees, no tolls and break through the vehicle;

• MTC vehicles entering the mixed toll lane normally park and pay before toll booths.

From the foregoing, we can see that the number of toll lanes should be determined by three factors: traffic volume, service time and service level.

We regard the lane configuration as the main axis of the program design, classify the different lane configurations, and use the traffic volume and lane utilization rate as the control variables for each category. Then we use the simulation to analyze and evaluate, which can be used as the basis of lane control, and expect to get the specific operation control scheme of toll station. Principles and methods of the design of the program described as follows:

• We take the exit and entrance of the six-lane toll station that designed before as a simulation object, as the closed-end charging system, the export charges longer than the entrance time, so the simulation of the export more than the entrance to illustrate the problem, in the experiment, export direction was selected for simulation.

• According to the results of ETC lane setup, the ETC lane is considered to be superior to out-of-direction amplification in its driving direction. Therefore, in the simulation of this paper, we do not analyze the situation of the amplification from the inside out, and the ETC lane is all set up from the inside to the outside. To preserve a common MTC lane as a prerequisite, where the simulation of the ETC lane set will be mixed lane and ETC lane combination of classified discussions.

• This simulation also analyzes the proportion of vehicles constituting ETC vehicles, which is ETC usage.

• For the variation of traffic flow into the toll station, the simulation developed several different traffic levels.

2. Model establishing

From the theory of traffic flow, the arrival distribution of the toll station obeys the Poisson distribution, and the headway is in line with the negative exponential distribution. The toll station simulation, the vehicle structure consists of four models: cars, cars (ETC), trucks, large passenger cars. We used VISSIM software default models. The traffic composition ratio is shown in Table 10.5.

Tab. 10.5 The Traffic Composition Ratio

Models	Small car	Middle-sized vehicle	Oversize vehicle
Vehicle proportion (%)	75%	20%	5%

The standard of the vehicle conversion coefficient of the toll station is a small car, and the conversion coefficient of each model is shown in Table 10.6.

Tab. 10.6 The Conversion Coefficient of Each Model

Vehicle model	Vehicle conversion factor
Small car	1.0
Middle-sized vehicle	1.3
Oversize vehicle	1.7

In order to unify it, we can use the matrix to calculate the relative length of all types of vehicles.

Tab. 10.7 Relative Length of All Types of Vehicles

Vehicle model	General vehicle length
Small car	4.3
Middle-sized vehicle	5.6
Oversize vehicle	9.8

From Table 10.7, we can know that $\frac{1}{3}(4.3 \quad 5.6 \quad 9.8)\begin{pmatrix}1.0\\1.3\\1.7\end{pmatrix}=9.413$. Obviously, when the small cars, midsize cars and large vehicles are traveling on the same road in the same time, the driver who is driving behind must feel this road is very blocked, and may give priority to other roads. We can see that the medium-sized vehicles have a larger impact on the traffic capacity. The higher the proportion is, the lower capacity of the lane should be. And the small car have a lower impact on the traffic capacity.

In the simulation process, it will also change the proportion settings according to the needs, to analyze the effect of different proportions on the charging system. The speed of preparation is shown in Table 10.8.

Tab. 10.8 The Speed of Preparation

Models	Small car	Middle-sized vehicle	Oversize vehicle
Speed (km/h)	30—60	40—50	30—50

3. Model solving

(1) The fixed parameters of simulation

• The simulation set up the simulation time of each program for 15 minutes, the simulation step using Time step[s]/Sim. sec., in order to make the simulation results are comparable, each simulation of the random number seed for the same value.

• Geometric design of the toll station: The length of the upstream section of the toll station, the length of the deceleration section, the length of the charging section, the length of the acceleration section, the length of the downstream section, and the number of the toll lanes are fixed values and specific geometric arrangement.

• In the simulation, the proportion of large, medium and small models is unchanged in the simulation, namely 5% for large cars, 20% for medium cars and 75% for small cars; global parameters in simulation is the same in the simulation, such as the driver reaction time, Business Hours, vehicle acceleration and deceleration etc.

(2) The control parameters of simulation

• ETC toll lanes: According to the simulation program, set from the inside out 1, 2, 3.

• Mixed toll lane number: the same under the program, set from the inside out of 1, 2 or 3.

• Simulation traffic: It can be divided into 400 vph/h, 600 vph/h, 800 vph/h, 1000 vph/h and 1200 vph/h, in total 5 series.

• ETC utilization rate: 10%, 20%, 30%, 40%, 50%, 60%, 70%, 80%, in total 8 series, the corresponding proportion of ordinary cars 70%, 60%,…, 0%, both the total amount of cars to maintain 80% of the total.

The average delay through toll stations was used as an indicator of the efficiency of service levels of the toll stations throughout the highway system, and the average queuing delay was used as an efficiency index to evaluate various lane service levels.

Each simulation program will correspond to five traffic series for simulation, and in accordance with the results of the simulation results obtained program table[16].

4. Result analysis

(1) ETC usage impact analysis

Using the simulation data to draw the different types of lane settings to delay time as the vertical axis, traffic flow for the horizontal axis of the curve comparison chart, in the chat, series 1 through Series 6 correspond to cases where the ETC usage is from 10% to 80%. The delay time under various traffic flows is shown in Table 10.9.

Tab. 10.9 The Delay Time under Various Traffic Flows

The traffic flow(veh/h)	Utilization rate of ETC	Drive Type 1	Drive Type 2	Drive Type 3	Drive Type 4	Drive Type 5	Drive Type 6
400	10%	17.5	165.4	245	152.4	152.4	29.6
	20%	12.5	95.2	206.4	78.2	77.6	18.4
	30%	9.4	21.4	163	24.1	20.4	11.5
	40%	6.1	9.6	115.4	11.5	11.5	8.4
	50%	5.7	6	42.6	7.5	7.2	6.4
	60%	4.6	4.3	6.5	5	5	5.5
	70%	4.6	3.2	3.5	4.6	4.6	3.4
	80%	4.3	2.8	2.4	3	3.2	2.7
600	10%	94.6	224.4	254.4	214.5	178.5	153.5
	20%	27.2	175.5	217.5	163.4	144.8	64.5
	30%	11.4	115.5	166.4	96.4	111.5	17.6
	40%	8.4	32.4	135.2	34.4	36.4	11.5
	50%	6.4	7.2	94.5	11.5	14.5	8.5
	60%	5.6	5.4	21.6	7.6	8.4	5.5
	70%	5.4	3.5	4.5	4.5	4.5	3.4
	80%	6.4	2.4	3.4	3.4	3.1	2.9
800	10%	171.5	238.5	257.6	221.4	181.5	175.5
	20%	118.6	197.5	221.5	182.2	154.6	139.4
	30%	28.6	152.5	181.6	134.5	123.5	59.5
	40%	10.5	102.2	146.5	86.5	92.5	16.5
	50%	7.4	14.6	103.5	21.5	24.5	10.5
	60%	6.4	5.8	63.5	10.5	10.5	6.8
	70%	7.2	3.5	6.5	5.9	5.4	4.8
	80%	13.5	3	3.5	3.4	4.5	2.6
1000	10%	187.5	245.3	265.4	231.5	186.6	186.5
	20%	146.5	204.6	224.6	188	152.6	153.1
	30%	95.4	165.2	186.5	142.2	127.5	114.5
	40%	17.5	124.5	152.1	108.5	102.5	33.5
	50%	10.5	58.6	114.6	48.5	64.2	15.5
	60%	8.6	7.6	75.6	15	13.5	8.5
	70%	15.2	4.2	10.6	7	7.2	5.6
	80%	33.5	3.6	3.5	4.2	4.5	3.7
1200	10%	194.6	249.1	246.5	241.5	185.6	190.5
	20%	164	221.5	228.5	195.5	154.6	157.6
	30%	129.5	174.6	118.5	142.5	135.5	128.4
	40%	56.2	132.4	154.1	117.4	105.2	85.6
	50%	11.5	82.6	117.5	82	72.5	25.1

When ETC usage is 40%, the relationship between delay time and traffic volume is shown in Figure 10.8.

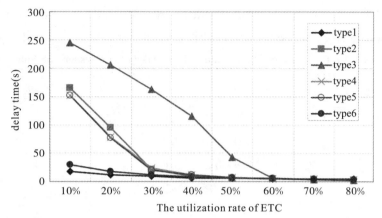

Fig. 10.8 The Relationship Between Delay Time and Traffic Volume

It can be seen that in the case of different lane settings, as the ETC usage increases, the delay time decreases; When ETC usage is large enough, delays in different lane settings are low and do not vary much.

(2) Effect analysis of mixed lanes

When the traffic flow is 400 veh/h, the relationship between the delay time and ETC is shown in Figure 10.9.

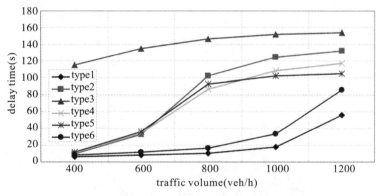

Fig. 10.9 The Relationship Between the Delay Time and ETC

From the further analysis, we can see that in the case of the same number of ordinary toll lanes, the more the number of ETC toll lanes change to mixed toll lanes, the more time can be saved accordingly; this toll station in operation, can be regarded as a very effective congestion mitigation measures, but also very good to adapt to ETC vehicles gradually increase the situation. Therefore, as more self-driving cars are added, increasing the percentage of ETC by increasing the number of small vehicles, ETC toll stations will increase and the toll rate for manual toll stations will decrease.

5. Model test

According to the literature, we can know that the relationship between lane and toll station is $B=[1.65L+0.9]^{[17]}$, for the different value added, we use the MATLAB software for sensitivity analysis, the results shown in Table 10.10 and Figure 10.10.

Tab. 10.10 Different Values and Values Under the Corresponding Value

B \ L	1	2	3	4	5	6
$B_1(a=1.65)$	1.65	2.55	4.2	5.85	7.5	9.15
$B_2(a=2)$	2	2.9	4.9	6.9	8.9	10.9
$B_3(a=3)$	3	3.9	6.9	9.9	12.9	15.9
$B_4(a=4)$	4	4.9	8.9	12.9	16.9	20.9
$B_5(a=5)$	5	5.9	10.9	15.9	20.9	25.9
$B_6(a=6)$	6	6.9	12.9	18.9	24.9	30.9

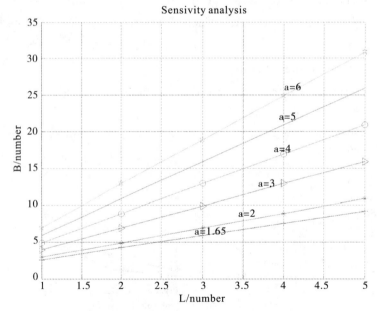

Fig. 10.10 Sensivity Analysis

According to the data in the Table 10.10, a sensitivity analysis of L can be obtained, as shown in Figure 10.10. From the sensitivity analysis we can see that when the value of a is constant, the value of B increases with the time, and the increasing amount is increasing. At the same time, for different values of L, B value of the beginning to grow slowly, and later grew faster.

10.5.5 Non-technical Letter

Dear sir/madam:

We are honored to be able to offer our own solutions for the traffic merge problem in the New Jersey area. I hope our recommendations will meet your requirements and increase the capacity of the road network.

First of all, we are strongly advised to use the ETC toll channel as much as possible due to traffic delays on the inter-continental highways during peak periods. In our model of different lane settings, with the ETC usage increasing, the delay time will go down. And when the ETC usage is large enough, delays in different lane settings are low and do not vary much.

In the case of the same number of common toll lanes, the greater the number of ETC lanes changed to mixed toll lanes, the more time we can save. This is a very effective congestion mitigation measures in toll station operation, and can be a good adaptation to the increasing tendency of ETC vehicle. Therefore, as more self-driving cars are added to increase the percentage of ETC for small vehicles, ETC toll stations will increase and the proportion of manual toll stations will decrease. Our model also applies to other interstate highways. We are happy to use our model to help you if you need.

Based on the size, shape and cost that you want to know about the toll plaza, we first consider the honeycomb model and it can save the material. Thus, the two "Y" types of the toll plaza are identified, and the traffic flow merging mode is "Y" type; Then we take into account the width of toll plaza on both sides and the width of the middle road is different, then we combined with the gradient of the side slope, using side slope incline angle and the charging square to simulate, when the tangent of the side slope inclination angle is 1/6, the width of the toll plaza is 65.2 m, the area of the toll plaza is 10,540 m². While the fees are equal to the cost of each part of the square, then we can measure the cost of charging Plaza by measuring the size of the toll plaza. We compared the road conditions in different proportions of self-driving. In order to find out the optimal proportion of vehicles in different time periods (traffic stationary period and traffic peak period). We use VISSIM simulation to get the following table.

The traffic flow(veh/h)	Utilization rate of ETC	Drive Type 1	Drive Type 2	Drive Type 3	Drive Type 4	Drive Type 5	Drive Type 6
400	10%	17.5	165.4	245	152.4	152.4	29.6
	20%	12.5	95.2	206.4	78.2	77.6	18.4
	30%	9.4	21.4	163	24.1	20.4	11.5
	40%	6.1	9.6	115.4	11.5	11.5	8.4
	50%	5.7	6	42.6	7.5	7.2	6.4
	60%	4.6	4.3	6.5	5	5	5.5
	70%	4.6	3.2	3.5	4.6	4.6	3.4
	80%	4.3	2.8	2.4	3	3.2	2.7
600	10%	94.6	224.4	254.4	214.5	178.5	153.5
	20%	27.2	175.5	217.5	163.4	144.8	64.5
	30%	11.4	115.5	166.4	96.4	111.5	17.6
	40%	8.4	32.4	135.2	34.4	36.4	11.5
	50%	6.4	7.2	94.5	11.5	14.5	8.5
	60%	5.6	5.4	21.6	7.6	8.4	5.5
	70%	5.4	3.5	4.5	4.5	4.5	3.4
	80%	6.4	2.4	3.4	3.4	3.1	2.9
800	10%	171.5	238.5	257.6	221.4	181.5	175.5
	20%	118.6	197.5	221.5	182.2	154.6	139.4
	30%	28.6	152.5	181.6	134.5	123.5	59.5
	40%	10.5	102.2	146.5	86.5	92.5	16.5
	50%	7.4	14.6	103.5	21.5	24.5	10.5
	60%	6.4	5.8	63.5	10.5	10.5	6.8
	70%	7.2	3.5	6.5	5.9	5.4	4.8
	80%	13.5	3	3.5	3.4	4.5	2.6
1 000	10%	187.5	245.3	265.4	231.5	186.6	186.5
	20%	146.5	204.6	224.6	188	152.6	153.1
	30%	95.4	165.2	186.5	142.2	127.5	114.5
	40%	17.5	124.5	152.1	108.5	102.5	33.5
	50%	10.5	58.6	114.6	48.5	64.2	15.5
	60%	8.6	7.6	75.6	15	13.5	8.5
	70%	15.2	4.2	10.6	7	7.2	5.6
	80%	33.5	3.6	3.5	4.2	4.5	3.7
1200	10%	194.6	249.1	246.5	241.5	185.6	190.5
	20%	164	221.5	228.5	195.5	154.6	157.6
	30%	129.5	174.6	118.5	142.5	135.5	128.4
	40%	56.2	132.4	154.1	117.4	105.2	85.6
	50%	11.5	82.6	117.5	82	72.5	25.1

Therefore, through the form of data and software analysis, we can get the optimal ETC utilization rate under different traffic flow.

Yours faithfully…
January, 2017

10.6 Evaluation and Spread of the Model

10.6.1 Evaluation of the Model

1. Strengths

In model I, We use the simulation to find the tangent of the slope angle and the image of the toll square width. In contrast, using LINGO to solve the linear programming is more complicated.

Using queuing theory legitimately. The delay time is used as the service level indicator, and VISSIM software is used to get the relationship between the different three charging methods and the delay time.

This article uses a lot of images and tables, simple and intuitive. It's easy to understand.

2. Weaknesses

In Model I, Model II and Model III, due to there are many complex factors, we can not do a comprehensive consideration, resulting in a certain inconsistent with the actual.

In Model I, the tollgates of the toll plaza entrance and exit are in a straight line, which looks simple and beautiful, but compared to entrances and entrances toll station, those waste a part of the area.

10.6.2 The Extension of Models

1. Model II uses delay time as an indicator of service level. This model can be applied to the queue of maintenance industry and the services of hairdressing industry. Using the barber's waiting time to determine the barber shop's service efficiency. When this model is applied to the outpatient service of a hospital specialist, the outpatient physician can be appropriately equipped with the waiting time of the outpatient service. It can be used as reference for highway accident treatment when it is applied to traffic flow of fluid mechanics.

2. Model III can be used for VISSIM simulation of highway toll under different charging methods and proportion of vehicles, it also can be applied to the cross section of the port traffic intersection optimization, then the simulation optimization model of the different sections' simulation can be established; and the model is mainly used to solve the traffic problems, what's more, all the traffic problems can be optimized using this model.

Reference

[1] Xiang Qiao, Huiru Lin. *Highways Overpass Planning and Design Practice* [M]. Beijing: People's Traffic Press, 2001.

[2] Junlong Cheng. *The Capacity Study of ETC and MTC Hybrid Toll Station* [D]. Chengdu:Southwest Jiaotong University, 2015.

[3] Weiwei Liu, Bo Lei. Queuing Theory Model in the Design of Expressway Toll Station [J]. *Communications Science and Technology Heilongjiang*, 2013, (04):184—185.

[4] Su Li. Study on Design of Expressway Toll Station [D]. Chang'an University, 2012.

[5] Chenchen Zhang, Yanhui Wang. Congestion Control Strategy of Freeway Toll Station [J]. *Chinese Journal of Highway*, 2013, (04):139—145.

[6] Guo Jia. Research on Operation Control Strategy of Expressway Toll Station [D]. Shenzhen: Institutes of Technology of South China, 2012.

[7] Pratelli, Schoen. Optimal Design of Motorway Toll Stations [J]. *European Transport Conference*, 2003(9), 8:1—33.

[8] Perry Ronald, Gupta Surendra M. Response Surface Methodology Applied to Toll Plaza Design for the Transition to Electronic Toll Collection [J]. *International Transactions in Operational Research*, 2001, 8(6):707—726.

[9] Dr. Marguerite L. Zarrillo. Capacity Calculation for Two Toll Facilities: Two Experiences in ETC Implementation[C]. The 79th Annual Meeting of the TRB, Washington, DC, USA, 2000.

[10] Klodzinski J., Al-Deek H. M. Proposed Level of Service Methodology for Toll Plazas[C]. Proceedings of the 81 th Annual Meeting of the TRB, Washington, DC, USA, 2002.

[11] KLODZINSKI J. Methodology for Evaluating the Level of Service (LOS) of Toll Plazas on a Toll Road Facility [D]. Florida: University of Central Florida, 2001.

[12] Xiaoyuan Wang, Xinyue Yang. *Microscopic Simulation of Traffic Flow and Driver Behavior Modeling Theory and Method* [M]. Beijing: Science Press, 2010.

[13] Li Li, Rui Jiang. *Modern Traffic Flow Theory and Application: Volume 1- freeway traffic flow*[M]. Beijing: Tsinghua University Press, 2011.

[14] Transportation Research Board. *High way Capacity Manual* 2000 [M]. Washington DC: National Research Council, 2000.

[15] Gonzalez Velez Enrique. Adaptation of VISSIM, A Dydamic Simulation

Model, to the Traffic Behavior at Intersections in Mayaguez, Puerto Rico [D]. University of Puerto Rico Mayaguez campus, 2006.

[16] The National ITS Architecture [S]. US: Department of Transportation, 2000.

[17] David Levinson, Elva Chang. A Modle for Optimizing Electronic Toll Collection Systems [J]. *Transportation Research* Part A, 2003, 37(4):293—314.

◆ 建模特色点评

❀论文特色❀

◆标题定位:"Design of Toll Plaza Based on Simulation"将"基于仿真的交通收费站设计",标题定位较好,融问题、方法和特色为一体,贴切、简洁、完美。

◆方法鉴赏:运用仿真来确定坡度角的正切和收费广场宽度的图像。运用 LINGO 来求解线性规划。以延迟时间作为服务水平指标,利用 VISSIM 软件分析了三种不同收费方式与延迟时间的关系。使用了很多图片和表格,简单直观。

◆写作评析:整体写作思路清晰,框图比较直观,运用图表结合法,整体效果好。建模思路、算法流程图、灵敏度分析等特色较鲜明。

❀不足之处❀

在三种模式中,均存在着诸多复杂因素,没有给出较全面的考虑,导致与实际存在一定的不一致。在模型 I 中,收费广场入口和出口的收费站是直线的,看起来简单漂亮,但与入口处和收费站相比,浪费了一部分区域。

第 11 篇
能源生产

◆ 竞赛原题再现

Problem C Energy Production

Background: Energy production and usage is a major portion of any economy. In the United States, many aspects of energy policy are decentralized to the state level. Additionally, the varying geographies and industries of different states affect energy usage and production. In 1970, 12 western states in the US formed the Western Interstate Energy Compact (WIEC), whose mission focused on fostering cooperation between these states for the development and management of nuclear energy technologies. An interstate compact is a contractual arrangement made between two or more states in which these states agree on a specific policy issue and either adopt a set of standards or cooperate with one another on a particular regional or national matter.

Problem: Along the US border with Mexico, there are four states-California (CA), Arizona (AZ), New Mexico (NM), and Texas (TX)-that wish to form a realistic new energy compact focused on increased usage of cleaner, renewable energy sources. Your team has been asked by the four governors of these states to perform data analysis and modeling to inform their development of a set of goals for their interstate energy compact.

The attached data file "ProblemCData.xlsx" provides in the first worksheet ("secedes") 50 years of data in 605 variables on each of these four states' energy production and consumption, along with some demographic and economic information. The 605 variable names used in this dataset are defined in the second worksheet ("miscodes").

Part Ⅰ:

A. Using the data provided, create an energy profile for each of the four states.

B. Develop a model to characterize how the energy profile of each of the four states has evolved from 1960 — 2009. Analyze and interpret the results of your model to address the four states' usage of cleaner, renewable energy sources in a way that is easily understood by the governors and helps them to understand the similarities and difference between the four states. Include in your discussion possible influential factors of the similarities and differences (e. g. geography, industry, population, and climate).

C. Determine which of the four states appeared to have the "best" profile for use of cleaner, renewable energy in 2009. Explain your criteria and choice.

D. Based on the historical evolution of energy use in these states, and your understanding of the differences between the state profiles you established, predict the energy profile of each state, as you have defined it, for 2025 and 2050 in the absence of any policy changes by each governor's office.

◆ 获奖论文精选

Renewable Energy Renews Your Life[①]

Summary: Energy usage targets are crucial to energy production and usage. Several states in the US formed compacts to cooperate and address the usage of clean, renewable energy resources. Therefore, we help the governors to form a realistic new energy compact.

Based on the multi-layer feed forward neural network model, we summarize the energy profile and evolution of four states—California, Arizona, New Mexico, and Texas. With the help of R language programming, we make a hierarchical clustering and decide which state has the "best" profile for use of cleaner, renewable energy in 2009. Then we apply a growth rate model and predict the energy profile. We identify and discuss several actions in order to find optimal solution to addressing the usage of cleaner, renewable energy. Finally, we further study the models and develop them.

For Task A of Part I, we select and aggregate variables based on the code rule. Using EXCEL, we draw stacking area maps to summarize the state profiles as of 2009.

For Task B of Part I, considering natural endowments and technologies, we build a multi-layer feed forward neural network model to characterize how the energy profile of each state has evolved from 1960 to 2009. We find that Arizona made the best usage of clean energy resources and New Mexico made the worst.

---

① 本文获 2018 年国际一等奖。队员：祁浩宇，陶冉，桂扬；指导教师：李琴。

For Task C of Part I, with hierarchical clustering method, we cluster the rest of the 46 states and make a discriminant analysis to compare the 4 states with 46 states. We make a conclusion that Texas has the "best" profile for use of cleaner, renewable energy in 2009.

For Task D of Part I, we determine the fixed growth rate of each state. With use of the multi-layer feed forward neural network model discussed above, we predict the energy profile of each state without any policy changes for 2025 and 2050.

Based on all of the models and indicators in Part I, we change the growth rate and fit the new growth model to data in order to determine renewable energy usage targets for 2025 and 2050. In this way, we complete the tasks in Part II and prepare a one-page memo to the group of governors.

At last, we further study the growth rate model for renewable energy usage targets. In this case, we try to establish an improved time series model to observe this model's performance.

Keywords: energy profile; renewable energy usage; neural network; growth rate model; hierarchical clustering; discriminant analysis

11.1 Introduction

In the United States, energy is the basic matter and dynamic resource for the development of economy and society. In 1970, 12 western states in the US formed the Western Interstate Energy Compact (WIEC), whose mission focused on fostering cooperation between these states for the development and management of nuclear energy technologies. Additionally, different environment and resources of different states affect energy usage and production. Therefore, the modeling and analysis of energy is one of the major elements of energy exploitation and strategy planning and layout.

The fundamental question is how we address the usage of according to different energy profiles of each state.

Our models summarize historical data and make projections. And they include cases with different assumptions regarding macroeconomic growth, world oil prices, technological progress, and energy policies.

11.2 Restatement of Problems

We are required to perform data analysis and modeling to inform development of a set of goals for their interstate energy compact of four states-California (CA), Arizona (AZ), New Mexico (NM), and Texas (TX).

By analysis, we decompose the problem into four sub-problems.

First, we select and aggregate data to create an energy profile.

Second, we build a model that can characterize how the energy profile of each of the four states has evolved from 1960 to 2009.

Third, we use the model to predict the energy profile of each state.

Fourth, we determine renewable energy usage goals for 2025 and 2050. Propose some actions for each of the states.

This is our flow chart.

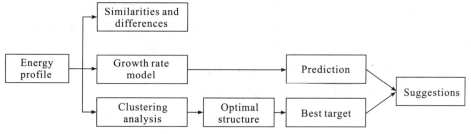

Fig. 11.1 Total Flow Chart

11.3 Assumptions and Justifications

- Assuming there is no policy change in predictions of the energy profile of each state.
- Assuming the missing data does not play an important part in the model.
- Assuming there is no abrupt development of technology that causes the variables to have a high growth rate.

11.4 Notations

Tab. 11.1 Notations

Symbol	Definition	Units
Z_i	Linear discriminant function i	—
x_i	Input variable j	—
F_n	Value for time n	—

Symbol	Definition	Units
w_{ij}	Weights between adjacent layers	—
θ_j	Neuron threshold j in hidden layer	—
a_i	Neuron i in input layer	—
g	Natural growth rate	%
d	Special growth rate	%
E_{jt}	Consumption of Energy j at time t	Btu

11.5 Depiction of Energy Profile

11.5.1 Preprocess the Variables

We collect data of the four states' energy production and consumption from data file "ProblemCData.xlsx". We consider the completeness of the data, judgment of experts and the meaning of indicators on official website then we make preliminary descriptive statistics. It is showed that the data structure of the given 583 variables is no equilibrium panel. The time span includes 30, 33, 40 and 50 years.

(1) Delete two variables GETXV and HYTXV that do not exist in the official data file.

(2) Considering the code rule of MSN as follows.

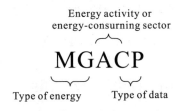

Fig. 11.2　Code Rule

We select variables according to the last character in MSN, which shows type of data. Table 11.2 shows different meaning of character 5.

Tab. 11.2　Meaning of Character 5

Character 5	Frequency	Meaning
B	40800	Data in British thermal units (Btu)
D	17920	Factor for converting data from physical units to Btu
K	2200	Other units
P	25200	Data in standardized physical units
R	132	Thousand Btu per chained (2000) dollar
S	160	Share or ratio expressed as a fraction
V	19200	Value, such as value of shipments
X	132	Million chained (2005) dollars

The deleted data includes variables of total quantity, price, costs and percentage. Because fixed conversion relation remains in British unit and other units, and also there is plenty of data in British units, we obtain variables with unit Btu when meaning of them is the same and delete variables with other units.

(3) We select variables which contain words "total", "total consumption", "produced" or "total production". Namely, we delete energy production and consumption by each sector.

(4) After process of selecting, we get 53 variables. From the description of the remaining 53 variables and similarities and differences, we make clear the hierarchy and inclusion relationships. We select 8 variables of clean energy and define a new variable as other renewable energy (ORTCB). Finally we get 9 variables that can describe energy consumption. Plus, the data of ORTCB is as follows:

$$ORTCB = RETCB-HYTCB-BMTCB-GETCB-SOTCB$$

A classification graph is shown as below Figure 11.3.

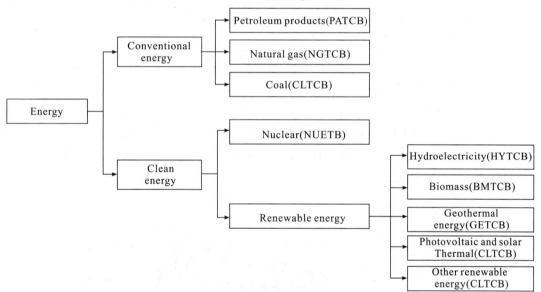

Fig. 11.3 Classification Graph

We do not take the usage of other energy into account for three reasons. First, other energy is too scarce. Second, it is hard to estimate the amount of them. The last reason is that technology cannot exploit some of the special energy.

We assume that the electricity produced from nuclear power and hydroelectricity total production is completely consumed. Table 11.3 shows the MSN code of 9 variables.

Tab. 11.3 MSN Code

MSN	Meaning	Unit
PATCB	All petroleum products total consumption.	Btu
NGTCB	Natural gas total consumption (including supplemental gaseous fuels).	Btu
CLTCB	Coal total consumption.	Btu
NUETB	Electricity produced from nuclear power.	Btu
HYTCB	Hydroelectricity total production.	Btu
BMTCB	Biomass total consumption.	Btu
GETCB	Geothermal energy total consumption.	Btu
SOTCB	Photovoltaic and solar thermal energy total consumption.	Btu
ORTCB	Other renewable energy total production.	Btu

Thus we can use the data of total consumption of the each kind of energy and create stacking area maps. These maps show the energy consumption and energy structure of each state from 1960 to 2009. For example, graph shows the energy consumption and graph shows the energy structure of Arizona.

11.5.2 Summarizing the Energy Profile

1. Energy Profile of Arizona

Most energy consumption increased during the period of 50 years with small amount of them fluctuating a little. PATCB was greatly used all the time. CLTCB and NUETB were greatly used during the last 10 years. ORTCB and BMTCB were used a little from 1960 to 2009.

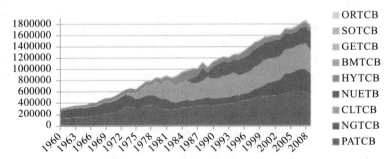

Fig. 11.4 Stacking Area Map of AZ

From the energy structure, from 1960 to 1970, PATCB and NGTCB account for the most of the energy consumption. Over time, the ratio of use of NGTCB decreases but that of PATCB remains steady. The use of CLTCB and NUETB suddenly increases from 1985 to 2009.

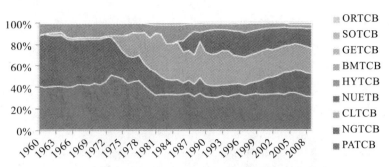

Fig. 11.5 Percentage Stacking Area Map of AZ

2. Energy Profile of California

Most of use of energy has slight fluctuation. PATCB and NGTCB make up a big portion. Other energy use makes a little portion.

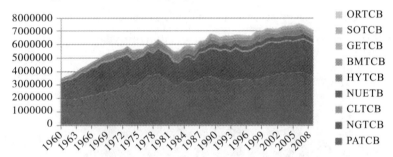

Fig. 11.6 Stacking Area Map of CA

From energy structure, NGTCB and PATCB account for a large portion. Over time the ratio of NGTCB decreases and PATCB remains steady. CLTCB and NUETB increase abruptly later.

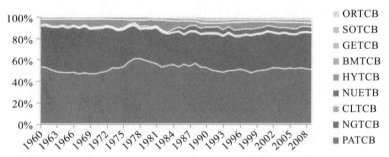

Fig. 11.7 Percentage Stacking Area Map of CA

3. Energy Profile of New Mexico

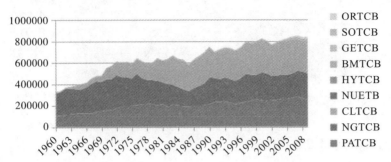

Fig. 11.8 Stacking Area Map of NM

New Mexico mainly uses three types of energy resources. The usage of each energy resource fluctuates a lot. Over time PATCH increases a little and NGTCB decreases at the beginning but increases later. CLTCB rises sharply year by year.

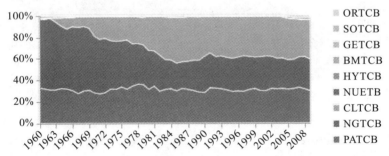

Fig. 11.9 Percentage Stacking Area Map of NM

From energy structure, the ratio of PATCB, NGTCB and CLTCB change a lot. From 1960 to 1962, NGTCB and PATCB make up nearly all the structure. Over time the ratio of NGTCB falls and keeps steady. PATCB remains steady during 50 years. The ratio of CLTCB rises sharply from 1962 and remains steady from 1985.

4. Energy Profile of Texas

The amount of PATCH rises every year, and that of NGTCB fluctuates a little. The amount of CLTCB increases from 1970 and reaches a large number.

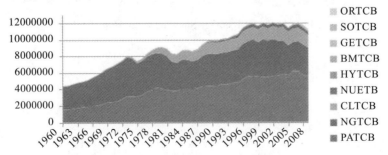

Fig. 11.10 Stacking Area Map of TX

From energy structure, the ratio of PATCB and NBTCB are a big portion. Over time, the ratio of PATCB rises a little, and percentage of NGTCB falls. CLTCB take up nearly 15% of the energy structure.

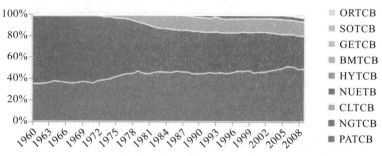

Fig. 11.11　Percentage Stacking Area Map of TX

11.6　Model for Evolution of Energy Profile

11.6.1　Determining Possible Influential Factors

We apply the multi-layer feedforward neural network model to characterize how the energy profile of each of the four states has evolved from 1960—2009 with restriction to natural endowments and technologies. Here we choose 7 criteria as input variables: TPOPP (Resident population including Armed Forces), GDPRX (Real gross domestic product), CLPRB (Coal production), NGMPB (Natural gas marketed production), PAPRB (Crude oil production), NUETB (Electricity produced from nuclear power) and HYTCB (Hydroelectricity total production). TPOPP reflects total consumption, GDPRX reflects generation capacity. The three types of conventional energy—CLPRB, NGMPB and PAPRB represent resource restriction. The two types of clean energy—NUETB and HYTCB represent climate and technology restriction. We choose 7 total consumption variables, which represent energy profile as output variables: PATCB, NGTCB, CLTCB, BMTCB, GETCB, SOTCB, and ORTCB. We make a neural network model for each output variable of each state. Namely, we need train the model for 28 times. These models have the same structure including input layer, hidden layer and output layer. Figure 11.12 shows the process of neural network model.

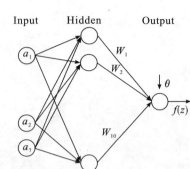

Fig. 11.12 Process of Neural Network

Selection scheme of important parameters are showed below table 11.4.

Tab. 11.4 Selection Scheme

Parameter	Selection Scheme
Number of input layer neurons	7
Number of hidden layer neurons	10
Number of output layer neurons	1
Maximum number of iterations	500
Weight of attenuation	0.01
The activation function	Sigmoid function

11.6.2 Specific Calculation Procedure

1. Calculate activation value b_j of each neuron on hidden layer.

$$b_j = \sum_{j=1}^{n} \sum_{i=1}^{n} w_{ij} a_i - \theta_j$$

w_{ij} is the weight of connection between output layer and hidden layer. θ_j is threshold value of hidden layer.

2. Calculate the output of unit j on hidden layer.

$$c_j = f(b_j) = 1/\{1 + \exp[-(\sum_{i=1}^{n} w_{ij} a_{ij} - \theta_j)]\}$$

3. Similarly, we calculate the activation value and output value.

4. Correct the error of each unit.

5. Loop memory training and get the result of regression.

When we build the neural network model, we lack 17 years' data of variable GDPRX, which has 33 years' data. We use the regression method to complete data. Then we apply completed data to training model. Here we only illustrate AZ, whose output variables are output variable. Put the data of AZ into neuron model and go through 500 iterations. When the number of iterations increases, iteration fitting error line and iteration test error line go falling down. It depicts that the algorithm converges and the error decreases. Figure 11.13 is iterative error graph.

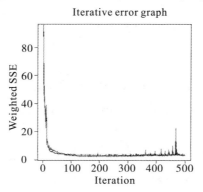

Fig. 11.13　Iterative Error Graph

The prediction of the model can show the predicted number of AZ from 1960 to 2009. Table show the predicted value and real value. Because of the limited space we only show data from 1960 to 1969.

Tab. 11.5　Predicted Value and Real Value from 1960 to 1969

Year	Predicted value	Real value
1960	3935.586438	4012.94849
1961	3826.893733	3836.9629
1962	3861.480711	3672.12878
1963	3979.83019	4027.53031
1964	4042.414785	4089.10173
1965	3840.209627	3695.21696
1966	3553.921324	3679.36818
1967	4534.417172	4180.13105
1968	4018.939984	4147.20373
1969	4078.548471	4389.1241

We use EXCEL to draw a line chart of the predicted value and real value. We excitedly find that they are very close to each other. Figure 11.14 is the line chart.

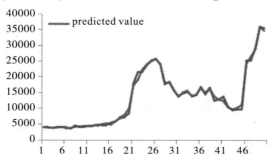

Fig. 11.14　Line Chart of Predicted Value and Real Value

The line chart vividly shows that the model has good simulation and high accuracy of prediction. Therefore, we use data form 1960 to 2009 to train neuron model of each state. When we input specific input variables, we can predict energy consumption variables.

11.6.3 Similarities and Differences

1. Similarities

The four states use a large portion of PATCB and NGTCB and use a little GETCB and ORTCB. And the ratio of NGTCB decreases over time.

2. Differences

CA consumes little CLTCB from 1990 to 2009. During this period CLTCB of the rest states has a great increase. AZ has complex energy structure in the later period but the rest states have single structures.

11.7 Model for the Best Energy Profile

First we use hierarchical clustering method to cluster the rest of the 46 states based on Euclidean distance and deviation square method. Then we cluster them according to the result of hierarchical clustering and reality. At last we make a multi-classification linear discriminant analysis and train the data of 46 states. Put the data of California (CA), Arizona (AZ), New Mexico (NM), and Texas (TX) into the trained discriminant model. The type of the four states can be decided in this way.

11.7.1 System Clustering

1. We standardize the data and cluster them. The hierarchical cluster graph is shown below Figure 11.15.

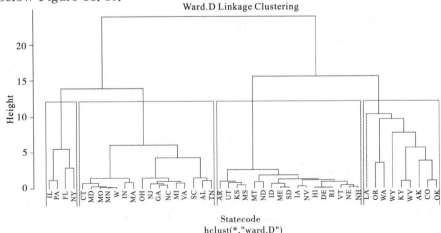

Fig. 11.15　Hierarchical Cluster Graph

2. According to the hierarchical cluster graph we combine 46 states into 4 types. Table shows the result of clustering. Table 11.6 shows the clustering result.

Tab. 11.6 Clustering Result

Class	State Code
1	AK CO KY LA OK OR WA WV WY
2	AL CT GA IN MA MD MI MN MO NC NJ OH SC TN VA WI
3	AR DE HI IA ID KS ME MS MT ND NE NH NV RI SD UT VT
4	FL IL NY PA

We refer to the reality of each state and find that Class 1 has high production of conventional energy, especially rich in natural gas and crude oil and poor in nuclear energy. Class 2 has higher GDP and poorer conventional energy, with full use of nuclear energy. Class 3 has small population and poor conventional energy and full use of hydroelectricity. Class 4 has denser population, higher GDP, poorer coal production and rich natural gas resource, with balanced production of other usage of energy. As for development, LA is the best in class 1, MA is the best in class 2, IA is the best in class 3, NY is the best in class 4.

11.7.2 System Clustering

1. According to the clustering result, we build multi-classification linear discriminant model considering the result of the above clustering. Namely, we use date in Table 11.7 to train the model. Since the table cannot show all the data, here we just show part of them as follows:

Tab. 11.7 Data of 46 States

	CLPRB	GDPRX	HYTCB	NGMPB	NUETB	PAPRB	TPOPP	Class
AK	29016	44595	12919.75	442657.4	0	1365900	694.69	1
AL	459528.9	152514	122345.2	451824.1	415431.5	42038.4	4707.496	2
AR	112.2001	91796	40920.81	691086.2	158677.8	33529.8	2887.331	3
CO	614581.1	232138	18404.67	1626658	0	164279.2	5015.155	1
CT	0	198394	4973.165	0	174236.2	0	3514.826	2
DE	0	54212	0	0	0	0	884.124	3
FL	0	660942	2032.052	263.425	304573	4036.8	18509.94	4
GA	0	356332	31814.51	0	331399.8	0	9813.588	2
HI	0	58682	1099.451	0	0	0	1288.285	3

Using R language we build the model and get three linear discriminant equations as follows:

$$Z_1 = 0.298 \times X_1 + 1.055 \times X_2 - 0.490 \times X_3 - 0.585 \times X_4$$
$$+ 1.183 \times X_5 + 0.081 \times X_6 + 1.267 \times X_7$$
$$Z_2 = -0.806 \times X_1 + 0.114 \times X_2 - 0.804 \times X_3 - 0.56 \times X_4$$
$$+ 0.207 \times X_5 - 0.513 \times X_6 - 0.643 \times X_7$$
$$Z_3 = -0.162 \times X_1 - 3.501 \times X_2 - 0.121 \times X_3 + 0.343 \times X_4$$
$$- 0.676 \times X_5 - 0.23 \times X_6 + 4.0718 \times X_7$$

Based on the three linear discriminant equations, we get the discriminant result of the trained data and calculate the accuracy of the trained model. The accuracy is 97.8%. As a result, the accuracy of the model is high.

2. Now we cluster the four target states. Table shows that AZ belongs to class 3, CA belongs to class 2, and TX belongs to Class 2.

Tab. 11.8 Classification of Four States

	Z_1	Z_2	Z_3	Class
AZ	-1.031	1.785	-0.388	3
CA	2.480	-0.572	0.068	2
NM	-3.211	0.517	0.258	3
TX	1.762	-1.730	0.062	2

By analysis of climate and location, the energy structure of MA is the best in class 2, and is similar to CA and TX. IA is the best in Class 3, and is similar to AZ, CA, NM and TX. The energy structure is showed in Table 11.9.

Tab. 11.9 Energy Structure of 6 States (Unit: Btu)

	AZ	NM	IA	TX	CA	MA
BMTCB	0.020185	0.020581	0.136647	0.013396	0.031738	0.040183
CLTCB	0.235553	0.364318	0.291374	0.135335	0.007403	0.075831
GETCB	0.000188	0.000377	0.00068	0.000186	0.018006	0.000558
HYTCB	0.035756	0.003147	0.006212	0.000907	0.038451	0.009656
NGTCB	0.2147	0.294061	0.208019	0.31282	0.337825	0.336574
NUETB	0.182808	0	0.032076	0.039219	0.046936	0.046494
ORTCB	0.000164	0.017963	0.047467	0	0.008052	4.62E-05
PATCB	0.307949	0.299217	0.27749	0.498062	0.507153	0.490315
SOTCB	0.002697	0.000336	3.39E-05	7.41E-05	0.004435	0.000343

We compare the energy structure of four target states AZ, CA, NM and TX with MA and IA according to different types. The specific procedure is: we calculate the

Euclidean distance of energy structure of every two states. The smaller the Euclidean distance is, the better the energy structure is. Table 11.10 shows the Euclidean distance and ranking.

Tab. 11.10 Euclidean Distance and Ranking

State	Euclidean distance	Ranking
TX	0.070797	1
CA	0.079079	2
NM	0.169045	3
AZ	0.208542	4

11.7.3 Conclusion

We do not consider the difference among four states, only compare each state with itself. The best state is the one which reaches its optimal condition. According to Table 10, the energy structure of TX is the best and that of AZ is the worst.

11.8 Model for Prediction of Energy Profile

11.8.1 Fixed Growth Model

We assume that there is no abrupt development of technology that cause the variables to have a high growth rate. Directly predicting the future energy usage and structure is not convenient, so we change the problem into the prediction of input variables of neural network in 2025 and 2050 with absence of policy changes. Later we put the data into the trained neural network to predict total consumption of each type of energy in 2025 and 2050. According to the total energy usage we calculate the energy structure.

Through modeling test, it is not suitable to use time series model and grey prediction model to predict temporal variables because of the too long time span. Consider establishing a more concise and realistic model.

Without policy changes, input variables can be considered as growing with a fixed growth rate. With fluctuation of each year, we set a random fluctuation term based on fixed ratio. Fixed term and random term make up the growth rate.

We determine different growth rate of variables of different states and predict input variables for 2025 and 2050. By analysis of data and estimating reality, we get fixed growth rate θ of each state.

Tab. 11.11　Fixed Growth Rate

	AZ	CA	NM	TX
GDPRX	0.004512	0.002812	0.000804	0.003955
CLPRB	−0.00096	0	−0.00017	−0.00043
HYTCB	0.01181	0.00948	0.0169	0.01201
NGMPB	0.0031	−0.00273	−0.00323	0.00325
NUETB	0.01952	0.01759	0	0.00676
PAPRB	−0.00125	−0.00304	−0.00094	−0.00096
TPOPP	0.00427	0.00145	0.00212	0.00402

We can use the fixed growth rate and earlier value to predict the later value. The formula is:

$$F_n = F_{n-1}(1 + \varphi + r_n)$$

where r_n is a random variable of uniform distribution in the interval $[-\varphi \times 10\%, \varphi \times 10\%]$.

11.8.2　Multi-layer Feedforward Neural Network Model

We get predictive value of the four states' independent variables in 2025 and 2050. Based on the multi-layer feedforward neural network model discussed above, we can also predict the energy profile of the four states and energy structure. Tables 11.12 to 11.13 show the energy profile of the four states and energy structure.

Tab. 11.12　Predicted Consumption of Each State in 2025

	AZ	CA	NM	TX
PATCB	501199.2	4254094.0	198374.9	2258986.0
NGTCB	698727.6	3937242.0	160324.2	1475554.8
CLTCB	456234.8	3015520.0	184546.1	2173588.0
NUETB	434786.9	439130.2	0.0	482895.3
HYTCB	76003.5	315899.6	3465.3	12145.4
BMTCB	35516.0	233915.5	17879.0	177580.8
GETCB	2933.6	20135.3	376.1	18547.2
SOTCB	29367.4	214730.5	8595.1	140823.6
ORTCB	7042.7	24562.3	1745.1	22470.2

Tab. 11.13　Predicted Consumption of Each State in 2050

	AZ	CA	NM	TX
PATCB	469080.0	4711630.0	197826.9	2334906.0
NGTCB	483498.2	3372906.0	34994.3	2288782.0
CLTCB	482221.4	3581904.0	136790.2	2226600.0
NUETB	703174.6	683375.5	0.0	570256.8
HYTCB	101973.1	401542.1	5287.1	16318.4
BMTCB	19258.4	216923.4	14086.3	190269.3
GETCB	3351.2	20699.3	194.3	20964.6
SOTCB	31710.9	186380.0	7919.1	147337.1
ORTCB	7693.1	24695.7	1896.2	24902.0

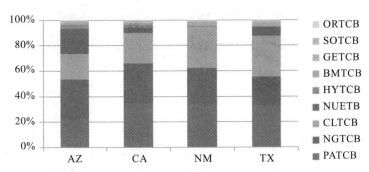

Fig. 11.16　Predicted Energy Structure of Each State in 2025

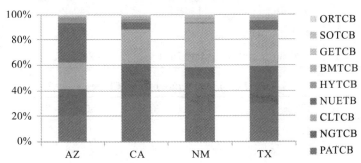

Fig. 11.17　Predicted Energy Structure of Each State in 2050

11.9　Goals of Compact

11.9.1　Determining Growth Rate

We determine natural growth rate g and special growth rate d. When we predict energy structure of MA and IA, we use g and d to illustrate different types of energy consumption in 2025 and 2050. Different consumption in different period adopts this prediction formula:

$$E_{jt+1} = E_{jt} \times (1+g)(1+d)$$

where E_{jt} represents consumption of energy j at time t.

Considering historical data, we determine all the growth rates as follows:

Tab. 11.14　Growth Rate

d	PATCB	NGTCB	CLTCB	NUETB	HYTCB	BMTCB	GETCB	SOTCB	ORTCB
MA	−0.18%	0.16%	−0.23%	0.39%	0.12%	0.75%	0.22%	1.66%	5.11%
IA	−0.10%	0.37%	−0.33%	0.14%	0.62%	1.09%	0.16%	2.79%	4.12%
g					0.05%				

According to the model we predict energy consumption and optimal structure of MA and IA in 2025 and 2050.

Tab. 11.15 Predicted Consumption Value of MA and IA

	2025(class 2)	2025(class 3)	2050(class 2)	2050(class 3)
PATCB	578309.9	416599.3	552840.4	406206.7
NGTCB	419402.3	336894.9	436875	369794.4
CLTCB	88698.8	421544.2	83697.08	387924.9
NUETB	60039.53	50024.63	66124.49	51765.02
HYTCB	11952.48	10466.35	12320.88	12219.83
BMTCB	54938.16	248163.8	66152.74	325788.2
GETCB	701.16	1064.79	740.28	1109.35
SOTCB	541.47	80.35	817.4	159.9
ORTCB	226.36	138175.4	2001.77	379127

Tab. 11.16 Structure of Two Classes

	2025(class 2)	2025(class 3)	2050(class 2)	2050(class 3)
PATCB	47.60%	25.67%	45.26%	21.00%
NGTCB	34.52%	20.76%	35.76%	19.12%
CLTCB	7.30%	25.97%	6.85%	20.06%
NUETB	4.94%	3.08%	5.41%	2.68%
HYTCB	0.98%	0.64%	1.01%	0.63%
BMTCB	4.52%	15.29%	5.42%	16.84%
GETCB	0.06%	0.07%	0.06%	0.06%
SOTCB	0.04%	0.00%	0.07%	0.01%
ORTCB	0.02%	8.51%	0.16%	19.60%

11.9.2 Conclusion

CA and TX should take the optimal energy structure of Class 2 as the best target. They should make full of nuclear energy and also consume petroleum and natural gas the most, and use less coal. The goal of CA and TX is to balance the usage of renewable energy and use less conventional energy.

AZ and NM should take the optimal energy structure of Class 3 as the best target. They should use petroleum, natural gas and coal equally and steadily generate hydroelectricity and make use of biomass energy.

11.10 Suggestions

Considering optimal energy structure and identify the states in Class 2 and Class 3, we discuss several actions the four states might take to meet their energy compact goals.

1. For California
Use less mineral resources and promote usage of biomass energy in agricultural sectors.

2. For Texas
Strengthen environmental awareness among residents and develop wind power generation.

3. For Arizona
Control the mining of petroleum and develop hydroelectricity generation.

4. For New Mexico
Build large water power facilities or solar energy installations. Make less of non-renewable energy. Make moderate use of nuclear power generation energy.

11.11　Error Analysis

1. When the energy profile is analyzed in PART1A, the non-uniqueness of energy classification can cause the emergence of a large number of new energy sources in the future.

2. There is a difference between the fitting value and the true value of GDP. It is very difficult to determine the number of neurons in the hidden layer of neural network, and it is more difficult to do corresponding function selection between input layer and output layer.

3. The growth rate of independent variables in PART1 and PART2 may vary in different years.

11.12　Sensitivity Analysis

For prediction of the energy profile of each state for 2025 and 2050, we base on the current value and fixed growth rate and get predicted value of variables. The equation is like this: $F_n = F_{n-1}(1+\theta+r_n)$

Because of the great number of states and variables, we just choose HYTCB of AZ to exemplify. In the paper, the fixed growth rate of HYTCB of AZ is 1.181%. In order to verify the rationality of the fixed growth rate we set, we take the fixed rate of growth θ as 1.1 times, 1.05 times, 0.95 times and 0.9 times. Using EXCEL software for sensitivity analysis, the results are shown in the Figure 11.17.

Figure 11.17 shows that with time going by, different fixed growth rates of HYTCB do not fluctuate a lot. Obviously, HYTCB value is not sensitive to the fixed growth rate so we set a reasonable growth rate.

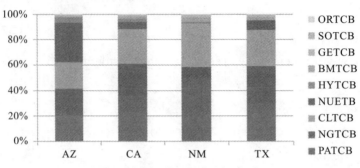

Fig. 11.17　Sensitivity Analysis

11.13　Development of the Model

The fixed growth rate model has some subjectivity and we ignore that it has characteristics of time series model such as periodic feature. However, we try to establish ARIMA, a time series model at the beginning. We find that the fitting is not good. Therefore, we improve the time series model. First, we use a differential equation to delete trend of non-stationary time series data. Then we take the Fourier transform. The formula is:

$$F(e^{jw}) = F(z)\mid_{z=e^{jwT}}$$
$$F(e^{jw}) = \sum_{-\infty}^{\infty} f(k) e^{-jkw}$$

Positive transformation: $F(e^{jw}) = DTFT\{f(k)\} = \sum_{-\infty}^{+\infty} F(e^{jw}) e^{jkw}$

Negative transformation: $f(k) = IDTFT\{F(e^{jw})\} = 1/2\pi \int_{-\pi}^{+\pi} F(e^{jw}) e^{jkw} dw$

According to the above transformation, the high-energy frequency points are extracted and fitted into Fourier series equation. The fitting result is much better than that of ARIMA. If the fitting effect is still not ideal, we can add wavelet analysis and other methods.

11.14　Strengths and Weaknesses

Like any model, our model has its own strengths and weaknesses. Some of the major points are presented below.

11.14.1 Strengths

1. The process of selecting variables starts with energy consumption. We consider the hierarchical structure of energy classification and it helps to reach a realistic result.

2. Offering growth rate model instead of time sequence model to reach a more realistic and operable result.

11.14.2 Weaknesses

1. We do not consider ways of production of energy consumption.

2. We do not consider the effect of energy price fluctuation on the selection of consumers.

3. We do not consider the great loss of energy and in the energy profile there is no real utilization.

4. We do not consider the export and import of each state.

5. We do not analyze interstate cooperation in quantity of energy resources.

Memo

To the group of Governors

During the period of 1960 to 2009, for Arizona and California, petroleum and natural gas accounted for the most of the energy consumption. Over time, the ratio of use of natural gas decreased but that of petroleum remained steady. The use of coal and nuclear of Arizona suddenly increased from 1985 to 2009 but that of California decreased sharply. New Mexico mainly used three types of energy resources and the consumption of coal rose sharply year by year. Texas mainly used petroleum and natural gas. Other energy resources were used over time.

In 2025 the conventional energy consumption of Arizona will take up 73.88%, nuclear energy will take up 19.39% and biomass energy will take up 1.58%. California will use conventional energy the most and hydroelectricity energy will account for 2.54%. The conventional energy consumption of New Mexico will take up 94.43% and biomass energy will reach 3.11%. Texas will use nuclear energy up to 7.14%.

In 2050, the consumption of nuclear energy of Arizona will account for 30.55%, which shows it has a bright prospect. California will make more use of hydroelectricity power. New Mexico will make more use of biomass energy. Texas will develop the technology of nuclear.

We discuss several actions the four states might take to meet their energy compact goals.

Note 1：For California

Use less mineral resources and promote usage of biomass energy in agricultural sectors.

Note 2：For Texas

Strengthen environmental awareness among residents and develop wind power generation.

Note 3：For Arizona

Control the mining of petroleum and develop hydroelectricity generation.

Note 4：For New Mexico

Build large water power facilities or solar energy installations. Make less use of non-renewable energy. Make moderate use of nuclear power generation energy.

Sincerely，Team♯81666

Reference

[1] *Yang Guiyuan. Mathematical modeling*[M]. Shanghai University of Finance and Economics，2015.

[2] http://www.eia.gov

[3] Yang Guiyuan, Zhu Jiaming. *Analysis of Outstanding Paper in Mathematical Contest in Modeling*[M]. University of Science and Technology of China. 2013.

[4] Pang Ningtan, Micheal Steinbach, Vipin Kumar. *Introduction to Data Mining*[M]. Posts & Telecom Press. 2011.

[5] Robert I. Kabacoff. *R in action*[M]. Posts & Telecom Press. 2016.

[6] Ma L., Liu P., Fu F., et al. Integrated energy strategy for the sustainable development of China[J]. *Energy*, 2011, 36(2)：1143—1154.

[7] Greenstone M., Looney A. A strategy for America's energy future：Illuminating energy's full costs[J]. *The Hamilton Project Strategy Paper*. Washington DC：Brookings，2011.

◆ 建模特色点评

❀论文特色❀

◆标题定位：标题 Renewable Energy Renews Your Life 即"可再生能源让你的生活焕然一新"紧扣问题，给予准确的定位，这样的标题生动活泼，作为赛题有创意。

◆方法鉴赏：文章处理较好，选择变量从能源消耗开始，运用聚类分析法构建能量分类的层次结构,有助于实现结果。所提出的增长率模型代替时间序列模型,以达到更为现实和可操作的结果。技术层面较好,多种软件的使用显示编程能力强,聚类分析与灵敏度分析显示创新能力较好。

◆写作评析:整体思路清晰,技术手段多样,写作流畅,图文并茂。

❅**不足之处**❅

建模处理较简单,不考虑能源消费的生产方式,没有考虑能源价格波动对消费者选择的影响。在能量分布中没有真正地利用浪费的能量,没有考虑每个国家的进出口,没有分析洲际能源合作案例的数量。

第3部分

全国研究生数学建模竞赛优秀论文评析

内容简介

一、全国研究生数学建模竞赛简介

全国研究生数学建模竞赛（National Post-Graduate Mathematical Contest in Modeling）是"全国研究生创新实践系列活动"的主题赛事之一，由教育部学位与研究生教育发展中心主办。该赛事发源于 2003 年东南大学发起并成功主办的"南京及周边地区高校研究生数学建模竞赛"；2013 年被纳入教育部学位中心"全国研究生创新实践系列活动"。其宗旨是为广大研究生探究实际问题、开展学术交流、培养团队意识搭建有效平台，培养研究生创新意识，提升研究生创新实践能力，进一步推动研究生培养机制改革和"研究生教育创新计划"的实施，促进研究生培养质量的提高。

二、安徽财经大学参赛获奖情况介绍

安徽财经大学从 2009 年起每年组织 4 组同学参加研究生竞赛，第一年获全国一等奖 1 项，全国二等奖 2 项，至今已持续 11 年，队伍在不继壮大中。限于篇幅，文中选出全国一等奖赛文 3 篇，二等奖 1 篇，分别是 2009 年全国一等奖、2010 年全国一等奖、2012 年全国一等奖、2013 年全国二等奖。

第 12 篇
我国就业人数或城镇登记失业率的数学建模

◆ **竞赛原题再现**

2009 年 A 题 我国就业人数或城镇登记失业率的数学建模

失业、经济增长和通货膨胀为宏观经济中特别重要的三个指标,就业(或者失业)是社会、国民经济中极其重要的问题。按照已有研究,就业可以定义为三个月内有稳定的收入或与用人单位有劳动聘用关系。失业的统计方法各国差异较大,我国采用城镇登记失业率,是指城镇登记失业人数同城镇从业人数与城镇登记失业人数之和的比。其中,城镇登记失业人员是指有非农业户口,在一定的劳动年龄内(16 岁以上及男 50 岁以下、女 45 岁以下),有劳动能力,无业而要求就业,并在当地就业服务机构进行求职登记的人员。但由于统计口径不同,存在一定的差异,有些历史数据也较难获得。

从经济学的角度,影响就业(或者失业)的因素很多。从宏观层面上,消费、投资、政府购买和进出口都是重要的因素;而从中观层面,不同地区、不同产业也会表现出不同的特征。当然,中央政府调整宏观经济政策(包括财政政策和货币政策),以及对不同地区和不同产业实行不同的扶持政策都会对就业产生巨大的影响。

就我国的现实情况,2008 年我国经济社会经受了历史罕见的考验,GDP 依然保持 9% 以上平稳较快增长,城镇新增就业 1113 万人,城镇登记失业率为 4.2%。2009 年我国就业面临更大的挑战,一是国际金融危机导致国际市场需求难以在短期内复苏;二是今年我国经济增速下滑;三是国内消费需求乏力;四是一些行业产能过剩与市场预期不确定导致企业投资不足,所以就业形势十分严峻。

为此,中央政府从 2008 年 10 月开始实施了 40000 亿元的投资计划,确定了十大产业振兴计划,采取扩大国内消费需求的措施,提高对外开放水平以增加出口。同时,中央财政拟投入 420 亿元资金实施积极的就业政策。2009 年我国在就业方面的目标:城镇新增就业 900 万人以上,城镇登记失业率控制在4.6%以内(以上数据取自温家宝总理的政府工作报告)。

请研究生参考就业问题的研究成果,利用近年来我国有关的统计数据并结合一年多来我国国民经济的运行数据(参见下面网站,也可以对比其他国家的统计数据)就我国就业人数或城镇登记失业率研究如下问题。

1. 对有关统计数据进行分析,寻找影响就业的主要因素或指标。

2. 建立城镇就业人数或城镇登记失业率与上述主要因素或指标之间联系的数学模型。

3. 对上述数学模型从包含主要的经济社会指标、分行业、分地区、分就业人群角度,尝试建立比较精确的数学模型。(由于时间限制,建议适度即可)

4. 利用所建立的关于城镇就业人数或城镇登记失业率的数学模型,根据国家的有关决策和规划对2009年及2010年上半年的我国就业前景进行仿真(可以根据模型的需要对未来的情况作适当的假设)。

5. 根据所建立的数学模型和仿真结果,对提高我国城镇就业人口数或减少城镇登记失业率提出你们的咨询建议。

希望利用下面网站上的数据,可以参考已有文献的数学模型,但更鼓励创新,使用翔实可靠的数据,在多方面对各项政策措施进行对比论证。

参考网站:

http://www.stats.gov.cn/tjsj/ndsj/

http://dlib.cnki.net/kns50/

http://www.drcnet.com.cn 点击宏观经济,再点击运行数据或数据平台或旧库数据平台

http://chinese.wsj.com/gb/index.asp 点击经济,再点击经济数据

http://news.hexun.com/ 点击宏观数据

附件中有部分逐月经济运行数据,供参考。

原题详见2009年全国研究生数学建模竞赛A题。

◆ 获奖论文精选

我国就业人数的主要影响因素分析及前景预测[①]

摘要: 本文针对社会与国民经济中极其重要的问题——就业问题进行分析研究,通过对有关统计数据进行分析,运用统计软件SPSS,采取主成分分析方法提取了影响就业的主要因素或指标,建立了我国就业人数的数学建模,分别从包含主要的经济社会指标、分行业、分地区、分就业人群角度,尝试建立比较精确的数学模型。根据国家的有关

---

① 本文获2009年全国一等奖。队员:王犁,童金萍,郭艳芳;指导教师:朱家明等

决策和规划对 2009 年及 2010 年上半年的我国就业前景进行仿真。最终结合模型与仿真结果,对提高我国城镇就业人口数或减少城镇登记失业率提出较好的建议。

针对问题 1,本文结合就业人数折线图,将我国就业分成两个阶段进行研究,综合运用了灰色关联度与相关系数两种方法,建立了指标排序模型,按所得值大小来选取影响就业的主要指标,依照 $\gamma \geqslant 0.81$ 的原则对主要因素或指标进行提取。

针对问题 2,本文分别利用主成分回归分析法与向量自回归分析法分段构建了城镇就业人数拟合模型,并依据所得到的向量自回归模型作出脉冲响应函数图以及方差分解图,得到每一个指标对城镇就业人数的冲击效应及方差贡献率,综合两阶段模型的结果可知对我国城镇就业人数影响最大的是城市化水平。

针对问题 3,结合上述数学模型从包含主要的经济社会指标,文章分别从东、中、西三大经济带,国有、集体和其他经济行业三个方面以及农民工、大学毕业生及其他三种不同人群的角度来构建相关的人口就业模型。

针对问题 4,利用神经网络预测的方法,结合国家的有关决策和规划,对 2009 年及 2010 年上半年我国城镇居民的就业前景进行仿真预测。文章最后根据所建立的数学模型的相关影响因素,我们从经济发展、财政政策、货币政策等 5 个角度出发,并结合社会实际,提出提高城镇就业水平的政策、建议。

关键词:就业;影响就业指标体系;灰色关联度;相关系数;神经网络预测

12.1 问题的重述

12.1.1 就业与失业统计法

1. 什么是就业

就业就是一定年龄段的人们所从事的为获取报酬或经营收入所进行的活动。细分则从三个方面进行界定:①就业条件,指一定的年龄;②收入条件,指获得一定的劳动报酬或经营收入;③时间条件,即每周的工作时间。根据已有研究,就业也可以定义为一定年龄阶段的人们在三个月内有稳定的收入或与用人单位有劳动聘用关系的活动。

2. 经济中地位

失业、经济增长和通货膨胀为宏观经济中特别重要的三个指标,就业(或者失业)是社会、国民经济中极其重要的问题。

3. 失业统计法

对失业的统计方法各国差异较大,我国采用城镇登记失业率,其定义如下。

$$城镇登记失业率 = \frac{城镇登记失业人数}{城镇从业人数 + 城镇登记失业人数}$$

其中,城镇登记失业人员是指有非农业户口,在一定的劳动年龄内(16 岁以上及男 50 岁以下、女 45 岁以下),有劳动能力,无业而要求就业,并在当地就业服务机构进行求职登记的人员。但由于统计口径不同,存在一定的差异,有些历史数据也较难获得。

12.1.2 就业的背景与对策

1. 影响因素分析

从经济学的角度,影响就业(或者失业)的因素很多。从宏观层面看,消费、投资、政府购买和进出口都是重要的因素;而从中观层面,不同地区、不同产业也会表现出不同的特征。当然,中央政府调整宏观经济政策(包括财政政策和货币政策),以及对不同地区和不同产业实行不同的扶持政策都会对就业产生巨大的影响。

2. 我国现实状况

2008 年我国经济社会经受了历史罕见的考验,GDP 依然保持 9% 以上平稳较快增长,城镇新增就业 1113 万人,城镇登记失业率为 4.2%。2009 年我国就业面临更大的挑战,一是国际金融危机导致国际市场需求难以在短期内复苏;二是今年我国经济增速下滑;三是国内消费需求乏力;四是一些行业产能过剩与市场预期不确定导致企业投资不足,所以就业形势十分严峻。

3. 政府就业对策

面对挑战,中央政府从 2008 年 10 月开始实施了 40000 亿元的投资计划,确定了十大产业振兴计划,采取扩大国内消费需求的措施,提高对外开放水平以增加出口。同时,中央财政拟投入 420 亿元资金实施积极的就业政策。2009 年我国在就业方面的目标:城镇新增就业 900 万人以上,城镇登记失业率控制在 4.6% 以内(以上数据取自温家宝总理的政府工作报告)。

12.1.3 数据获取的途径

1. 参考网站

(1)http://www.stats.gov.cn/tjsj/ndsj/;

(2)http://dlib.cnki.net/kns50/;

(3)http://www.drcnet.com.cn 点击宏观经济,再点击运行数据或数据平台或旧库数据平台;

(4)http://chinese.wsj.com/gb/index.asp 点击经济,再点击经济数据;

(5)http://news.hexun.com/ 点击宏观数据。

2. 相关附件

原题附件中有部分逐月经济运行数据,供参考,具体如下。

(1)附件一:WORD 文档"A 题经济数据";

(2)附件二:WORD 文档"A 题新闻"。

3. 统计资料

《新中国 55 年统计资料汇编》《2005—2008 年中国统计年鉴》《1996—2008 中国劳动

力统计年鉴》《2005—2008年中国人口统计年鉴》等统计资料。

12.1.4 要解决的具体问题

请研究生参考就业问题的研究成果,利用近年来我国有关的统计数据,并结合一年多来我国国民经济的运行数据(参见上面网站,也可以对比其他国家的统计数据),就我国就业人数或城镇登记失业率研究如下问题。

问题1 对有关统计数据进行分析,寻找影响就业的主要因素或指标;

问题2 建立城镇就业人数或城镇登记失业率与上述主要因素或指标之间联系的数学模型;

问题3 对上述数学模型从包含主要的经济社会指标、分行业、分地区、分就业人群角度,尝试建立比较精确的数学模型(由于时间限制,建议适度即可);

问题4 利用所建立的关于城镇就业人数或城镇登记失业率的数学模型,根据国家的有关决策和规划对2009年及2010年上半年的我国就业前景进行仿真(可以根据模型的需要对未来的情况作适当的假设);

问题5 根据所建立的数学模型和仿真结果,对提高我国城镇就业人口数或减少城镇登记失业率提出你们的咨询建议。

12.2 问题的分析

12.2.1 问题的总体分析

就业(失业)是社会、国民经济中极其重要的问题,本文要求利用近年来我国有关的统计数据,并结合一年多来我国国民经济的运行数据进行统计分析,寻找影响就业的主要因素或指标。

研究对象为我国近年来经济发展中的一系列时间序列分析,采用方法为主成分分析、关联度分析,找出与就业问题相关的主要因素或指标,并结合各类具体数据,运用灰色预测法对未来一两年内的就业问题进行预测。

本问题的难点是涉及数据量大,涉及指标太多,且国家政策等不确定的因素混合在内,因此合理指标体系的构建是本文的重中之重。构建指标体系时,指标的个数不可太少,也不可太多。为此,首先我们结合经济学知识从我国经济相关时间序列数据中查找尽可能多的相关因素;其次我们可运用基本统计方法将相关数据进行分类筛选,采取同类合并、化多为少、取重舍次等方法,最终构建出较为合理、方便操作的"适度"指标以便于研究。

12.2.2 对具体问题的分析

1. 对问题 1 的分析

问题 1 要求对有关统计数据进行分析,寻找影响就业的主要因素或指标,这要求我们寻找与就业紧密相关的各类因素,然后将这些因素进行分门别类,形成三级指标体系。然后,我们运用多元统计分析法中的主成分分析法与灰色关联度方法,由此排序可找出影响就业的主要因素。

2. 对问题 2 的分析

首先根据对问题 1 提出来的影响就业的主要指标,结合历年来我国城镇就业人数的变化规律,可考虑运用主成分回归分析法与向量自回归分析法来建立城镇就业人数或城镇登记失业率与上述主要因素或指标之间联系的数学模型。

3. 对问题 3 的分析

前面建立的模型是全盘考虑的多元回归分析,问题 3 要求因素较多,从包含主要的经济社会指标,然后从分行业、分地区、分就业人群三个角度,尝试建立比较精确的数学模型。对分行业,我们考虑到行业过多,由于时间限制,我们选取了几个容易获取数据且影响就业的行业:国有经济企业、乡镇集体经济和其他经济类型等。对分地区,从简单考虑,我们分东部、中部、西部三个地区进行建模研究,至于更细的可类似处理。对分就业人群,我们选择几个主要就业人群:大中专毕业生、农民工、其他人群(如高中待业、转业复员军人、下岗工人)等几个主要就业人群。

4. 对问题 4 的分析

为了科学合理地对未来城镇就业前景预测,减小预测模型的误差,本文利用问题 2 中已建立的数学模型预测结果与神经网络预测结果,得出城镇就业的综合预测值。第一种预测方法我们首先利用神经网络方法模拟仿真出不同模型中各影响因素的数值,将其代入我们得到的数学模型,得到预测结果,第二种方法我们直接利用神经网络得到仿真预测结果。最后我们进行细化,结合国家的有关决策和规划对我国不同行业的 2009 年及 2010 年上半年的就业前景进行仿真预测。

5. 对问题 5 的分析

问题 5 要求根据所建立的数学模型和仿真结果,对提高我国城镇就业人口数或减少城镇登记失业率提出咨询建议。对此,我们要结合问题 2 所建立的模型及问题 4 的仿真结果,分别依照所选的六个主要影响结果进行相关的政策建议,为有关政府部门提供政策,进而有效应对当前金融危机下我国高校毕业生以及农民工就业难的困境,在危机中抢抓机遇,从而实现我国经济的稳定可持续健康快速发展。

12.3 模型的假设

(1) 我国城市劳动力处于供过于求的状态,没有达到最佳的就业状态;
(2) 假设男女就业(获取工作)的机会相同,即不考虑性别因素;
(3) 规定年龄在 15～64 岁之间的人为我国的劳动人口,不考虑不具有劳动能力的人;
(4) 规定失业人口不仅包括城镇登记失业人口,还包括城镇未登记的、非自愿失业人口和农村剩余劳动力;
(5) 不考虑因自然灾害(如汶川地震)等原因对就业产生的意外影响。

12.4 符号说明

序号	符号	符号说明
1	P	就业人数
2	GDP	国内生产总值
3	K	固定资产投资
4	ρ	可比价格表示的固定资产投资额的平均增长率
5	δ	折旧率
6	mar	市场化指数
7	str	产业结构调整水平
8	zr	人口自然增长率
9	csh	城市化水平
10	lb	劳动力资源总数占总人口比例
11	cxc	城乡收入差距
12	czs	财政收入
13	czz	财政支出
14	ne	净出口
15	M2	广义货币供给量
16	tax	税收
17	kzf	科学研究占财政支出比重
18	eng	恩格尔系数
19	com	居民消费水平
20	$huma$	人力资本
21	d_1	小学人口所占比重
22	d_2	初中人口所占比重
23	d_3	高中人口所占比重
24	d_4	高中以上人口所占比重
25	pr	平均货币工资水平

12.5 模型的准备

12.5.1 数据的采集与指标的构建

1. 数据的采集

根据问题要求,我们首先通过经济意义的分析,初步选取30个因素作为影响就业人数的主要统计数据:就业人数、国内生产总值、固定资产投资总额、财政收入、财政支出、进口总额、出口总额、货币供给量M2、税收总收入、科学研究占财政支出比重、职工工资总额、劳动者报酬、城镇非国有单位从业人员、城镇从业人员、外商固定资产投资、第一产业增加值、第二产业增加值、第三产业增加值、人口自然增长率、城镇人口数、总人口数、15~65岁人口数、城镇居民可支配收入、农村居民纯收入、恩格尔系数、居民消费、小学在校学生数、中学在校人数、大专及以上在校人数、平均货币工资等。

数据来源于《新中国55年统计资料汇编》《2005－2008年中国统计年鉴》《1996－2008中国劳动力统计年鉴》《2005－2008年中国人口统计年鉴》等统计资料。

2. 指标体系的说明

根据上述统计数据资料,为了便于进一步建模分析,以下我们将30个因素相关的数据进行同类合并、化多为少、取重舍次等方法,构建影响就业的18个主要指标。

(1)就业人数(P):在本文模型中,我们选取每年就业人数来衡量我国就业情况,考虑到城镇登记失业率在统计上存在一定的系统性误差,没有将既未从业又未参加登记的人群计算在内,故不能反映出真实的就业状况。

(2)国内生产总值(GDP):从宏观调控角度看,GDP的增加是创造新增就业机会的主要来源,所以我们初步选取GDP作为影响就业人数的主要因素。

(3)固定资产投资(K):固定资产是社会扩大再生产的基本手段,是实现国民经济持续、快速、健康发展的原动力,它通过拉动经济、促使产业结构升级来扩大就业,因此可选取固定资产投资作为一个影响因素。

对于资本投入量的测算,按照惯例文章采取资本存量并通过永续盘存法来对资本投入量进行测算,即用上一年的资本净存量(资本存量减折旧)加上当年的投资来表示,其计算公式为:

$$K = K_{t-1} + I_t - \delta K_{t-1}$$

其中,初始年份的资本存量可以采取Hall-Jones(1999)的方法[1]来估算,其计算公式为:

$$K_0 = I_0/(\rho + \delta)$$

其中ρ为计算期内可比价格表示的固定资产投资额的平均增长率,δ代表折旧率,一般对于我国固定资产折旧率取值为10%(Hall和Jones在模拟世界上127个国家的资本存量时折旧率取值为6%)。

(4)财政收入(czs):增加税收会对人民就业产生影响,财政收入从某种程度上减少了人们的收入,降低了就业的积极性。因此我们引入财政收入作为影响就业人数的主要因素。

(5)财政支出(czz):能够给予社会补贴,实施特别培训计划,支持企业吸纳就业、支持相关机构提供就业服务,扶持职业中介机构,并通过大规模增加政府投资,吸纳更多的人就业,实现就业增长等。

(6)净出口(ne):净出口额是出口总额与进口总额的差值,其计算公式为:
$$净出口额=出口总额-进口总额$$
净出口额的增加,在某种程度上会刺激国内经济发展,增加就业机会。

(7)货币供给量(M2):货币供给量是指一国在某一时期内为社会经济运转服务的货币存量,它由包括中央银行在内的金融机构供应的存款货币和现金货币两部分构成。我们一般用广义货币量 M2 来代替。

M2=流通中现金+企业活期存款+机关团体部队存款+农村存款+个人持有的信用卡类存款+城乡居民储蓄存款+企业存款中具有定期性质的存款+信托类

(8)税收(tax):税收是国家为实现其职能,凭借政治权力,按照法律规定,通过税收工具强制地、无偿地参与国民收入和社会产品的分配和再分配,取得财政收入的一种形式。税收对就业的增加,可以分两方面考虑:一方面,税收增加,人们的可支配收入减少,人们为了维持原有的收入水平,从而多劳动,劳动供给增加。另一方面,税收增加,劳动力报酬减少,打击劳动者的工作积极性,劳动供给减少。

(9)科研支出率(kzf):科技经费支出额占 GDP 比例(反映对科技活动的投入强度和重视程度)。由于政府对科技活动的投入强度和重视程度的加大,将会直接导致全社会科技创新水平的显著提高,进而带动产业结构升级等一系列现象的发生。因此,本文构建科学研究占财政支出比重指标来反映这种现象,其计算公式为:
$$科学研究占财政支出比重=科技活动的投入/财政支出$$

(10)市场化指数(mar):对于市场化指数的测量,由于观察角度及使用的标准不同,许多学者提出了各自的方法论。国家计委课题组(1996 年)是从商品市场(包括生产环节和流通环节)的市场化和要素市场(包括劳动力市场和资金市场)的市场化程度入手进行测算的。陈宗胜(1999 年)认为,对经济体制市场化程度的测度,要按企业、政府、市场三方面来考虑。徐明华(1999 年)则从所有制结构、政府职能转变和政府效率、投资的市场化、商品市场发育、要素市场发育、对外开放、经济活动频度、人的观念 8 个方面进行了测算。樊纲、王小鲁(2003 年)则以政府与市场的关系、非国有经济的发展、产品市场的发育程度、要素市场的发育程度、市场中介组织发育和法律制度环境 5 个方面,23 个指标为基础,运用"主因素分析法"进行市场化指数的测量。

对于市场化指数的影响,可以分为正向指标和逆向指标。正向指标越大,逆向指标越小,市场化指数越大;反之市场化指数越小。本文根据以往研究经验,经过正逆指标的调整(劳动力市场中城镇国有单位从业人员/城镇从业人员为逆指标,而城镇非国有单位从业人员/城镇从业人员则为正指标),从正向指标的方向考虑,选取分别代表收入分配、劳动力市场和资本市场的各一项指标,运用 SPSS 的主成分分析方法对市场化指数进行计算。

(11)产业结构调整水平(str):对于影响就业结构的产业而言,第二、三产业行业多,门类广,劳动密集、资本密集、技术密集行业并存,具有吸纳各类劳动力就业的独特优势和作用。因此,为了更好地研究产业结构演进和就业数量之间的关系,文章参照张雷(2008)设定的用来反映区域产业结构多元化演进程度的指标 str,其计算公式为:

$$str = \sum (P/P, S/P, T/P) \quad (1 \to \infty)$$

其中,P 为第一产业产出,S 为第二产业产出,T 为第三产业产出,并且 str 的值域可以从 1 到无穷大,其值越大代表产业结构的多元化程度越高,否则反之。

(12)人口自然增长率(zr):人口自然增长率是反映人口发展速度和制定人口计划的重要指标,也是计划生育统计中的一个重要指标,它表明人口自然增长的程度和趋势。它反映了人口再生产活动的能力。在数值上,它是人口出生率与人口死亡率的差值,其计算公式为:

$$人口自然增长率 = 人口出生率 - 人口死亡率$$

根据分析,认为人口自然增长率的增加是就业人数增加的原因之一。

(13)城市化水平(csh):又称城市化率,是衡量城市化发展程度的数量指标,一般用一定地域内城市人口占总人口的比例来表示。

$$城市化水平 = 城镇人口数/农村居民人口数。$$

(14)劳动资源率(lb):劳动资源率即劳动力资源总数占总人口比例。劳动力的定义分广义和狭义,广义上的劳动力指全部人口,狭义上的劳动力则指具有劳动能力的人口。本文取统计年鉴中 15 岁至 65 岁的人口为广义劳动力人口。

$$劳动资源率 = 广义劳动力人口/总人口数。$$

(15)恩格尔系数(eng):恩格尔系数是根据恩格尔定律得出的比例数,是表示生活水平高低的一个指标,主要表述的是食品支出占总消费支出的比例随收入变化而变化的一定趋势,其计算公式为:

$$恩格尔系数 = 食品支出总品/家庭或个人消费额 \times 100\%$$

其中恩格尔系数越大,代表该地区或家庭生活越贫困,越小则反之。

(16)居民消费水平(com):居民消费水平是指居民在物质产品和劳务的消费过程中,对满足人们生存、发展和享受需要方面所达到的程度。通过消费的物质产品和劳务的数量和质量反映出来。

(17)人力资本($huma$):人力资本是指人们花费在教育、健康、训练、迁移和信息获取等方面的开支所形成的资本,之所以称为人力资本是因为无法将其同它的载体分离开。目前,关于人力资本的测量有四种方法:一是以人们已接受学校的教育年限来分类确定;二是用平均每万人在校中学生人数来衡量(Barro, Lee, 1993);三是用入学率作为替代变量来衡量人力资本的教育程度(Barro,Lee,1996;Barro,1997,2001);四是用教育经费占 GDP 或财政支出的比重来衡量(沈利生,朱运法,1997)。

目前学者们普遍使用人均受教育年数来度量,因此,文章对于人力资本的测算,通过考虑小学、中学、大学等不同层次的教育对劳动者素质的提高起到的作用,即平均受教育年数来代表影响能源使用效率的人力资本,其计算公式为:$H=6d_1+9d_2+12d_3+16d_4$,其中 $d_i(i=1,2,3,4)$ 表示在 6 岁及 6 岁以上人口中文化程度是小学、初中、高中及高中以上人口所占的比重。

(18)城乡收入差距(cxc):城乡收入差距指的是我国城镇居民收入与农村居民收入差值。其计算公式为:$cxc=Ic-In$,其中,Ic 代表城市居民收入水平,In 代表农村居民收入水平。

城乡收入差距越来越大,使得大量农村居民外出务工,在一定程度上导致就业人数增加。城乡收入差距变化的背后原因是,城市经济部门对劳动力需求的增加速度与农村劳动力供给的速度的不对称现象,并且当前者大于后者时即出现了所谓的"刘易斯转折点"。

12.5.2　指标体系模型的建立

下面我们将上述 18 个影响就业的指标分成四大类,宏观经济(国内生产总值、固定资产投资、净出口、市场化指数、产业结构调整水平),调控政策(财政收入、财政支出、税收、货币供给量、科研支出率),人口状况(就业人数、人口自然增长率、城市化水平、劳动资源率、人力资本),生活水平(城乡收入差距、恩格尔系数、居民消费水平)。为了便于直观理解,我们将以上 18 个影响就业指标体系结合分类,构建出影响就业指标体系的模型(如图 12.1)。

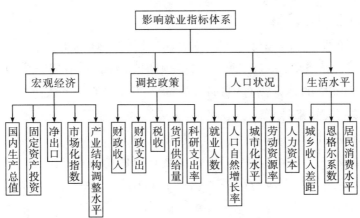

图 12.1　影响就业指标体系的模型

12.5.2 初始数据提取与分析

1. 就业问题研究阶段的划分

以下利用 EXCEL 软件对我国历年来城镇就业人数作折线图(见图 12.2),从折线图可以发现:图形大致呈折线形变化,其中城镇就业人数在 1989 年至 1990 年发生较大幅度的浮动。因此,就模型的准确性与合理性方面考虑,本文可将我国城镇就业人数的增长模型分两个阶段构建:即第一阶段 1978～1989 年,我国改革开放初期阶段;第二阶段 1990～2007 年,是我国在经济上进一步深化改革带来就业大幅增加的阶段。

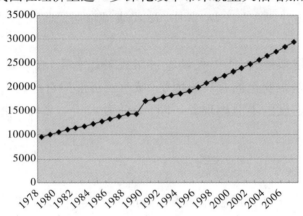

图 12.2　全国城镇就业人数折线图

2. 数据的补充及算法

由于对数据来源的可获得性以及统计资料的完备性均受到限制,甚至某些指标出现少许缺失值,出于对模型准确与合理方面的考虑,我们以下利用 Newton 插值法[2]对原始数据进行插值,补全缺失数据。

(1)牛顿插值法基本思路:

给定插值点序列$(x_i, f(x_i))$, $i=0,1,\cdots,n$。构造牛顿插值多项式 $N_n(u)$。输入要计算的函数点 x,并计算 $N_n(x)$ 的值,利用牛顿插值公式,当增加一个节点时,只需在后面多计算一项,而前面的计算仍有用;另一方面 $N_n(x)$ 的各项系数恰好又是各阶差商,而各阶差商可用差商公式来计算。

(2)牛顿插值法计算步骤:

①输入 n 值及$(x_i, f(x_i))$, $i=0,1,\cdots,n$;要计算的函数点 x。

②对给定的 x,由 $N_n(x) = f(x_0) + (x-x_0)f[x_0,x_1] + (x-x_0)(x-x_1)f[x_0,x_1,x_2] + \cdots + (x-x_0)(x-x_1)\cdots(x-x_{n-1})f[x_0,x_1,\cdots,x_n]$ 计算 $N_n(x)$ 的值。

③输出 $N_n(x)$。

3. 实际价格的计算

为了消除我们收集的原始数据中价格因素的影响,对于全国统计数据,我们利用全国居民消费价格指数对原始数据进行价格平减,即用实际价格除以以 1978 年为基期的居民消费价格指数来近似代替。对于各省份之间的价格指数计算我们利用各省份的居民价格指数进行平减。

12.6 模型的建立与求解

12.6.1 问题1:指标排序模型

1. 建模思路

问题一要求对有关统计数据进行分析,寻找影响就业的主要因素或指标,我们接下来对上一部分所构建的18个指标分别与就业人数进行关联性分析,我们分别运用灰色关联度与相关系数两种方法进行分析,然后将它们综合成排序指标模型,按所得值大小来选取影响就业的主要指标。

2. 理论准备

(1)灰色关联度[3]。

灰色系统理论用灰色关联度顺序来描述因素间关系的强弱、大小、次序。其基本思想是:以因素的数据列为依据,用数学的方法研究因素间的几何对应关系。GRA实际上是动态指标的量化分析,充分体现了动态意义。

$$\gamma_{关联度} = \frac{\min\limits_{i,k}|x_{0(k)}{'}-x_{i(k)}{'}|+\xi\cdot\max\limits_{i,k}|x_{0(k)}{'}-x_{i(k)}{'}|}{\Delta_{0ij}+\xi\cdot\max\limits_{i,k}|x_{0(k)}{'}-x_{i(k)}{'}|}$$

若关联度 $\gamma_{关联度}$ 最大,则说明 $x_i(k)$ 与最优指标 $x_0(k)$ 最接近,即第 i 个被评价对象优于其他被评价对象,据此可以排出各被评价对象的优劣次序。

(2)相关系数。

相关系数是描述变量之间相关程度的指标,用 $\gamma_{相关系数}$ 表示,取值范围为 $[-1,1]$。$|\gamma_{相关系数}|$ 值越大,误差 Q 越小,变量之间的线性相关程度越高;$|\gamma_{相关系数}|$ 值越接近 0,Q 越大,变量之间的线性相关程度越低。

$$\gamma_{相关系数} = \frac{\sum\limits_{i=1}^{n}(x_i-\bar{x})(y_i-\bar{y})}{\sqrt{\sum\limits_{i=1}^{n}(x_i-\bar{x})^2\cdot\sum\limits_{i=1}^{n}(y_i-\bar{y})^2}}$$

3. 模型建立

灰色系统理论的关联度分析与数理统计学的相关系数是不同的,两者的区别主要在于以下三个方面。

(1)它们的理论基础不同。前者是基于灰色系统的灰色过程,而后者则是基于概率论的随机过程。

(2)分析法不同。前者是进行因素间时间序列的比较,而后者是因素间数组的比较。

(3)数据量要求不同。前者不要求数据太多,而后者则需有足够的数据。

因此，为了更加全面地反映各项指标与就业人数之间的相关性，本文综合考虑灰色关联 $\gamma_{\text{关联度}}$ 及相关系数 $\gamma_{\text{相关系数}}$ 这两方面，构建排序指标模型如下：

$$\lambda = 1/2 \times (\gamma_{\text{关联度}} + \gamma_{\text{相关系数}})。$$

可根据此排序指标值的大小对各个指标进行排序，依此选取影响就业的主要指标。

4. 模型求解

根据对初始就业人数数据的分析，本文认为对该问题的分析应分为两个阶段：第一阶段为 1978—1989 年，第二阶段为 1990—2007 年。分别对这两个阶段影响我国就业人数的指标进行排序，具体结果如表 12.1 和表 12.2 所示。

表 12.1 1978—1989 年影响就业人数因素的 r 值

名次	1	2	3	4	5	6	7	8	9
指标	M2	csh	czz	GDP	com	K	pr	inc	tax
r 值	0.85	0.84	0.83	0.82	0.80	0.80	0.79	0.74	0.73
名次	10	11	12	13	14	15	16	17	18
指标	lb	czs	str	zr	eng	ne	mar	kzf	huma
r 值	0.69	0.67	0.62	0.61	0.50	0.40	0.31	0.30	0.23

表 12.2 1990—2007 年影响就业人数因素的 r 值

名次	1	2	3	4	5	6	7	8	9
指标	K	csh	inc	com	M2	GDP	pr	czz	czs
r 值	0.86	0.83	0.83	0.82	0.82	0.81	0.80	0.78	0.76
名次	10	11	12	13	14	15	16	17	18
指标	huma	tax	str	mar	ne	zr	eng	lb	kzf
r 值	0.75	0.74	0.71	0.69	0.60	0.60	0.59	0.55	0.52

5. 指标选取

根据表 12.1 及表 12.2 我们可以清楚地观察到，不同时期不同指标对就业人数的关联性大小及排序，可以发现各指标与就业人数都有一定的关联性。然而对主要指标的选取原则，却具有一定程度的主观性。此处，结合数据分析所受到的限制及建模结果，本文按照 $\gamma \geqslant 0.81$ 的原则对主要因素或指标进行提取。1978—1989 年，影响就业人数的有货币供给量 M2、城市化水平、财政支出、国内生产总值四个指标。而在 1990—2007 年这个阶段，影响就业人数的有固定资产投资总额、城市化水平、城乡收入差距、居民消费水平、货币供给量 M2、国内生产总值这六个指标。

12.6.2 问题 2：城镇就业人数拟合模型

1. 建模思路

从上面的分析得知，各个阶段影响城镇就业人数的指标是不同的，可以运用多元线性回归方法衡量这些因素对我国城镇就业人数影响的大小。为了避免回归时产生多重共线性，我们对不同阶段所选取出来的影响指标提取主成分，利用主成分进行回归，同时，用各阶段影响指标建立 VAR 模型，通过观察其脉冲响应函数及方差贡献率图研究各阶段的影响指标对我国城镇就业人数变动的长期影响及短期影响。

2. 理论准备

(1) 主成分回归分析[5]。

主成分回归分析是将回归模型中有严重多重共线性的变量进行因子提取,得到正交的因子变量,然后对因子变量进行回归模型的建立,从而可以解决由于共线性而造成伪回归方程的问题。其中主成分是一种通过降维技术把多个指标约化为少数的几个综合指标,这些称为主成分的综合指标能够反映出原始指标的绝大部分信息。

(2) 向量自回归分析。

向量自回归(VAR)是基于数据的统计性质建立模型,VAR 模型把系统中每一个内生变量作为系统中所有内生变量的滞后值的函数来构造模型,从而将单变量自回归模型推广到由多元时间序列变量组成的"向量"自回归模型。脉冲响应函数描述的是 VAR 模型中的一个内生变量的冲击给其他内生变量所带来的影响,它是指是随着时间的推移,观察模型中的各变量对于冲击是如何反应的,然而对于只是要简单地说明变量间的影响关系又稍稍过细了一些。方差分解是通过分析每一个结构冲击对内生变量变化(通常用方差来度量)的贡献度,进一步评价不同结构冲击的重要性。因此,方差分解给出对 VAR 模型中的变量产生影响的每个随机扰动的相对重要性的信息。也就是说相对方差贡献率是根据第 j 个变量基于冲击的方差对 y_i 的方差的相对贡献度来观测第 j 个变量对第 i 个变量的影响。

3. 变量检验

建立 VAR 模型的前提条件是各个变量同阶单整,在建立 VAR 模型前,必须对各个变量进行平稳性检验。检验结果如表 12.3、表 12.4。

表 12.3　相关变量的平稳性检验(1978—1989)

变量名	检验值	临界值	滞后期数	显著性水平
$\Delta \ln p$	−2.852	−2.748	0.000	0.100
$\Delta^2 \ln GDP$	−1.951	−1.600	0.000	0.100
$\Delta \ln czz$	−4.095	−3.321	2.000	0.010
$\Delta \ln M2$	−3.690	−3.213	0.000	0.050
$\Delta \ln csh$	−7.798	−5.835	1.000	0.010

表 12.4　相关变量的平稳性检验(1990—2007)

变量名	检验值	临界值	滞后期数	显著性水平
$\Delta^2 \ln P$	−5.622	−3.959	2.000	0.010
$\Delta \ln GDP$	−5.584	−4.800	2.000	0.010
$\Delta \ln K$	−2.277	−1.966	1.000	0.050
$\Delta \ln M2$	−4.566	−3.920	0.000	0.010
$\Delta^2 \ln csh$	−3.806	−3.081	0.000	0.050
$\Delta \ln cxc$	−2.945	−2.681	0.000	0.100
$\Delta \ln com$	−4.078	−3.920	0.000	0.010

由表 12.3、表 12.4 可知,在 1978—1989 年间,$\Delta \ln P$、$\Delta^2 \ln GDP$、$\Delta \ln czz$、$\Delta \ln M2$、$\Delta \ln csh$ 是平稳的;在 1990—2007 年这一阶段,$\Delta^2 \ln P$、$\Delta \ln GDP$、$\Delta \ln K$、$\Delta^2 \ln csh$、$\Delta \ln cxc$、$\Delta \ln M2$、$\Delta \ln com$ 是平稳的。

4. 模型的建立与求解

(1)主成分回归模型。

①1978—1989 年阶段。对于 1978—1989 年这一阶段,我们利用已提取出的四个因素:国内生产总值 GDP,财政支出 czz,广义货币供给量 M2,城市化水平 csh,对就业人数进行拟合。若直接利用最小二乘法对数据进行拟合,容易产生多重共线性等问题,为了避免这种情况的发生,我们首先对各个指标提取主成分:第一个主成分的方差贡献率为 97.844%,$\lambda_1 = 3.915$,这里我们提取一个主成分,表达式为:

$$F = 0.4938 \times \ln GDP + 04933 \times \ln czz + 0.5003 \times \ln M2 + 0.5024 \times \ln csh$$

这里得到一个关于主成分的时间序列,用这个序列对我国 1978—1989 年城镇就业人数进行拟合,结果为:$\ln P_{78} = 8.5518 + 0.1636 \times F$,其中 $R^2 = 0.9885$ 拟合效果较好,且综合指标的系数显著。

因此我们可以进一步求出就业人数拟合模型:

$$\ln P_{78} = 8.5518 + 0.0808 \times \ln GDP + 0.0808 \times \ln czz + 0.0819 \times \ln M2 + 0.0822 \times \ln csh$$

②1990—2007 年阶段。

对于 1990—2007 年这一阶段,根据前面所说的指标选取原则,这里,我们主要选取国内生产总值 GDP、固定资产投资 K、广义货币供给量 M2、城市化水平 csh、城乡收入差距 cxc 以及居民消费水平 com,就业人数记为 P,为了便于分析,我们先对各项指标取对数,分别记为 $\ln GDP$、$\ln czz$、$\ln M2$、$\ln csh$、$\ln cxc$、$\ln com$、$\ln p$。

首先来分析我们选取出来的此阶段的各项指标与就业人数的计量模型。由于选取的是六个指标,运用最小二乘法对各个指标进行拟合会存在多重共线性的问题,为了避免这种情况的发生,我们首先运用主成分分析法提取这六个指标的主成分,然后用所提取出来的主成分对我国城镇就业人数进行回归,然后再展开,可以得到各个指标的系数情况。

主成分回归的相关结果:此时累计方差贡献率大于 85%,特征值 $\lambda = 2.07494$ 选取一个主成分,表达式为:$F = 0.688 \times \ln GDP + 0.684 \times \ln K + 0.690 \times \ln M2 + 0.679 \times \ln csh + 0.683 \times \ln cxc + 0.686 \times \ln com$,这里得到一个关于主成分的时间序列,用这个序列对我国 1990—2007 年城镇就业人数进行拟合,结果为:$\ln P_{90} = 10.3305 + 0.0253 \times F$,其中 $R^2 = 0.9899$ 拟合效果较好,且综合指标的系数显著。

因此我们可以进一步求出城镇就业人数拟合模型:

$$\ln P_{90} = 10.331 + 0.0174 \times \ln GDP + 0.0171 \times \ln czz$$
$$+ 0.018 \times \ln M2 + 0.017 \times \ln csh + 0.017 \times \ln cxc + 0.017 \times \ln com$$

(2)向量自回归模型。

①1978—1989 年阶段。

下面我们利用 1978—1990 年的主要影响指标变动对我国城镇就业人数变动的影响建立向量自回归模型。根据表 12.3 的平稳性检验结果，$\Delta \ln p$，$\Delta^2 \ln \mathrm{GDP}$，$\Delta \ln czz$，$\Delta \ln M2$，$\Delta \ln csh$ 是平稳的，可以建立 VAR 模型。

$$\begin{pmatrix} \Delta \ln p_t \\ \Delta^2 \ln \mathrm{GDP}_t \\ \Delta \ln cee_t \\ \Delta \ln M2_t \\ \Delta \ln csh_t \end{pmatrix} = \begin{pmatrix} 3.3417 \\ 16.6871 \\ -47.7658 \\ -35.3568 \\ -6.8870 \end{pmatrix} + \begin{pmatrix} 0.0115 & -0.0167 & 0.0229 & 0.2536 & 0.6193 \\ 0.7052 & -0.1774 & 0.3928 & 0.0801 & -1.7377 \\ 1.6617 & 0.5889 & 0.0905 & -4.2356 & 5.8674 \\ 1.3537 & -0.1541 & 0.2413 & -0.0706 & 3.9683 \\ 0.2120 & 0.2197 & 0.1238 & 0.2563 & 0.9512 \end{pmatrix} \times$$

$$\begin{pmatrix} \Delta \ln p_{t-1} \\ \Delta^2 \ln \mathrm{GDP}_{t-1} \\ \Delta \ln cee_{t-1} \\ \Delta \ln M2_{t-1} \\ \Delta \ln csh_{t-1} \end{pmatrix} + \begin{pmatrix} \varepsilon_{1t} \\ \varepsilon_{2t} \\ \varepsilon_{3t} \\ \varepsilon_{4t} \\ \varepsilon_{5t} \end{pmatrix}$$

三个方程的拟合优度分别是 $R_P^2 = 0.998$，$R_{\mathrm{GDP}}^2 = 0.785$，$R_{cee}^2 = 0.979$，$R_{M2}^2 = 0.979$，$R_{csh}^2 = 0.996$，由此可以看出，拟合的效果比较好。

②1990—2007 年阶段。

而对于 1978—1990 年的主要影响指标变动对我国城镇就业人数变动的影响，建立向量自回归模型。从表 12.4 可以看出，$\Delta^2 \ln P$、$\Delta \ln \mathrm{GDP}$、$\Delta \ln K$、$\Delta^2 \ln csh$、$\Delta \ln cxc$、$\Delta \ln M2$、$\Delta \ln com$ 是平稳的，可以建立 VAR 模型。

$$\begin{pmatrix} \Delta^2 \ln p_t \\ \Delta^2 \ln csh_t \\ \Delta \ln com_t \\ \Delta \ln cxc_t \\ \Delta \ln \mathrm{GDP}_t \\ \Delta \ln K_t \\ \Delta \ln M2_t \end{pmatrix} = \begin{pmatrix} -0.6912 & 0.0592 & -0.0094 & -0.0183 & 0.021 & 0.0011 & 0.0127 \\ 0.1016 & -0.3686 & -0.0147 & -0.0781 & -0.010 & 0.0294 & 0.1201 \\ 2.3825 & -2.629 & 0.03117 & -0.3953 & 0.4783 & 0.037 & -0.1028 \\ 12.5463 & -2.1424 & 0.3287 & 0.6705 & -0.0796 & -0.0572 & 0.1025 \\ 0.2454 & -0.258 & 0.1290 & 0.102 & 0.5097 & -0.048 & 0.1885 \\ -0.1379 & -0.0369 & 0.0252 & -0.0117 & 0.0561 & 0.6554 & 0.0221 \\ -0.2666 & -0.7129 & 0.0415 & -0.0481 & -102246 & 0.1227 & -0.5189 \end{pmatrix} \times$$

$$\begin{pmatrix} \Delta^2 \ln p_{t-1} \\ \Delta^2 \ln csh_{t-1} \\ \Delta \ln com_{t-1} \\ \Delta \ln cxc_{t-1} \\ \Delta \ln \mathrm{GDP}_{t-1} \\ \Delta \ln K_{t-1} \\ \Delta \ln M2_{t-1} \end{pmatrix} + \begin{pmatrix} \varepsilon_{t1} \\ \varepsilon_{t2} \\ \varepsilon_{t3} \\ \varepsilon_{t4} \\ \varepsilon_{t5} \\ \varepsilon_{t6} \\ \varepsilon_{t7} \end{pmatrix}$$

5. 模型结果的分析

(1) 主成分回归的结果。

由1978—1989年的主成分回归模型可以看出，国内生产总值、财政支出、广义货币供给量、城市化水平对我国城镇就业人数的弹性大小相差不大；同样，对于1990—2007年的主成分回归模型，国内生产总值、城市化水平、固定资产投资、广义货币供给量或城乡收入差距以及居民消费水平都变动1%，相比之下，1978—1989年的弹性系数大一些，国内生产总值、财政支出、货币供给量或城市化水平变动1%，将会引起城镇就业人数约0.081%的变动。

(2) 向量自回归模型的结果分析。

①1978—1989年阶段。

观察1978—1989年这一阶段每一指数对我国城镇就业人数的脉冲响应函数图形以及方差分析图。

[GDP对就业的影响]

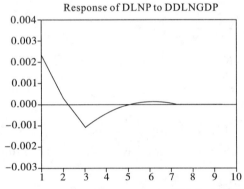
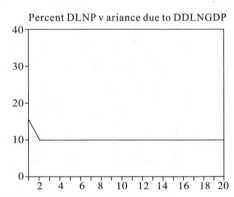

图12.3　GDP对就业人数的脉冲响应函数图形及方差分析图

图12.3左图是GDP增长率变动冲击引起的我国城镇就业人数的增长率变化的脉冲响应函数。从图中可以看出，当在本期GDP增长率变动受到一个冲击后，在前两期对我国城镇就业人数增长有正向作用，但其滞后作用力度呈下降趋势，在第二年达到最小，短暂恢复到原有水平，在第五年左右恢复到原有的均衡水平，此时达到稳定。这说明，给GDP增长变动率一个冲击，可以引起我国城镇就业人数增长率增加的滞后期为两年，在第二年末，正向冲击很小，几乎为零。从方差贡献率图中也可以看到这种情况。不考虑就业人数本身的贡献率，GDP增长速度对城镇就业人数增长的贡献在当期最大，达15%，之后呈下降趋势，两期后稳定在10%左右。

[城市化水平对就业的影响]

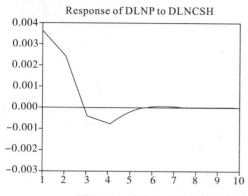
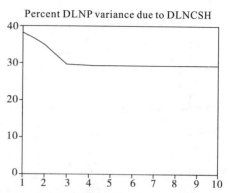

图 12.4　城市化水平对就业人数的脉冲响应函数图形及方差分析图

图 12.4 左图是城市化的变化冲击对我国城镇就业人数增长率的影响，从图中可以看出，本期给城市化一个正冲击后，前两期对我国城镇就业人数的增加有正向冲击，但逐渐减少，受到冲击后我国城镇就业人数增长率在五期左右恢复到原来水平。从城市化对就业人数增加的方差贡献率来看，当期最高，达 38%，之后虽有下降，但下降幅度并不大，且在三期后达到均衡，维持在 30% 的水平。

[货币供给量对就业的影响]

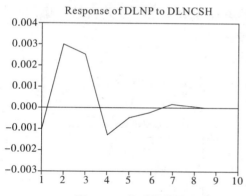
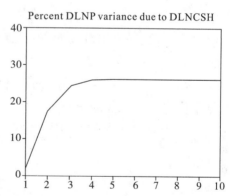

图 12.5　货币供给量对就业人数的脉冲响应函数图形及方差分析图

从图 12.5 左边的脉冲响应函数图中可以看出，给货币供给量增长率一个冲击后，当期对就业的影响不大，其作用在两期后达到最大，这一点从货币供给量对我国城镇就业人数变动的方差贡献率中也可以看出，当期的方差贡献率小于 5%。但其滞后效应比较大，可以引起我国城镇就业人数增长率增加的滞后期为未来三年且增长幅度比较大。从其方差贡献率中可以看出，货币供应量对城镇就业人数的贡献率是呈指数函数形式递增的，在第三期达到稳定状态，约为 27%。

[财政支出对就业的影响]

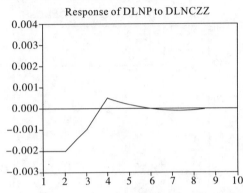
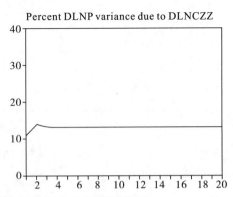

图12.6 财政支出对就业人数的脉冲响应函数图形及方差分析图

从图12.6的左边财政支出增长率对城镇就业人数增长率的脉冲图形中看到,当期给财政支出增长率一个反向冲击后,两期之后通过各路径逐步反映到就业市场,同时,它对城镇就业人数变动的贡献率波动不大,在13%左右。

从以上图中可以看出,当期给各个影响指标一个冲击,其对城镇就业人数的滞后效应一般在三期左右,影响波动最大的是城市化指数。不考虑城镇就业人数自身的贡献率,各个指标对城镇就业人数变化的贡献率情况是:GDP的贡献率最小,约为10%,其次是财政支出,城市化指数与货币供给量对城镇就业人数变动的方差贡献率稳定时相差不大,但总体上说,城市化指数的方差贡献率要大些。

② 1990—2007年阶段。

观察1990—2007年这一阶段的脉冲响应函数图形及方差分解图。

[城市化水平对就业的影响]

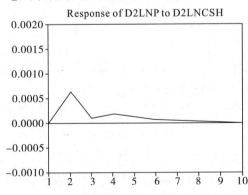
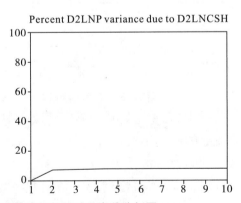

图12.7 城市化水平对就业人数的脉冲响应函数图形及方差分析图

图12.7的左边是对城市化的冲击引起的城镇就业人数增加率变化的脉冲响应函数图。从图中我们可以看出,当在本期给城市化一个正冲击后,城镇就业人数在第2期达到最高点;从第4期以后开始趋于稳定。这表明城市化受外部条件的某一冲击后,经过市场传递给就业人数,给就业带来同向的冲击,冲击效应在第2期时达到最大,即城市化的正向冲击对就业具有显著的促进作用,并且这一显著促进作用具有较长的持续效应。

[居民消费水平对就业的影响]

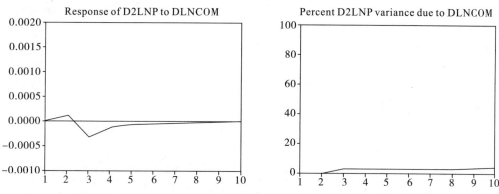

图 12.8　居民消费水平对就业人数的脉冲响应函数图形及方差分析图

从图 12.8 的左边的脉冲响应曲线可以看到,当期居民消费水平的变动受到一个单位的冲击后,对就业的作用不是很明显,而且滞后影响持续时间不长,且对就业人数变动的方差贡献率很低。

[GDP 对就业的影响]

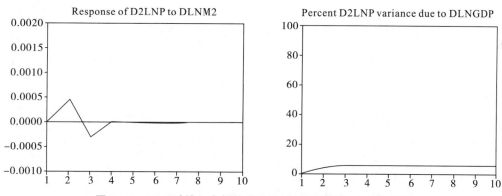

图 12.9　GDP 对就业人数的脉冲响应函数图形及方差分析图

从图 12.9 的左边的脉冲响应图形可以看出,国内生产总值变动率当期受到一个冲击后,对我国就业人数变动的滞后影响在前两期其作用力度呈上升趋势,在第二年达到最大,然后滞后影响开始下降,经过两期恢复到原有水平。从方差贡献率来看,国内生产总值变动率最多能解释大约 8% 的方差。

[货币供给量对就业的影响]

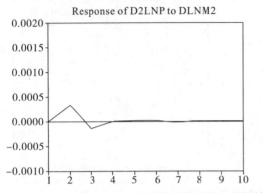
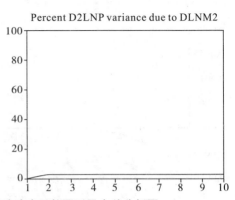

图12.10　货币供给量对就业人数的脉冲响应函数图形及方差分析图

结合图12.8、图12.9、图12.10，可以看出这三者的脉冲响应函数图具有一定的相似性。从图形上看，三者对影响就业人数变动具有相似的作用模式，滞后期不长，且波动趋势相似，其中影响幅度波动最大的是国民生产总值，其中货币供给量的方差贡献率最小。

[城乡收入差距对就业的影响]

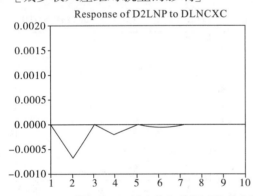
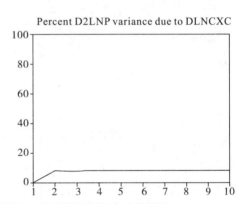

图12.11　城乡收入差距对就业人数的脉冲响应函数图形及方差分析图

根据图12.11左图，当期城乡收入差距受到一个单位的正向冲击后，在一定时间内对城镇就业人数的增加有阻碍作用，这种影响将持续大约5期，其方差贡献率在2期后维持在10%左右的水平。

[固定资产投资总额对就业的影响]

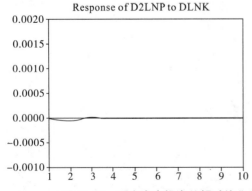
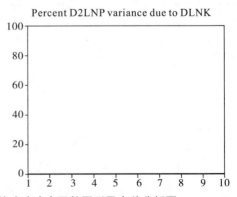

图12.12　固定资产投资总额对就业人数的脉冲响应函数图形及方差分析图

从图 12.12 中可以看出,固定资产投资总额的变动受到一个冲击后,滞后作用不明显,且其方差贡献率很低,几乎为零。

从上面的几个图中我们可以看出,不考虑城镇就业人数自身的贡献率,城乡收入差距及城市化对就业人数的贡献率较大,在 8% 左右,其次是 GDP,固定资产对城镇就业人数的贡献率最小。从中还可以看出以下几个方面的特点:居民消费虽然对城镇就业人数的贡献较小,但它对城镇就业的拉动作用将逐渐加强,这一判断来自于居民消费结构、消费观念正发生着相当大的变化。对家庭而言,消费结构的变化可能会引起以下两种后果:第一,家庭负债额可能增大;第二,家庭将寻找增加收入的门路,如兼职、加班、主妇就业等,这两种后果都会在一定程度上使城镇就业人数上升。

综合两个阶段的特点,我们发现,城市化水平在 1978—1989 年及 1990—2007 年这两个阶段对我国城镇就业人数的变化有较大影响,究其原因,城市化水平的提高将急剧扩大消费需求,有力地拉动经济的快速增长,扩张城镇就业总量。城市化使城市人口规模不断地扩大,同时也会使生活消费和生产消费的规模不断扩大,这使许多行业能够获取一定规模的收益,而且也将催生一些新的产业和行业。城市化促进社会分工和职业细化,从而广开就业门路,扩张城镇就业总量,同时社会分工的深化和职业的细化,有力地带动了各行各业的产生和发展,从而提供大量的就业机会。此外,城市化程度的进一步提高还可以提高就业质量、优化产业结构、增强就业弹性、完善就业体系等,这些也间接促进了城镇就业。

12.6.3 问题 3:具体分类就业人口模型

1. 分区域的城镇就业人口模型

(1)建模思路。

本文参考上面的城镇就业人口估计模型,并由于受到数据的限制,选取各个省份影响城镇就业人数的固定资产投资总额、城市化水平、城乡收入差距、居民消费水平、货币供给量 M2、国内生产总值这六个指标,进行面板数据分析,进而从中观察不同区域,构建分区域的就业人口模型。

(2)建模准备。

①我国区域的划分。

对于我国区域的划分,本文按照国家统计局的划分方法,我国 31 个省、直辖市、自治区(香港、澳门、台湾除外)从地理位置上可分为东、中、西部三个经济带,其中东部经济带包括北京、上海、天津、辽宁、河北、山东、江苏、浙江、福建、广东、海南、广西 12 个省、市、自治区;中部经济带包括黑龙江、吉林、山西、安徽、江西、河南、湖北、湖南、内蒙古 9 个省、市、自治区;西部经济带包括新疆、四川、重庆、西藏、云南、青海、甘肃、宁夏、陕西、贵州 10 个省、市、自治区。而西部大开发战略中所定义的西部地区包括内蒙古、广西、宁夏、西藏、新疆 5 个少数民族自治区和陕西、甘肃、青海、云南、贵州、四川、重庆 7 个省区。

并且为了便于分析,我们将重庆数据一概并入四川来进行计算。

②指标数据的选取。

由于受到数据的限制,本文主要对1995—2007年我国的省际面板数据进行分析,并且这段时间属于我国就业人口变化1990—2007的这个阶段。因此,本文认为在这个阶段的基础上,文章应选取各个省份固定资产投资总额、城市化水平、城乡收入差距、居民消费水平、货币供给量M2、国内生产总值这六个指标,来对城镇就业人口进行分析。

(3)模型的建立与求解。

①模型的初步建立与分析。

要根据我国省际面板数据来构建精确的分区域就业人口模型。我们先根据固定资产投资总额、城市化水平、城乡收入差距、居民消费水平、货币供给量M2、国内生产总值这六个指标,对我国东、中、西三大区域进行初步的估算,进而分析各变量的影响程度。此处我们采用固定效应来对面板数据进行分析,三大区域的初步估算模型如表12.5所示。

表12.5 三大区域面板数据的初步回归建立结果

	东部	中部	西部
M2	0.3412*	0.0751*	0.2725*
	(7.8514)	(2.1323)	(8.2583)
czz	−0.0578	−0.2556*	0.0600
	(−1.4576)	(−3.7122)	(1.5247)
com	−0.8868*	−0.3384*	−1.6073*
	(−8.3712)	(−2.2251)	(−7.2795)
csh	−0.2509*	−0.8774*	−0.6758*
	(−2.4797)	(−10.3981)	(−7.4980)
cxc	−0.0797	0.0422	−0.1960*
	(−0.8073)	(0.4130)	(−1.7159)
K	0.7258*	0.3775*	0.8452*
	(12.0981)	(5.8208)	(14.9040)
常数	7.9146*	9.3568*	11.9288*
	(12.4248)	(10.7923)	(12.5449)
$Adj\text{-}R^2$	0.9121	0.7933	0.9146
N	12	9	9

注:表中括号内的数字为t检验值。*代表显著性水平为10%

②模型的最终建立与求解。

根据表12.5中三大区域城镇就业人口模型的初步估算,我们剔除不显著的变量,进而可以得到更加精确的估计模型,如表12.6所示。

表 12.6 三大区域面板数据的最终回归建立结果

	东部	中部	西部
ln$M2$	0.3825*	0.0712*	0.2895*
	(9.4504)	(2.1075)	(7.2012)
lnczz		−0.2409*	
		(−4.1017)	
lncom	−0.9759*	−0.33824*	−1.5055*
	(−9.9181)	(−2.2052)	(−7.7505)
lncsh	−0.1672*	−0.8766*	−0.5949*
	(−1.7326)	(−10.4379)	(−4.4199)
lncxc			−0.2231*
			(−1.8406)
ln$K$	0.6893*	0.3841*	0.8549*
	(11.7709)	(6.1374)	(13.4679)
常数	7.3064*	9.4089*	12.2991*
	(12.2829)	(11.0178)	(11.0702)
$Adj\text{-}R^2$	0.9095	0.7951	0.9304
N	12	9	9

注:表中括号内的数字为 t 检验值。*代表显著性水平为 10%

其中具体分区域的城镇就业人口模型为:

东部地区:$\text{Ln}P = 7.3064 + 0.3825 \times \ln M2 - 0.9759 \times \ln com - 0.1672 \times \ln csh + 0.6893 \times \ln K$;

中部地区:$\text{Ln}P = 9.4089 + 0.0712 \times \ln M2 - 0.2409 \times \ln czz - 0.3382 \times \ln com - 0.8766 \times \ln csh$;

西部地区:$\text{Ln}P = 12.2991 + 0.2895 \times \ln M2 - 1.5055 \times \ln com - 0.5949 \times \ln csh - 0.2231 \times \ln inc + 0.8549 \times \ln K$。

2. 分行业的城镇就业人口模型

(1)建模思路。

对于分行业建立城镇就业人口模型,由于我国行业划分较多,且不同年份有关指标变量的统计口径不一致等原因,我们从国有经济行业、集体经济行业和其他经济行业三方面,并参照之前的有关指标来进行简要的分析,然后运用主成分回归的方法来构建分行业的城镇就业人口模型。

(2)建模准备。

本文着重从国有经济行业、集体经济行业和其他经济行业这三方面来进行分析,而对于影响所选取行业的指标,我们进行了以下划分。

对于国有经济行业,本文认为影响国有经济行业就业人数的因素主要有:国有经济行业职工工资 x_1、国有经济行业固定资产投资 x_2、广义货币供给量 M2、城市化水平 csh、城乡收入差距 cxc 以及居民消费水平 com。

对于集体经济行业,本文认为影响集体经济行业就业人数的因素主要有以下几种:集体经济行业职工工资 y_1、国有经济行业固定资产投资 y_2、广义货币供给量 M2、城市化水平 csh、城乡收入差距 cxc 以及居民消费水平 com。

对于其他经济行业,本文认为影响其他经济行业就业人数的因素主要有以下几种:其他经济行业职工工资 z_1、国有经济行业固定资产投资 z_2、广义货币供给量 M2、城市化水平 csh、城乡收入差距 cxc 以及居民消费水平 com。

而对于时间区间范围的考虑,本文根据指标数据的可得性以及准确性将其指定为1984—2007年。

(3)模型的建立与求解。

为了研究分行业对就业数量的影响,我先对各类企业人员的时间序列进行分析。

表12.7 历年来各企业就业人数

年份	1984	1985	1986	1987	1988	1989	1990	1991	1992	1993	1994	1995
国有	8637	8990	9333	9651	9984	10108	10346	10664	10889	10920	10890	10995
集体	811	967	1092	1207	1426	1557	1681	1866	2109	2592	3245	3931
其他	37	41	55	72	97	132	164	216	282	536	748	877
年份	1996	1997	1998	1999	2000	2001	2002	2003	2004	2005	2006	2007
国有	10949	10766	8809	8262	7715	7168	6621	6621	6438	6232	6170	6148
集体	4302	4512	5331	5774	6262	6867	7667	8678	9814	11283	13014	15595
其他	942	1085	1628	1951	2274	2597	2920	2920	3287	3849	4264	4595

注:数据来源于国家统计局网站

①国有经济行业就业人口模型。通过观察国有经济行业就业人数趋势图(见图12.13),我们以1995年为界把国有经济行业就业人口时期划分为1984—1995年和1996—2007年两个时期,再分别进行主成分回归分析,得出相应模型。

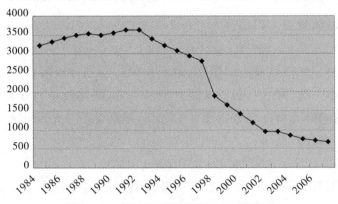

图12.13 国有经济行业就业人口趋势图

本文对国有经济行业选取的六个指标提取主成分:第一个主成分的方差贡献率为98.55%,$\lambda_1=5.913$,这里我们提取一个主成分,即为影响国有经济行业就业人口数量因素的综合值,表达式为:$F=0.416\times\ln x_1+0.404\times\ln x_2+0.410\times\ln x_3+0.407\times\ln x_4+$

$0.409 \times \ln x_5 + 0.409 \times \ln x_6$。

我们运用这个综合值序列分别对国有经济行业所划分不同阶段的就业人数进行拟合,其结果为:
$$\ln P_{84} = 7.8857 + 0.0815 \times F$$
$$\ln P_{95} = 13.4496 - 0.2338 \times F$$

其中 $R_{84}^2 = 0.869, R_{95}^2 = 0.878$ 拟合效果较好,且综合指标的系数显著。

因此我们可以进一步求出不同时期国有经济行业就业人口模型:
$$\ln P_{84} = 7.8857 + 0.0335 \times \ln x_1 + 0.0329 \times \ln x_2 + 0.0334 \times \ln x_3 + 0.0332 \times \ln x_4$$
$$+ 0.0334 \times \ln x_5 + 0.0333 \times \ln x_6$$
$$\ln P_{95} = 13.4496 + 0.096 \times \ln x_1 + 0.0945 \times \ln x_2 + 0.0959 \times \ln x_3 + 0.0951 \times \ln x_4$$
$$+ 0.0957 \times \ln x_5 + 0.0955 \times \ln x_6$$

②集体经济行业就业人口模型。

通过观察集体经济行业就业人数时序图,我们可以发现其大体上呈现出逐年递减的趋势。因此,我们可以直接进行主成分回归分析,并得出相应模型。

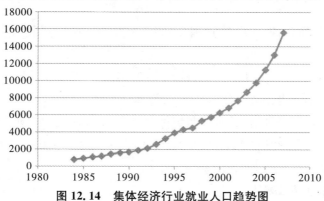

图 12.14 集体经济行业就业人口趋势图

本文对国有经济行业选取的六个指标提取主成分:第一个主成分的方差贡献率为 96.42%,$\lambda_1 = 5.785$,这里我们提取一个主成分,即为影响集体经济行业就业人口数量因素的综合值,表达式为:
$$F = -0.389 \times \ln x_1 + 0.413 \times \ln x_2 + 0.406 \times \ln x_3 + 414 \times \ln x_4 + 0.414 \times \ln x_5 + 0.413 \times \ln x_6$$

我们运用这个综合值序列对我国 1984—2007 年集体经济就业人数进行拟合,结果为:
$$\ln P = 13.0229 - 0.0516 \times F$$

其中 $R_{84}^2 = 0.803$,拟合效果较好,且综合指标的系数显著。

因此我们可以进一步求出不同时期集体经济行业就业人口模型如下:
$$\ln P = 13.0229 + 0.02 \times \ln x_1 - 0.02 \times \ln x_2 - 0.02 \times \ln x_3 - 0.021 \times \ln x_4$$
$$- 0.021 \times \ln x_5 - 0.021 \times \ln x_6$$

③其他经济行业就业人口模型。

通过观察其他经济行业就业人数时序图,我们可以发现其大体上呈现出逐年递增的趋势。因此，我们可以直接进行主成分回归分析，并得出相应模型。

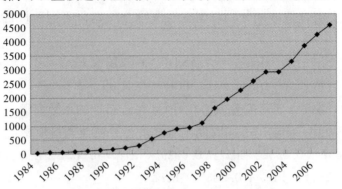

图 12.15 其他经济行业就业人口趋势图

本文对其他经济行业选取的六个指标提取主成分：第一个主成分的方差贡献率为 98.63%，$\lambda_1=5.918$，这里我们提取一个主成分，即为影响其他经济行业就业人口数量因素的综合值，表达式亦为：

$$F = -0.389 \times \ln x_1 + 0.413 \times \ln x_2 + 0.406 \times \ln x_3 + 414 \times \ln x_4 + 0.414 \times \ln x_5 + 0.413 \times \ln x_6$$

我们运用这个综合值序列对我国 1984—2007 年其他经济就业人数进行拟合，结果为：

$$\ln P = 7.8884 + 0.8137 \times F$$

其中 $R_{84}^2 = 0.96$，拟合效果较好，且综合指标的系数显著。

因此我们可以进一步求出不同时期集体经济行业就业人口模型如下：

$$\ln P = 7.8884 + 0.328 \times \ln x_1 + 0.331 \times \ln x_2 + 0.332 \times \ln x_3 + 0.329 \times \ln x_4 + 0.331 \times \ln x_5 + 0.331 \times \ln x_6 。$$

3. 分人群的就业人口模型

(1) 模型建立的思路。

对于分人群建立就业人口模型，由于我国就业人群有多种，且不同的分类有着不同的标准，现阶段对于就业的关注主要体现在农民工与大学毕业生这两大人群上。这里我们讨论影响这两大人群的就业情况的指标，我们将人群分为三类：农民工、大学毕业生及其他。

(2) 模型的准备。

根据产业就业互动理论，产业结构调整必然会使劳动力结构和技术结构出现一系列变化，促使劳动力就业产生新组合，由于农民工在很大程度上受产业结构的影响，分析影响农民工人数的因素就要考虑产业结构这一指标。选取 2001—2007 年外出务工的农民工人数，在之前所选取的主要影响因素的基础上引入产业结构、城乡收入差距两个指标，计算各项指标与我国就业人数的灰色关联度 $\gamma_{关联度}$ 及相关系数 $\gamma_{相关系数}$，根据灰色关联度及相关系数构建排序指标 γ，然后根据这个排序指标的大小对各个指标进行排序，依此选取影响我国就业人数的指标。

(3)模型的建立与求解。

①农民工就业模型。

根据排序结果,选取5个指标:净出口 lnnet、产业结构 str、城市化水平 csh、城乡收入差距 lncxc、居民消费 lncom 等。对选取的5个指标提取主成分:第一个主成分的方差贡献率为96.275%,且 $\lambda=4.184$,提取一个主成分,表达式为:

$$F = 0.439 \times \ln net_1 + 0.445 \times str + 0.448 \times csh + 0.449 \times lncxc + 0.454 \times lncom$$

运用这个综合值序列对我国2001—2007年农民工人数进行拟合,结果为:

$$\ln p_{农民工} = 10.899 + 0.01 \times F$$

其中 $R^2=0.9744$,拟合效果较好,且综合指标的系数显著。

因此我们可以进一步求出城镇就业人口模型如下。

$$\ln p_{农民工} = 10.899 + 0.0044 \times \ln net_1 + 0.0045 \times str + 0.0045 \times csh + 0.0045 \times lncxc + 0.0045 \times lncom$$

②大学毕业生就业人数模型。

根据排序结果,选取4个指标:净出口 lnnet、产业结构 str、城市化水平 csh、城乡收入差距 lncxc 等。

本文对影响大学毕业生就业人数的四个指标提取主成分:第一个主成分的方差贡献率为95.602%,$\lambda_1=3.842$,这里我们提取一个主成分,表达式为:

$$F = 1.878 \times \ln net + 1.917 \times str + 0.921 \times csh + 1.931 \times lncxc$$

我们运用这个综合值序列对我国2001—2007年大学毕业生就业人数进行拟合,结果为:

$$\ln P_{毕业生} = 1.525 + 0.0035 \times F$$

其中 $R^2=0.988$,拟合效果较好,且综合指标的系数显著。

因此我们可以进一步求大学毕业生就业人数模型如下。

$$\ln P_{毕业生} = 1.525 + 0.0066 \times \ln net + 0.0067 \times str + 0.0067 \times csh + 0.0068 \times lncxc$$

③其他人群就业人数模型。

根据排序结果,选取6个指标:国内生产总值 GDP、城乡收入差距 lncxc、居民消费 lncom、净出口 lnnet、广义货币供给量 M2 以及固定资产投资 K 等。对选取的6个指标提取主成分:第一个主成分的方差贡献率为85.768%,且 $\lambda=5.164$,提取一个主成分,表达式为:

$$F = 0.436 \times \ln GDP + 0.429 \times lncxc + 0.431 \times \ln M2 + 0.436 \times \ln K - 0.249 \times str + 0.428 \times \ln net$$

运用这个综合序列对其他人群就业人数进行拟合,得到:

$$\ln p_{其他} = 8.748 + 0.049 \times F$$

其中 $R^2=0.784$,拟合效果较好,且综合指标的系数显著。

因此我们可以进一步求其他类就业人数模型如下。

$$\ln P_{其他} = 8.748 + 0.021 \times \ln GDP + 0.021 \times \ln cxc + 0.021 \times \ln M2$$
$$+ 0.021 \times \ln K - 0.012 \times \ln str + 0.021 \times \ln net$$

12.6.4　问题4:就业前景仿真模型

1. 建模思路

为了科学合理地对未来城镇就业前景进行预测,减小预测模型的误差,本文利用问题二中已建立的数学模型预测结果与神经网络预测结果,得出城镇就业的综合预测值。第一种预测方法我们首先利用神经网络方法模拟仿真出不同模型中各影响因素的数值,将其代入我们得到的数学模型,得到预测结果,第二种方法我们直接利用神经网络得到仿真预测结果。最后我们进行细化,结合国家的有关决策和规划对我国不同行业的2009年及2010年上半年的就业前景进行仿真预测。

2. 理论的准备

BP网络[4],是1974年 *P. Werbos* 在其博士论文中提出第一个适合多层网络的学习算法,但当时该算法并未受到足够的重视被广泛地应用,直到20世纪80年代中期,美国加利福尼亚的PDP小组(Parallel Distributed Procession)于1986年发表了专著——*Parallel Distributed Processing*,将该算法应用于神经网络,才使之成为迄今为止最著名的多层网络学习算法——BP算法。由此算法训练的网络称为BP神经网络[6]。其网络拓扑结构见图12.16。

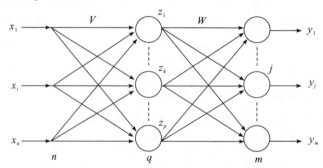

图12.16　具有单隐层的BP网络结构

设BP网络的输入层有 n 个节点,隐层有 q 个节点,输出层有 m 个节点,输入层与隐层之间的权值为 v_{kj},隐层与输出层之间的权值为 w_{jk},如图12.16所示。隐层的传递函数为 $f_1(\cdot)$,输出层的传递函数为 $f_2(\cdot)$,则隐层节点的输出为(将阈值写入求和项中):

$$z_k = f_1\left(\sum_{i=0}^{n} v_{ki} x_i\right) \quad k = 1, 2, \cdots, q$$

输出层节点的输出为:

$$y_i = f_2\left(\sum_{k=0}^{q} w_{jk} z_k\right) \quad j = 1, 2, \cdots, m$$

BP网络完成 n 维空间向量对 m 维空间向量的映射。

根据以上原理,可以将人工神经网络计算过程归纳为如下几步。

(1) 初始值选择 $w(0)$;
(2) 前向计算,求出所有神经元的输出: $a^k(t)$;
(3) 对输出层计算 $\delta: \delta_j = (t_j - a_j)a_j(1-a_j)$;
(4) 从后向前计算各隐层 $\delta: \delta_j = a_j(1-a_j)\sum_i w_{ji}\delta_i$;
(5) 计算并保存各权值修正量: $\Delta w_{ij} = -\eta\delta_j a_i$;
(6) 修正权值: $w_{ij}(t+1) = w_{ij}(t) + \Delta w_{ij}$;
(7) 判断是否收敛,如果收敛则结束,不收敛则转至步骤(2)。

3. 仿真模型的建立与求解

(1) 间接仿真法。

首先,我们利用神经网络方法预测出影响城镇就业人口的6个因素的数值,结果见表12.8。

表12.8 影响城镇就业人口的因素预测值

年份	GDP	czz	M2	csh	cxc	com
2008	55238.3	58400.89	102638.97	47.69	2010.97	1493.83
2009	61195.3	68847.27	118415.53	49.42	2194.05	1612.51
2010	67794.71	81162.25	136617.09	51.22	2393.8	1740.63

然后根据上述预测值,并结合问题二中得到的城镇就业人口模型:

$$\ln\hat{p} = 0.3305 + 0.0174 \times \ln GDP + 0.0173 \times \ln czz + 0.0175 \times \ln M2 \\ + 0.0172 \times \ln csh + 0.0173 \times \ln cxc + 0.0174 \times \ln com$$

得到城镇就业人口的预测值,见表12.9。

表12.9 城镇就业人口预测值(万人)

年份	2008	2009	2010
$P_{城}$	34525.72	34910.53	35299.57

(2) 直接仿真法。

利用神经网络方法对城镇就业人口进行预测,得到预测结果见表12.10。

表12.10 神经网络城镇就业人口预测值(万人)

年份	2008	2009	2010
$P_{城}$	30299.72	31336.73	32409.24

为了更加准确地对就业人口进行预测,减小预测误差,增加预测精确度,以下利用组合预测原理综合考虑上述两种预测方法,将两种方法的预测结果进行加权平均,得到组合预测结果见表12.11。

表12.11 组合预测城镇就业人口预测值(万人)

年份	2008	2009	2010
$P_{城}$	32412.72	33123.63	33854.41

12.6.5 问题5:相关建议方案

从本文的分析中可知,影响我国城镇就业人口数量的指标主要是城市化水平、城乡居民收入差距、居民消费、货币供给量、国内生产总值和固定资产投资额六项指标。以下我们针对这几方面逐一进行讨论。

1. 经济发展与就业

首先,根据上面问题分析得出的模型,GDP对就业人数变动的弹性是 0.0174,我们认为经济增长是解决就业的根本出路,努力实现促进经济增长与扩大就业的良性互动。

农民工与我国的产业结构与区域结构发展有关。在珠三角、长三角以及一些沿海城市,它们是农民工聚集的富裕城市。经济的发展促进了企业的成长,劳动密集型企业的成长吸纳了劳动力,劳动力的增加成就了城市的发展,在经济高速发展中农民工就业不是问题。正如深圳为代表的珠三角地区,在中国经济高速发展的时候,该地区的外向型企业劳动密集型企业中缺少的最多的就是劳动力。但当经济发展形势开始出现转变与放缓后,中国的农民工就业问题再次浮出水面,如 2008 年我国出现农民工提前返乡潮现象,主要就是受到国际金融危机对我国经济发展的冲击和影响。

从经济发展的角度出发,我们提出如下建议:经济危机发生之后,农民工的就业方向应该在党中央提出的中部崛起、西部大开发和现在的 4 万亿经济刺激方案下逐步向中西部欠发达地区转移。因为在内陆欠发达地区,GDP 增长 1 个百分点比沿海发达地区增长 1 个百分点所增加的劳动力要多 20% 左右,同时在内陆城市可以减少企业的用工成本和工业用地成本,还可以得到当地政府更多的支持。这些城市的企业吸纳周边的劳动力,带动当地的经济发展与建设,这样就逐步带动了现在中国经济中最困难的内需不足问题,同时在带动内需的时候又增加了就业的机会。

经济发展对大学生就业也有重大影响,经济的发展使企业需要更多的高度专业化和高度技术型人才,我们建议学校应该在了解、分析市场和行业后有计划地开设相应的课程。这既有利于学生的就业率,也有利于企业的技术进步和产能提升,以进一步扩大企业规模、增加就业。

2. 财政政策与就业

积极的财政政策对完善和实施积极就业政策具有重要意义,从我国国情看,财政支出政策在很大程度上缓解了就业压力,我们认为可以通过财政拨款兴修水利、建设基础设施,创造大量的就业机会;增加再就业培训、职业介绍服务等方面的财政支出,缩短工人寻找工作的时间,增强工人适应不同岗位的能力,等等。

3. 货币政策与就业

根据我国实际状况,当前最佳战略选择是发展中小企业来吸纳劳动力,而这就需要在制度上保证中小企业的经营环境,从而需要货币政策和就业政策共同发挥作用的最佳组合空间内寻求到较好的解决途径。我们认为实现低失业率不能单纯依靠货币供应量的增加和财政支出的扩大,我们提出如下建议:开辟适合中小企业发展的专门融资渠道;开拓新的融资手段,为中小企业提供必要的信贷支持;加快中小企业直接融资步伐,让中小企业进入资本市场。

4. 城市化与就业

城市化进程是城市在国家经济和社会生活中的主导作用不断增强的过程,它使人口进一步向大城市集中,这一规律不仅表现在大城市数量和规模的更快增长上。它对就业的影响一方面使人们的消费观念发生变化,使一些新兴产业得到进一步发展的机会,从而提供更多的就业机会;另一方面,它会使农村剩余劳动力发生转移,从而增加了就业人数。

我国的城市化水平不仅低于世界平均水平,而且与我国的经济发展及工业化水平不一致。造成这一现象的原因主要是片面发展重工业的产业政策、落后的农业生产率以及我国城市发展政策对城市化的制约。解决这个问题的相关建议是,在政策上,要为农村人口向城市的转移扫除障碍;同时,在增加城市数量,扩大城市规模的同时,要注意提高城市化的质量水平,在模式上,要使大中城市和小城镇协调发展。

5. 城乡收入差距与就业

城乡收入差距对就业的影响一方面是通过影响消费来影响就业,另一方面是通过影响城市化进度来影响就业。我国城乡居民收入差距呈现出扩大的趋势,导致这种现象的重要原因是二元经济结构、人力资本及物质资本的差异。二元经济结构的核心问题是农村剩余劳动力的转移问题,通过向城市工业转移来消除农村劳动力剩余,并引起产业结构的调整而提高农村的劳动生产率以减小城乡收入差距;提高农业生产的技术水平,缩小城乡生产的技术水平差异;加强农村人力资本投资,缩小城乡的人力资本差异;加强物质投入,提高农村居民的平均产出。

12.7 模型的误差分析、检验及进一步讨论

12.7.1 误差分析

1. 考虑影响因素的选取方面

在指标选取模型中,在国家财政政策影响因素的确定上,考虑到在数据,利率因素的确定并非是按照年度时间制定的。因此并未将此因素引入模型中。但是我们通过定性分析可知利率的影响对就业人数是有一定作用的,因此在模型的建立过程中会产生一定的误差。

当利率发生变动时,可能会间接导致就业人数变化,因为利率与人们手中握有的实际财富量密切相关,会引起人们就业取向变化。对于不同行业、不同就业人群等,其作用机制与作用大小是不同的。

2. 考虑缺省值

在数据的选取上,因为某些指标部分年份数据缺省,我们利用牛顿插值法进行差值补全数据,我们可以通过相应时刻拟合模型的拟合值来判断误差大小。

3. 考虑我国区域的划分

在问题三不同地区的就业人口模型分为中部、东部和西部三地区,没有考虑东北三省老工业基地,模型的精度可能会受影响。

12.7.2 模型的检验

本文在解决四个问题的过程中,多元线性回归模型、主成分回归模型、向量自回归模型、灰色预测模型和神经网络预测模型在建立的过程中通过了相应软件的检验,具有一定的合理性。

12.7.3 模型的进一步讨论

本文对模型指标的选取,只是通过关联系数选取了固定资产投资总额、城市化水平、城乡收入差距、居民消费水平、货币供给量M2、国内生产总值六个指标,然而影响就业水平的因素还有很多,本文所构建的影响就业指标体系就包括了18个相关指标,诸如:产业结构、市场化指数、知识层次,等等,都会对城镇就业人口产生影响。此外,利率、房地产价格、股票等一系列金融指标也同样会对城镇就业人口的变化起到一定的作用。同时,不同行业、地区、人群都存在着不同的特征,其相关影响因素也不同,从而使其相应的就业人口数量也产生不同。

12.8 模型的评价与推广

12.8.1 模型的优缺点

1. 模型的优点

(1)对于全国就业量的研究,本文以1990年为界,对就业人口的影响因素进行了分段考虑。因而,能够从不同时间阶段考虑对就业量的主要影响因素。

(2)本文利用主成分回归方法,有效避免变量间的多重共线性问题,进而提高了模型的估计精度。

(3)对于影响全国就业量的因素,本文从不同角度,选取了大量指标进行了分析。

(4)本文运用SPSS、DPS、EVIEWS等多种数学软件进行计算,取长补短,使计算结果更加准确。

2. 模型的缺点

(1)在对影响就业量主要因素的选取过程中,主观性较大,极易漏掉其他重要变量,从而不能全面地的分析就业量。

(2) 建立就业量模型时，对影响就业量的指标我们设定的是一种对数形式，然而现实生活中，其与就业量的相关性并不一定遵守这个规律，因而会产生一定的偏差。

12.8.2 模型的推广

依据主成分回归的方法我们可以进一步从不同的角度来对就业量的估计进行推广。此处，文章选取农业、工业、建筑业、房地产业相关数据进行进一步分析。

1. 农业行业就业人口模型

本文对农业行业选取的六个指标：支农财政支出 x_1、农业各税 x_2、农业总产值 x_3、农村居民消费水平 x_4、农村居民恩格尔系数 x_5、农村居民纯收入 x_6 六项指标，提取主成分：第一个主成分的方差贡献率为 94.41%，$\lambda_1 = 5.665$，这里我们提取一个主成分，即影响农业行业就业人口数量因素的综合值，表达式为：

$$F = 0.398 \times \ln x_1 + 0.409 \times \ln x_2 + 0.410 \times \ln x_3 + 0.416 \times \ln x_4 - 0.399 \times \ln x_5 + 0.417 \times \ln x_6$$

我们运用这个综合值序列对我国 1990—2007 年就业人数进行拟合，结果为：

$$\ln P_{\text{农}} = 11.907 + 0.189 \times F$$

其中，$R^2 = 0.746$ 拟合效果较好，且综合指标的系数显著，因此我们可以进一步求出农业行业就业人口模型：

$$\ln P_{\text{农}} = 11.907 + 0.0752 \times \ln x_1 + 0.0773 \times \ln x_2 + 0.0775 \times \ln x_3 + 0.0786 \times \ln x_4 - 0.0754 \times \ln x_5 + 0.0788 \times \ln x_6$$

2. 工业行业就业人口模型

本文对影响工业行业就业人数的因素：工业企业个数 y_1、工业总产值 y_2、利润总额 y_3、工业贷款 y_4 四项指标提取主成分：第一个主成分的方差贡献率为 95.185%，$\lambda_1 = 3.807$，这里我们提取一个主成分，即为影响工业行业就业人口数量因素的综合值，表达式为：

$$F = 0.5012 \ln y_1 + 0.5079 \ln y_2 + 0.4941 \ln y_3 + 0.4945 \ln y_4$$

我们运用这个综合值序列对我国 1998—2007 年就业人数进行拟合，结果为：

$$\ln P_{\text{工}} = 4.4489 + 0.2693 \times F$$

其中，$R^2 = 0.669$，拟合效果较好，且综合指标的系数显著，因此我们可以进一步求出工业行业就业人口模型：

$$\ln P_{\text{工}} = 4.4489 + 0.135 \ln y_1 + 0.1368 \ln y_2 + 0.1331 \ln y_3 + 0.1332 \ln y_4$$

3. 房地产行业就业人口模型

本文对房地产业选取的四个指标：房地产企业个数 z_1、职工平均工资 z_2、商品房销售额 z_3、房地产业贷款 z_4，提取主成分：第一个主成分的方差贡献率为 97.75%，$\lambda_1 = 3.91$，这里我们提取一个主成分，即为影响房地产行业就业人口数量因素的综合值，表达式为：

$$F = 0.5193\ln z_1 + 0.5017\ln z_2 + 0.5037\ln z_3 + 0.5023\ln z_4$$

我们运用这个综合值序列对我国1997—2007年就业人数进行拟合,结果为:

$$\ln P_{房} = 4.2618 + 0.0621 \times F$$

其中,$R^2 = 0.9642$,拟合效果较好,且综合指标的系数显著,因此我们可以进一步求出房地产业行业就业人口模型:

$$\ln P_{房} = 4.2618 + 0.0322\ln z_1 + 0.0312\ln z_2 + 0.0313\ln z_3 + 0.0312\ln z_4$$

4. 建筑业行业就业人口模型

本文对建筑业行业选取的四个指标:建筑业生产总值 e_1、建筑业企业单位数 e_2、建筑业技术装备率 e_3、建筑业贷款 e_4,提取主成分:第一个主成分的方差贡献率为85.97%,$\lambda_1 = 3.439$,这里我们提取一个主成分,表达式为:

$$F = 0.5193 \times \ln e_1 + 0.4287 \times \ln e_2 + 0.5134 \times \ln e_3 + 0.5322 \times \ln e_4$$

我们运用这个综合值序列对我国1995—2007年就业人数进行拟合,结果为:

$$\ln P_{建} = 4.751 + 0.1874 \times F$$

其中,$R^2 = 0.781$,拟合效果较好,且综合指标的系数显著,因此我们可以进一步求出建筑业行业就业人口模型如下:

$$\ln P_{建} = 4.751 + 0.0973 \times \ln e_1 + 0.0803 \times \ln e_2 + 0.0962 \times \ln e_3 + 0.0997 \times \ln e_4$$

根据以上模型,我们可以进一步对未来几年的数值进行仿真预测,其结果如表12.12。

表12.12 各行业就业人数预测值

年份	2008	2009	2010
$P_{农}$	30 186.09	31 564.21	33 004.97
$P_{工}$	13 394.76	14 736.04	16 211.82
$P_{房}$	458.0703	468.9532	480.0946
$P_{建}$	4 755.009	4 957.898	5 169.442

由以上分析,此处仅是对模型的初步推广,更深层次更精确的研究有待进一步讨论,希望能够为政府相关部门提供一定的借鉴。

参考文献

[1] Hall R., Jone C., Why Do Some Countries Produce So Much More Output Per Worker than Others? [J], *Quarterly Journal of Econnormics*, 1999, 114: 83—116.

[2] 易大义,沈云宝,李有法. 计算方法(第二版)[M]. 杭州:浙江大学出版社,2002.

[3] 邓聚龙. 灰色系统理论教程[M]. 武汉:华中科技大学出版社,1990.

[4] 韩力群. 人工神经网络理论、设计及应用[M]. 北京:化学工业出版社,2001.

[5] 高铁梅. 计量经济分析方法与建模[M]. 北京:清华大学出版社,2006.

[6] 吴建国. 数学建模案例精编[M]. 北京:中国水利水电出版社,2005.

[7] 范金,朱强,王艳. 中级宏观经济学[M]. 北京:经济管理出版社,2007.

[8] 王振龙. 应用时间序列分析[M]. 北京:科学出版社,2007.

[9] 许毓坤. 股指及宏观经济变量的神经网络仿真建模和预测[J]. 宁波职业技术学院学报,11(5):91—94,2007.

[10] 朱加凤. 我国财政货币政策就业效应的实证分析[J]. 学术交流,6(6):91—94,2009.

[11] 石为人,冯治恒. 给予灰色理论与BP算法的宏观经济预测模型研究[J]. 计算机与数字工程,35:8—10,2006.

建模特色点评

❋论文特色❋

◆标题分析:"我国就业人数的主要影响因素分析及前景预测",涉及研究对象"我国就业人数"预测,研究内容"主要影响因素分析及前景预测"两个方面,内容紧扣主题,定位准确、贴切、完美。

◆方法鉴赏:以1990年为界,对于全国就业人口的影响因素进行分段考虑。利用主成分回归方法,有效避免变量间的多重共线性问题,进而提高了模型的估计精度。对于影响全国就业量的因素,分别从分地区、分行业、分人群等不同角度选取指标进行分析。运用SPSS、DPS、EVIEWS等多种数学软件进行计算,取长补短,使计算结果更加准确。

◆写作评析:文章写作语言流畅,内容结构条理性好,框图及神经网络图直观,响应曲线反映技术面好。论文写作条理性好,层次分明,图文并茂。

◆特别之处:该文获全国研究赛C题最高分,并收录在数学实践与认识上发表。

❋不足之处❋

不足相对较少,个别图从网上截取,缺少灵敏度分析。

第13篇
确定肿瘤的重要基因信息

◆ 竞赛原题再现

2010年A题 确定肿瘤的重要信息——提取基因图谱信息方法的研究

癌症起源于正常组织在物理或化学致癌物的诱导下,基因组发生的突变,即基因在结构上发生碱基对的组成或排列顺序的改变,因而改变了基因原来的正常分布(即所包含基因的种类和各类基因以该基因转录的mRNA的多少来衡量的表达水平)。所以探讨基因分布的改变与癌症发生之间的关系具有深远的意义。

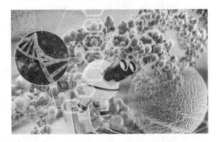

DNA微阵列(DNA microarray),也叫基因芯片,是最近数年发展起来的一种能快速、高效检测DNA片段序列、基因表达水平的新技术。它将数目从几百个到上百万个不等的称之为探针的核苷酸序列固定在小的(约1 cm^2)玻璃或硅片等固体基片或膜上,该固定有探针的基片就称之为DNA微阵列。根据核苷酸分子在形成双链时遵循碱基互补原则,就可以检测出样本中与探针阵列中互补的核苷酸片段,从而得到样本中关于基因表达的信息,这就是基因表达谱,因此基因表达谱可以用一个矩阵或一个向量来表示,矩阵或向量元素的数值大小即该基因的表达水平。

随着大规模基因表达谱(Gene expression profile或称为基因表达分布图)技术的发展,人类各种组织的正常的基因表达已经获得,各类病人的基因表达分布图都有了参考的基准,因此基因表达数据的分析与建模已经成为生物信息学研究领域中的重要课题。如果可以在分子水平上利用基因表达分布图准确地进行肿瘤亚型的识别,对诊断和治疗肿瘤具有重要意义。因为每一种肿瘤都有其基因的特征表达谱。从DNA芯片所测量的成千上万个基因中,找出决定样本类别的一组基因"标签",即"信息基因"(informative genes)是正确识别肿瘤类型、给出可靠诊断和简化实验分析的关键所在,同时也为抗癌药物的研制提供了捷径。

通常由于基因数目很大,在判断肿瘤基因标签的过程中,需要剔除掉大量"无关基因",从而大大缩小需要搜索的致癌基因范围。事实上,在基因表达谱中,一些基因的表达水平在所有样本中都非常接近。例如,不少基因在急性白血病亚型(ALL,AML)两个类别中的分布无论其均值还是方差均无明显差别,可以认为这些基因与样本类别无关,没有对样本类型的判别提供有用信息,反而增加信息基因搜索的计算复杂度。因此,必须对这些"无关基因"进行剔除。1999 年《Science》发表了 Golub 等针对上述急性白血病亚型识别与信息基因选取问题的研究结果[1]。Golub 等以"信噪比"(Signal to noise ratio)指标作为衡量基因对样本分类贡献大小的量度,采用加权投票的方法进行亚型的识别,仅根据 72 个样本就从 7 129 个基因中选出了 50 个可能与亚型分类相关的信息基因。Golub 的工作大大缩小了决定急性白血病亚型差异的基因范围,给出了亚型识别的基因依据,富有创造性。Guyon 等则利用支持向量机的方法再从中选出了 8 个可能的信息基因[2]。

但信噪比肯定不是衡量基因对样本分类贡献大小的唯一标准,肿瘤是致癌基因、抑癌基因、促癌基因和蛋白质通过多种方式作用的结果,在确定某种肿瘤的基因标签时,应该设法充分利用其他有价值的信息。有专家认为[3]在基因分类研究中忽略基因低水平表达、差异不大的表达的倾向应该被纠正,与临床问题相关的主要生理学信息(见问题 4)应该融合到基因分类研究中。

面对提取基因图谱信息这样前沿性课题,命题人根据自己科学研究的经历和思考,猜测以下几点是解决前沿性课题的有价值的工作。这种猜测是科学研究中的重要环节,当然猜测不会总是可行的,更不一定总是正确的。但不探索就不能前进,如果能够通过数学建模,得到的部分结果可以佐证你们的猜测或为新探索提供若干依据,就很有价值。我们的目的只是给研究生以启发,鼓励研究生培养这样的创造性发现的能力。所以研究生完全可以独立设计自己的技术路线,只要能够有效提取附件的基因图谱信息就行。

(1)由于基因表示之间存在着很强的相关性,所以对于某种特定的肿瘤,似乎会有大量的基因都与该肿瘤类型识别相关,但一般认为与一种肿瘤直接相关的突变基因数目很少。对于给定的数据,如何从上述观点出发,选择最好的分类因素?

(2)相对于基因数目,样本往往很小,如果直接用于分类会造成小样本的学习问题,如何减少用于分类识别的基因特征是分类问题的核心,事实上只有当这种特征较少时,分类的效果才更好些。对于给定的结肠癌数据如何从分类的角度确定相应的基因"标签"?

(3)基因表达谱中不可避免地含有噪声(见 1999 年 Golub 在《Science》发表的文章),有的噪声强度甚至较大,对含有噪声的基因表达谱提取信息时会产生偏差。通过建立噪声模型,分析给定数据中的噪声能否对确定基因标签产生有利的影响?

(4)在肿瘤研究领域通常会已知若干个信息基因与某种癌症的关系密切,建立融入了这些有助于诊断肿瘤信息的确定基因"标签"的数学模型。比如临床有下面的生理学信息:大约 90%结肠癌在早期有 5 号染色体长臂 APC 基因的失活,而只有 40%~50%的 ras 相关基因突变。

原题详见 2010 年全国研究生数学建模竞赛 A 题。

获奖论文精选

结肠癌基因标签的识别模型[①]

摘要：本文针对结肠癌基因标签的识别问题,首先对正常组织样本和癌症样本两组数据剔除了无关基因,在高度相关基因中选取代表基因,使用小波信号去噪等方法提高了基因标签识别率,并通过加入临床中确定性致病基因,运用贝叶斯判别法,建立了基因标签的识别模型。使用 Matlab 软件编程,得到比较合理的结肠癌基因标签识别结果。

对于问题 1,我们建立了致病基因与正常基因的识别模型,首先利用 Wilcoxon 秩和检验,剔除在正常组织和癌症之间无显著差异的基因,然后对余下的致病基因进行 K 均值聚类,最后在各个类别中采用稳健的基因选择方法,测度各个类别中的基因鉴别能力,选择出鉴别能力较强的作为分类因素。结果初步认为 590 个基因与癌症相关,聚为六类并选择了 55 个代表基因。

对于问题 2,在问题 1 的基础上使用支持向量机法进行降维和分类,主要是从代表基因中寻找基因标签,从而降低判别正常样本和癌症样本的误判率。我们将样本划分为训练集和测试集,结果表明 55 个代表基因在 22 个测试集中出现了 3 次误判,而选取的 16 个基因标签仅出现一次误判,共 4 次误判都是将癌症样本误判为正常样本。

对于问题 3,在之前降维和去冗余的基础上,进一步考虑用小波分析方法对基因表达谱数据进行去噪处理,然后对去噪后的数据,重复问题 1、2 的步骤分析。结果表明去噪后选取的 55 个代表基因没有出现误判,而进一步筛选的 3 个基因标签中出现二次误判,均是将正常样本误认为癌症样本。

对于问题 4,在上述问题得到的基因标签中加入临床确定性致病基因,运用贝叶斯判别分析对已知的 62 个样本进行重新的判别检验,以验证前述模型的正确性,同时说明已知若干个信息基因和本文选取的基因标签与结肠癌均有密切关系。结果表明 22 个正常组织样本中有 1 个被误判为肿瘤组织;40 个肿瘤组织中有 4 个被误判为正常组织,贝叶斯判别的平均误判率为 0.0797。最后运用灵敏度分析,选出使癌症组织误判为正常组织可能性最低的先验概率。

本文最后对模型作出分析、评价和改进。本文的特点是综合使用了 SPSS,Matlab 等软件,运用了 wilcoxon 非参数检验,聚类分析,小波分析,支持向量机以及贝叶斯判别,使得识别模型不断改进,达到最优情况。

关键词:基因标签;聚类分析;支持向量机;小波分析;贝叶斯判别;最优识别

[①] 本文获 2010 年全国一等奖。队员:宋峰,李小庆,黄蓓;指导教师:朱家明等。

13.1 问题的重述

13.1.1 背景知识

癌症起源于正常组织在物理或化学致癌物的诱导下,基因组发生的突变,即基因在结构上发生碱基对的组成或排列顺序的改变,因而改变了基因原来的正常分布(即所包含基因的种类和各类基因以该基因转录的 mRNA 的多少来衡量的表达水平)。所以探讨基因分布的改变与癌症发生之间的关系具有深远的意义。

1. DNA 微阵列

DNA 微阵列(DNA microarray),也叫基因芯片,是最近数年发展起来的一种能快速、高效检测 DNA 片段序列、基因表达水平的新技术。它将数目从几百个到上百万个不等的称之为探针的核苷酸序列固定在小的(约 1 cm^2)玻璃或硅片等固体基片或膜上,该固定有探针的基片就称之为 DNA 微阵列。

2. 基因表达谱

根据核苷酸分子在形成双链时遵循碱基互补原则,就可以检测出样本中与探针阵列中互补的核苷酸片段,从而得到样本中关于基因表达的信息,这就是基因表达谱,因此基因表达谱可以用一个矩阵或一个向量来表示,矩阵或向量元素的数值大小即该基因的表达水平。

随着大规模基因表达谱(*gene expression profile*,或称为基因表达分布图)技术的发展,人类各种组织的正常的基因表达已经获得,各类病人的基因表达分布图都有了参考的基准,因此基因表达数据的分析与建模已经成为生物信息学研究领域中的重要课题。如果可以在分子水平上利用基因表达分布图准确地进行肿瘤亚型的识别,对诊断和治疗肿瘤具有重要意义。因为每一种肿瘤都有其基因的特征表达谱。从 DNA 芯片所测量的成千上万个基因中,找出决定样本类别的一组基因标签,即"信息基因"(*informative genes*)是正确识别肿瘤类型、给出可靠诊断和简化实验分析的关键所在,同时也为抗癌药物的研制提供了捷径。

13.1.2 研究现状

通常由于基因数目很大,在判断肿瘤基因标签的过程中,需要剔除掉大量"无关基因",从而大大缩小需要搜索的致癌基因范围。事实上,在基因表达谱中,一些基因的表达水平在所有样本中都非常接近。例如,不少基因在急性白血病亚型(ALL,AML)两个类别中的分布无论其均值还是方差均无明显差别,可以认为这些基因与样本类别无关,没有对样本类型的判别提供有用信息,反而增加信息基因搜索的计算复杂度。因此,必须对这些"无关基因"进行剔除。1999 年 Science 发表了 Golub 等针对上述急性白血病亚型识别与信息基因选取问题的研究结果[1]。Golub 等以"信噪比"(Signal to noise ratio)指标作为衡量基因对样本分类贡献大小的量度,采用加权投票的方法进行亚型的识别,仅根据 72 个样本就从 7 129 个基因中选出了 50 个可能与亚型分类相关的信息基因。Golub 的工作大大缩小了决定急性白血病亚型差异的基因范围,给出了亚型识别的基因依据,富有创造性。Guyon 等则利用支持向量机的方法再从中选出了 8 个可能的信息基因[2]。

但信噪比肯定不是衡量基因对样本分类贡献大小的唯一标准,肿瘤是致癌基因、抑癌基因、促癌基因和蛋白质通过多种方式作用的结果,在确定某种肿瘤的基因标签时,应该设法充分利用其他有价值的信息。有专家认为[3]在基因分类研究中忽略基因低水平表达、差异不大的表达的倾向应该被纠正,与临床问题相关的主要生理学信息应该融合到基因分类研究中。

13.1.3 具体试验数据

本文的实验数据来自 Alon 公布的结肠癌基因表达谱数据集。Project_data1.txt 的数据集包括 40 个结肠癌组织样本和 22 个正常组织样本,每个样本包含 2000 个基因的表达数据。

13.1.4 要解决的具体问题

问题 1 由于基因表示之间存在着很强的相关性,所以对于某种特定的肿瘤,似乎会有大量的基因都与该肿瘤类型识别相关,但一般认为与一种肿瘤直接相关的突变基因数目很少。对于给定的数据,如何从上述观点出发,选择最好的分类因素?

问题 2 相对于基因数目,样本往往很小,如果直接用于分类会造成小样本的学习问题,如何减少用于分类识别的基因特征是分类问题的核心,事实上只有当这种特征较少时,分类的效果才更好些。对于给定的结肠癌数据如何从分类的角度确定相应的基因"标签"?

问题 3 基因表达谱中不可避免地含有噪声(见 1999 年 Golub 在 *Science* 发表的文章),有的噪声强度甚至较大,对含有噪声的基因表达谱提取信息时会产生偏差。通过建立噪声模型,分析给定数据中的噪声能否对确定基因标签产生有利的影响?

问题 4 在肿瘤研究领域通常会已知若干个信息基因与某种癌症的关系密切,建立融入了这些有助于诊断肿瘤信息的确定基因"标签"的数学模型。比如临床有下面的生理学信息:大约 90%结肠癌在早期有 5 号染色体长臂 APC 基因的失活,而只有 40%—50%的 ras 相关基因突变。

13.2 问题的分析

13.2.1 问题的总体分析

结肠癌基因标签的识别问题是当前医学中肿瘤病症诊断的热点之一。由于该问题涉及因素较多,比如从基因表达谱中寻求基因标签用来识别肿瘤特征,而基因表达谱中基因数目通常很大,所以需要剔除掉大量"无关基因"。这些无关基因除了包括重复的基因项外,还有某些基因在癌症患者和正常人之间的表达数据无显著性差异,另外还包括相关性很强的同类基因。当扣除这些"无关基因"后,再通过聚类分析将剩余基因聚为若干类代表基因,以正确判别正常人和癌症患者为原则,从代表基因中选取若干基因标签。由于原始数据存在噪声因素,在数据处理中,就要考虑去除噪声影响,从而增加数据的可信度。若已知某种癌症发生时,有确定性的基因会随之变动,即已知先验信息,将这些先验信息加入到基因标签中,可以增强模型的识别能力。

13.2.2 对具体问题的分析

1. 对问题 1 的分析

肿瘤基因表达谱中往往包含大量的基因,但大部分基因与分类无关,这些冗余信息的存在,人为地增加了分类计算的成本,也阻碍了分类精确性的提高。问题 1 基于此种考虑,首先对原始数据中重复的基因项,取均值合并成一个基因;然后对 normal 和 cancer 中的同一种基因表达数据有无显著性差异进行检验,我们认为如果检验没有显著性差异,则说明该种基因不会导致癌症也不会抑制癌症的发生,最后对剔除无显著性差异后的基因表达数据,使用 K 均值聚类算法进行分类,并在各类中使用基因选择理论方法选择代表基因。

2. 对问题 2 的分析

问题 1 对与癌症有关的基因进行了分类,而问题 2 要解决的则是找到能够区分正常人和癌症患者的基因"标签"。由于相对于样本,基因数目仍然较多,而 SVM 集成了降维和分类两个任务,一个最大的优点就是在小样本情况下依然可以保持很好的泛化性能,并且本文所研究的也符合 SVM 只能用于两类别分类问题的要求,所以我们针对问题一选择出的代表基因,运用支持向量机法进行基因表达数据的降维和识别。

3. 对问题 3 的分析

基因表达谱是进行癌症检测成为癌症研究的重点之一,但其数据具有高维性、高噪声、高冗余等特点。问题 1、2 已对数据进行降维,去冗余,问题 3 在此基础上进一步去噪。本文的解决方法考虑在去无关基因之前,采用在时域和频域都具有局部化特征的小波分析方法对基因表达谱数据进行去噪处理,从而能在降低噪声的同时保证较好的信号和细节特征。然后再运用解决问题1、2的方法处理去噪的数据,跟之前的结果进行对比,从而分析出给定数据中的噪声能否对确定基因标签产生有利的影响。

4. 对问题4的分析

在问题4中,已知结肠癌与5号染色体长臂APC基因的失活以及40%~50%的ras相关基因突变有关。因此,可以综合题目附带的基因信息和问题3中提取出的基因标签,运用贝叶斯判别分析对已知的62个样本进行重新的判别检验,以验证前述模型的正确性,同时说明已知若干个信息基因和本文选取的基因标签与结肠癌均有密切关系。

综上所述:问题研究的总体流程图如图13.1所示。

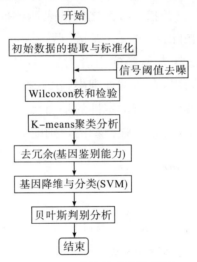

图13.1 总体思路的流程图

13.3 模型的假设

(1)对于重复出现的ESTs,用其均值代表其平均水平;
(2)假设训练集和测试集样本容量近似为2∶1,且选取满足随机性;
(3)假设基因表达数据值为定量数据,可以比较大小;
(4)假设噪声由基因信息或信号传递时产生;
(5)所有数据均为原始数据,来源真实可靠。

13.4 名词解释与符号说明

13.4.1 名词解释

1. 基因表达水平

基因表达就是基因转录及翻译的过程。在一定调节机制控制下,大多数基因经历基因激活、转录及翻译等过程,产生具有特异生物学功能的蛋白质分子。rRNA、tRNA

编码基因转录合成RNA的过程也属于基因表达。基因表达水平即转录翻译水平,包括mRNA转录和蛋白表达两方面。

2. 特征基因(基因标签)

用于确定相应肿瘤的某些特定基因,称为基因"标签",也即特征基因。

3. 基因失活

由于调控元件的突变,基因移位至异染色质部位,或编码序列出现突变、移框等因素导致基因不能正常表达的现象。

13.4.2 符号说明

表 13.1 符号说明

序号	符号	符号说明
1	normal	正常组织样本
2	cancel	癌症样本
3	a_{ij}	样本 i 上基因 j 的表达值
4	$\bar{a}_{kj}$	第 k 类样本第 j 个基因的表达值均值
5	$\hat{a}_j$	癌症样本和正常组织样本中第 j 个基因均值的平均
6	L	样本个数,在本文中 $L=2$
7	$Scatter(j)$	第 j 个基因类间差别的测度值
8	x_{ij}	第 k 类样本中基因 j 在各个样本上的表达值与其均值的绝对值
9	$\bar{x}_{kj}$	第 k 类样本第 j 个基因 x_{ij} 的均值
10	$\mu(j)$	癌症样本和正常组织样本中第 j 个基因 x_{ij} 均值的平均
11	$Compact(j)$	第 j 个基因类内变化的测度值
12	$score(j)$	第 j 个基因鉴别能力的测度值
13	D	后验概率之差

13.5 模型的建立与求解

13.5.1 问题1的分析与求解

1. 建模思路

问题1要求对给定的数据进行分析,选择最好的分类因素。本文从题中所述观点出发,首先利用Wilcoxon秩和检验对经提取和标准化的数据进一步剔除在正常组织和癌症之间无显著差异的基因,再对筛选出的基因进行K均值聚类,最后在各个类别中采用稳健的基因选择方法,测度各个类别中的基因鉴别能力,选择出鉴别能力较强的作为分类因素。

2. 理论准备

(1) Wilcoxon 秩和检验。

Wilcoxon(或称 Mann-Whitney)秩和检验常用来检验两总体的中位数是否相等,假定两个样本分别有 m 个和 n 个观测值。把两个样本混合后把这 $m+n$ 个观测值升幂排序,记下每个观测值在混合排序下面的秩。之后分别把两个样本所得到的秩相加。记第一个样本观测值的秩的和为 W_x 而第二个样本秩的和为 W_y。这两个值可以互相推算,称为 Wilcoxon 统计量。该统计量的分布和两个总体分布无关。由此分布可以得到 p 值。直观上看,如果 W_x 与 W_y 之中有一个显著地大,则可以选择拒绝零假设。该检验需要的唯一假定就是两个总体的分布有类似的形状(不一定对称)。

(2) K 均值聚类算法。

K-means 聚类是最常用于表达数据的聚类分析之一。该算法根据聚类中的均值进行划分,主要思想是:首先从 n 个数据对象中任意选择 k 个作为初始聚类中心,根据剩下的对象与这些聚类中心的相似度,分别分配给最相似的类;然后计算每个新类的聚类中心。不断重复上述过程直到标准测度函数开始收敛为止。一般采用均方差作为标准测度函数,公式如下:

$$E = \sum_{i=1}^{k} \sum_{x \in C_i} |x - m_i|^2$$

其中,x 为样本对象空间中的一个点,m_i 为聚类 C_i 的均值,k 为聚类个数。

(3) 基因选择理论方法。[①]

一个理想的鉴别基因一定是类间差别较大,而类内变化较小。其中,类间差别是指基因在不同类别的不同样本中表达值之间的大小差距,记为 $scatter$;类内变化是指把基因在属于相同类别的不同样本中表达值之间的大小差距,记为 $compact$。基因 j 的类间差别、类内变化和鉴别能力(score)公式分别为:

$$scatter(j) = \sqrt{\frac{1}{L} \sum_{k=1}^{L} (\bar{a}_{ij} - \hat{a}_j)^2} + \frac{1}{2} \min_{w \neq v} |\bar{a}_{vj} - \bar{a}_{wj}|$$

其中 $\bar{a}_{kj} = \frac{1}{n_k} \sum_{i \in C_k} a_{ij}, \hat{a}_j = \frac{1}{L} \sum_{k=1}^{L} \bar{a}_{kj}$,$a_{ij}$ 为样本 i 上基因 j 的表达值,L 为 n 个样本来自不相交的类别数,n_k 为第 k 类的样本数。

$$compact(j) = \sqrt{\frac{1}{L} \sum_{k=1}^{L} \left(\frac{1}{n_k} \sum_{i \in C_k} x_{ij}^2\right)} + \sqrt{\frac{1}{L} \sum_{k=1}^{L} \left(\frac{1}{n_k} \sum_{i \in C_k} (x_{ij} - \mu(j))^2\right)}$$

其中,$x_{ij} = |a_{ij} - \bar{a}_{kj}|, \bar{x}_{kj} = \frac{1}{n_k} \sum_{i \in C_k} x_{ij}, \mu(j) = \frac{1}{L} \sum_{k=1}^{L} \bar{x}_{kj} = \hat{x}_j$。

$score(j) = compact(j)/scatter(j)$。

对任意一个基因 j,其获得的分数值 $score(j)$ 越小,表明这个基因与类别的相关程度

[①] 李建中等. 考虑样本不平衡的模型无关的基因选择方法[J]. 软件学报. 2006(7):1485-1492.

越高,鉴别能力越强,以基因 j 为特征的基因表达数据的可分性越好。在基因表达数据中,该种方法能够合理的处理类别中样本不平衡的问题,并强调了基因选择方法对样本数目的稳健性,使方法更稳健,适应性更强,适于本文的研究问题使用。

3. 初始数据的提取与标准化

本文采用的是题目附带的包含 2000 个人类基因表达数据的 40 个癌症样本和 22 个正常组织的样本。一方面,若数据有大量的冗余会导致计算复杂度增加,并影响运算结果的准确度;另一方面,结合题目观点综合考虑,本文首先运用 Excel 中的函数 IF(COUNTIF(A:A,A1)>1,"重复")标记出初始的 2000 个基因中 ETSs 重复的基因,再对相应基因分别各个样本的数据取均值,经过处理,基因数目削减为 1727 个。再对此数据利用下列公式进行标准化:

$$a_{ij}' = \frac{a_{ij} - \mu_j}{\sigma_j} \quad i = 1, 2, \cdots, n,$$

其中,方差 $\sigma_j = \sqrt{\frac{1}{n}\sum_{i=1}^{n}(a_{ij}-\mu_j)^2}$,$\mu_j$ 为均值。经过处理的数据均值为 0,方差为 1。

4. 去除无关基因并聚类

在上一步初始数据处理的基础上,本文采用 Wilcoxon 秩和检验进一步去除无关基因,即在正常组织和癌症间无显著差异的基因。运用 matlab 软件编程实现。这里,显著性水平取 0.05,若输出 p 值小于等于 0.05,表明拒绝原假设,即两样本存在显著差异,选中符合该条件的基因进行聚类,根据筛选结果,进入下一步研究的基因数为 590 个。

聚类分析使用的数据为经过这 590 个基因在癌症样本中的原始数据,利用 SPSS 软件重新对其标准化,再进行 K-means 聚类。鉴于缺乏先验知识,所以 k 并不能够事先确定,但在该分类中确定类别数 k 是个很关键的问题,针对此种情况,本文先利用分层聚类生成冰柱图及树形图,依据图形判断 k 取 6 时分类效果最佳。

5. 基因的选择

根据 K-means 聚类获得六大类的基因,整理出各类基因的肿瘤样本和正常组织样本的原始数据,由于第一类只包含一种基因,所以直接选入我们的样本中,而对第二到六类则利用 matlab 软件编程,分类计算出 590 个基因的鉴别能力并从小到大排序,本文计划按约 10% 的基因数来选择,同时考虑到各类基因鉴别能力 score 的得分数值大小情况,因此确定常数 1.6,选择 $score(j) \leqslant 1.6$ 的基因来进行进一步分析,以降低基因表达数据的维度,减少数据分析的复杂性。过滤了鉴别能力差的基因,有理由认为选择出的该类基因是理想的分类因素。表 13.2 整理出入选基因的 54 个基因得分数值。

表 13.2 选择的各类基因鉴别能力的得分值

GenBank Acc No	score	类别	两类样本基因均值差值	GenBank Acc No	score	类别	两类样本基因均值差值
M26697	1.3015	2	−0.4347	U22055	1.5786	3	−0.2970
X55715	1.3596	2	−0.3908	X12671	1.0968	4	−0.6408
T95018	1.4203	2	−0.3392	M63391	1.1346	4	0.7640
T51023	1.4471	2	−0.3902	M22382	1.1973	4	−0.5206
T57619	1.4511	2	−0.2880	H40095	1.2965	4	−0.5477
T58861	1.5680	2	−0.3234	X14958	1.3328	4	−0.4183
T61609	1.5715	2	−0.3038	T71025	1.3570	4	0.4230
H55758	1.5872	2	−0.3860	R84411	1.3866	4	−0.4134
R87126	0.9538	3	0.6959	T47377	1.4033	4	−0.8096
M26383	1.0920	3	−0.9805	X12466	1.4516	4	−0.5154
R36977	1.0945	3	−0.5242	D31885	1.4542	4	−0.4159
Z50753	1.1346	3	0.4588	M36981	1.4553	4	−0.4269
H43887	1.2004	3	0.8068	T51571	1.5292	4	−0.5356
M76378	1.2830	3	0.5961	X70326	1.5392	4	−0.4593
U09564	1.3023	3	−0.4542	H20426	1.5444	4	−0.3387
T86473	1.3259	3	−0.5560	R08183	1.5447	4	−0.6148
X54942	1.3696	3	−0.5550	J02854	1.1260	5	0.6659
U26312	1.3967	3	−0.5089	H08393	1.1265	5	−0.4431
U30825	1.3978	3	−0.3297	J05032	1.1770	5	−0.5657
T56604	1.4476	3	−0.4744	X63629	1.2152	5	−0.5848
T92451	1.4542	3	0.8060	T86749	1.4279	5	−0.4368
X56597	1.4626	3	−0.4672	T62947	1.4311	5	−0.3892
R54097	1.4888	3	−0.7189	U17899	1.4431	5	−0.7604
T83368	1.5273	3	−0.4198	M36634	1.2345	6	0.3935
T51261	1.5480	3	−0.3906	X86693	1.3688	6	0.8252
R42501	1.5484	3	−0.3596	H06524	1.3818	6	0.4621
H23544	1.5703	3	−0.5302	U32519	1.5731	6	−0.3389

计算两类样本均值差值，差值为正表明该项基因在癌症样本里面表达数据值更高，我们认为该指标能够促使癌症的发生，如表 13.2 中的 2,3,4,5 类。差值为负则为抑癌基因。如表 13.2 中的第 6 类。

13.5.2 问题 2 的分析与求解

1. 建模思路

模型中,相对于基因数目,样本往往很小,如果直接用于分类会造成小样本的学习问题,所以我们使用支持向量机算法基于问题 1 中基因选择后的数据进行降维和分类,进一步缩减基因数目,找出能够区分正常人和癌症患者的基因"标签"。

2. 理论准备

支持向量机(Support Vector Machine, SVM)借助于最优化方法解决机器学习问题的新工具,是数据挖掘中的一项新技术。SVM 集成了降维和分类两个任务,一个最大的优点就是在小样本情况下依然可以保持很好的泛化性能,是传统的机器学习方法所不具备的,但只能用于两类别分类问题。对于支持向量机用于两类线性数据的分类问题就是要找到一个可计算的线性分类函数。

假设对给定线性可分样本集为 $(x_1, y_1), (x_2, y_2), \cdots, (x_n, y_n)$,其中,$x \in R^n$,$y_i \in \{-1, 1\}$;一般来说,其 n 维空间中线性判别函数的形式为:$g(x) = wx + b, x \in R^n$。

对应的分类面方程为:$wx + b = 0$。

若分类面方程满足两类所有样本能正确分类,则称其为最优分类面。按照支持向量机的理论,对于最优分类面问题我们可以转化为如下的二次优化问题:$mim\Phi(w) = 1/2 \|w\|^2$。

满足约束条件:$y_i(wx_i + b) \geqslant 1, i = 1, 2, 3, \cdots, n$。

在线性可分的情况下,原始二次最优化问题的对偶问题是:

$$Max: M(a) = -\frac{1}{2} \sum_{i=1}^{N} a_i \cdot a_j \cdot y_i y_j (x_i, x_j) + \sum_{i=1}^{N} a_i,$$

$$s.t., a_i \geqslant 0, i = 1, 2, \cdots, N,$$

$$\sum_{i=1}^{N} a_i y_i = 0。$$

通过解上面的优化问题即可获得相应的分类函数为:

$$f(x) = sign\left[\sum_{i=1}^{N} a_i \cdot y_i (x \cdot x_i) + b\right]$$

支持向量机的步骤可以概括为:首先将输入空间变换到一个新的空间,然后在这个新空间中求取最优线性分类面。支持向量机方法的一个重要的优点是所获得的分类器的复杂度可以采用支持向量的个数,而不是变换空间的维数来刻画。因此,SVM 往往不像一些别的方法一样容易产生过拟合现象。

3. 模型的建立与求解

基于问题 1 的分类数据,和上述 SVM 的理论分析,首先把样本数据集按照大约 2∶1 的比例随机分配到训练集和测试集中,见表 13.3。

表 13.3　结肠癌基于表达数据集

样本分组	训练集	测试集
normal	14	8
cancer	26	14

我们使用 matlab7.0 编程实现 SVM 算法，找到两类训练样本 55 个代表基因的最优线性分类面，寻求支持向量并且对样本进行测试，测试结果发现，22 个测试集中，8 个正常样本的测试表明全部是正常的，判别正确，然而 14 个癌症患者的测试中有 3 个结果出现了误判，即我们的分类方法使得 3 个癌症患者判别为正常人，这种结果在现实中是比较危险的。所以我们有理由怀疑 55 个代表基因的代表性，由于聚类分析后我们选取了 6 类，但是不能排除类与类之间的基因不存在高度的相关性，所以我们进一步剔除 55 个各类中得分靠后的基因表达数据。对剔除后的数据再次进行 SVM 分类，循环下去，直到找到最优的分类基因指标，见表 13.4。

表 13.4　基因标签

GenBank Acc No	score	类别
M27190	1	
M26697	1.3015	2
T58861	1.5680	2
R87126	0.9538	3
M26383	1.0920	3
H43887	1.2004	3
R54097	1.4888	3
X12671	1.0968	4
M63391	1.1346	4
J02854	1.1260	5
H08393	1.1265	5
J05032	1.1770	5
T62947	1.4311	5
M36634	1.2345	6
X86693	1.3688	6
H06524	1.3818	6

4. 模型结果的分析

对 55 个代表基因进行 SVM 分类识别，识别率为 86.36%，而且出现了比较严重的误判，即把癌症患者判别为正常人，这种后果非常严重，尽管对正常人的判别没有出现错误，我们仍然对模型采取了修正。经过循环进行 SVM 算法分类，得到最优的分类基因集包含 16 类基因，对正常人的分类没有错误，而对 14 个癌症患者的判别仍然有 1 个出现误判，误判率为 4.55%，说明模型识别的准确率得到了提高，图 13.2 中给出了两类基因在两种样本情况下的坐标图，从图中可以明显看到正常样本和癌症样本集中在两个不同的区域,被最优超平面划分开来。

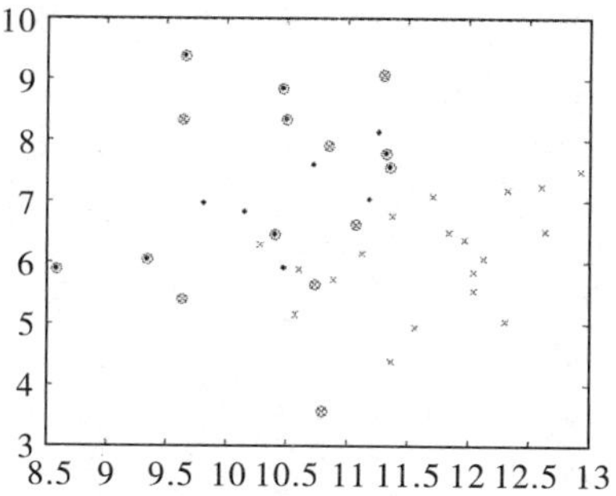

图 13.2　SVM 分类识别图

13.5.3　问题 3 的分析与求解

1. 建模思路

问题 3 要求建立噪声模型,分析给定数据中的噪声能否对确定基因标签产生有利影响。我们认为噪声主要是附加在原始数据中,包括原始数据的搜集和统计带来的误差。本文考虑到传统去噪方法使信号变换后的熵增高、无法刻画信号的非平稳特性并且无法得到信号的相关性,因此采用小波变换解决信号去噪问题。首先按第 1 问中初始数据的提取方法来整理原始数据,再用信号去噪进一步处理,最后按解决第 1 问的方法提取出鉴别力高的基因,用第 2 问中支持向量机的算法对选出基因的识别能力进行检验,以观测效果是否较第 2 问更好。

2. 理论准备

阈值去噪方法是一种实现简单、效果较好的小波去噪方法,该方法思想就是对小波分解后的各层系数中模大于和小于某阈值的系数分别处理,然后对处理完的小波系数再进行反变换,重构出经过去噪后的信号。基本步骤主要包括信号的小波分解;小波分解高频系数的阈值化以及小波重构。它针对多为独立正态变量联合分布,在维数趋向无穷时得出的结论,在最小最大估计的限制下得出的最优阈值。阈值的选择满足:$T = \sigma_n \sqrt{2\ln N}$,其中 σ_n 是噪声标准方差,N 是信号的长度。

3. 模型的建立与求解

重复第一步初始数据的提取方法,同样选出 1727 个基因,对此数据运用 Matlab 软件实现信号的阈值去噪,进而采用和第 1、2 问相同的方法去除无关基因并聚类,然后分类对 590 个基因的鉴别能力计算出 $score$ 的得分数值并从小到大排序,一方面,同样考虑到各类基因鉴别能力 $score$ 的得分数值大小情况,确定常数 1,选择 $score(j) \leqslant 1$ 的基因;另一方面,为了与第 1 题提取的基因进行比较,对选择基因的 $score$ 值从小到大排序选取 55 个,具体见表 13.5。

表 13.5　选择的各类基因鉴别能力的得分值

GenBank Acc No	SCORE	类别	GenBank Acc No	SCORE	类别
T57619	0.7238	1	X12466	0.8541	3
T58861	0.7586	1	R42501	0.8590	3
T61609	0.8285	1	T79152	0.9261	3
T95018	0.8347	1	X15183	0.9283	3
R87126	0.4973	2	R10066	0.9528	3
R36977	0.6217	2	H20426	0.9530	3
M26383	0.6725	2	T47377	0.9605	3
T56604	0.6736	2	D00860	0.7530	4
Z50753	0.6921	2	M84721	0.7600	4
U09564	0.7032	2	M36634	0.8204	4
T86473	0.7106	2	X86693	0.9070	4
M76378	0.7272	2	H06524	0.9167	4
U30825	0.7485	2	H08393	0.4640	5
T51261	0.7918	2	T62947	0.5679	5
U26312	0.8074	2	J02854	0.6101	5
R97912	0.8161	2	X63629	0.6605	5
H43887	0.8482	2	J05032	0.7331	5
T70062	0.8485	2	U17899	0.8509	5
X56597	0.8618	2	H55758	0.6633	6
U22055	0.8776	2	M22382	0.6825	6
M63391	0.6369	3	X55715	0.7677	6
X12671	0.6790	3	T51529	0.7842	6
H40095	0.7181	3	M36981	0.8188	6
X14958	0.7213	3	M26697	0.8334	6
D31885	0.7398	3	T52185	0.8492	6
X70326	0.7693	3	T51023	0.9220	6
R08183	0.7887	3	T71025	0.9304	6
R84411	0.8410	3			

从表 13.5 可以看出,利用同样的方法对经过小波去噪后的基因数据进行处理,与表 13.2 相比较可知,选出的基因判别能力显然更高。

与问题 2 实现算法相同,观察测试结果发现,22 个测试集中,8 个正常人和 14 个癌症患者的测试全部判别正确,即识别率达到 100%。但是我们考虑当基因代表性过高时,可能存在比某些基因不作为或者搭便车的情况,也即可能存在比较少的基因标签就能够达到很高的识别率。所以我们也进一步剔除了 55 个各类中得分靠后的基因表达数据。对剔除后的数据再次进行 SVM 分类,循环下去,发现只需 4 个基因表达数据即可实现百分百的准确判别,见表 13.6。而其中的 3 个基因表达数据识别率也已经达到了

90.91,说明去噪后的数据比去噪前,提取同样的基因标签识别率更高。而且在 3 个基因标签识别中,M26383 基因在映射结肠癌复杂(分子)病理机制的基因网络中起到一个中心枢纽的作用。分子生物学实验的证据表明 MONAP 在多种人类肿瘤细胞系中总是高表达的,而其余两类基因标签与相关文献得出的结论一致。从图 13.3 中可以明显看到正常样本和癌症样本被最优超平面划分为两个不同的区域。

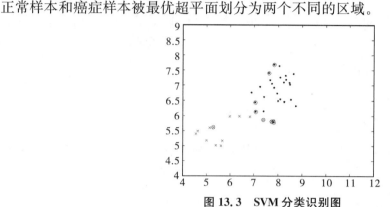

图 13.3　SVM 分类识别图

表 13.6　基因标签

GenBank Acc No	score	类别
H08393	0.4640	5
H36634	0.8204	4
M26383	0.6725	2

13.5.4　问题 4 的分析与求解

1. 建模思路

问题 4 要求建立融入有助于诊断肿瘤信息的确定基因"标签"的数学模型。解题思路为综合问题 3 筛选出的三个基因标签、5 号染色体长臂 APC 基因以及 ras 相关基因,整理出附件中所给的 40 个结肠癌组织样本和 22 个正常组织样本中这些基因的数据,运用贝叶斯判别法对 62 个样本重新判别分类,并计算平均误判概率。

2. 理论的准备

(1) 贝叶斯判别分析的基本思想。

贝叶斯判别法的基本思想是假定对所研究的对象已经有一定的认识,常用先验概率来描述。已知 $G_1,G_2,\cdots,G_n$ 是 n 个 p 维总体,每个总体 G_i 可认为是属于 G_i 的指标 $X=(X_1,X_2,\cdots,X_p)^T$ 的取值全体,它们分别具有互不相同的 p 维概率密度函数 $f_1(x)$,$f_2(x)\cdots f_n(x)$,对于任一给定的新样品关于指标 X 的观测值 $x=(x_1,x_2,\cdots,x_n)^T$,要判断该样品属于这 n 个总体中的哪一个,因而判别分析是根据所研究个体的某些指标的观测值来推断该个体所属类型的一种统计方法。

(2) 贝叶斯判别法求解步骤如下:

① 判断总体的协方差矩阵是否相等,一般情况下,可构造服从卡方分布的 Q 检验统

计量进行检验,表达式为: $Q_i = (n_i-1)[\ln|S|-\ln|S_i|-p+tr(S^{-1}S_i)] \sim \chi^2(p(p+1)/2), i=1,2$, 其中, p 是向量维数;

② 判断总体是否服从正态分布。多元正态分布的检验方法有多种,如主成分方法、QQ 图方法等;

③ 利用按比例分配方法估计两个总体的先验概率, $p_1 = \dfrac{n_1}{n_1+n_2}$, $p_1 = \dfrac{n_2}{n_1+n_2}$, 其中 n_1, n_2 为两样本个数;

④ 假设两个 p 元正态分布密度函数为:

$$f_j(x) = \dfrac{1}{(2\pi)^{\frac{p}{2}}|\sum|^{\frac{1}{2}}} \exp\{-\dfrac{1}{2}(x-\mu_j)^T \sum\nolimits^{-1}(x-\mu_j)\}, j=1,2,$$

其中, μ_j 是第 j 个总体的均值向量, $\sum$ 为总体的协方差矩阵。上式两边取自然对数,得:

$$\ln f_j(x) = -\dfrac{p}{2}\ln(2\pi) - \dfrac{1}{2}\ln|\sum| - \dfrac{1}{2}(x-\mu_j)^T \sum\nolimits^{-1}(x-\mu_j)。$$

因而有 $\{p_1 f_1 \leq p_2 f_2\} \sim \{\ln p_1 - 0.5 d(x, G_1) \leq \ln p_2 - 0.5 d(x, G_2)\}$, 其中 $d(x, G_j)$ 为 x 到总体 $G_j(j=1,2)$ 的马氏距离平方。

3. 模型的建立与求解

(1) 贝叶斯判别分析的条件检验。

在运用贝叶斯判别分析对 62 个样本进行判别检验之前,先检验 40 个结肠癌组织样本和 22 个正常组织样本的协方差矩阵是否相等以及它们的总体是否服从二元正态分布。

① 检验两总体的协方差矩阵是否相等。

利用 matlab 软件求得, Q_1, Q_2 对应的概率值 p 分别为 0.1296 和 1.0000, 所以接受原假设,认为两总体的协方差矩阵相等。

② 总体的正态性检验。

利用 matlab 软件运用数据作出总体正态分布的 QQ 图检验如图 13.4 所示。

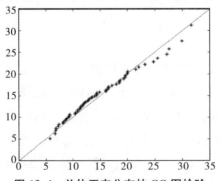

图 13.4 总体正态分布的 QQ 图检验

从图 13.4 可以看出,总体正态分布的 QQ 图中的点散布在一条通过原点,斜率为 1 的直线上,即表示接受数据来自多元正态总体的假设。

综上所述，两总体的协方差矩阵相等，且总体服从多元正态分布，因此满足贝叶斯判别条件。

(2) 贝叶斯判别。

选取 5 号染色体长臂 APC 基因 $L35545$，11 个 ras 相关基因和问题 3 中筛选出的 3 个基因标签，运用 matlab 软件编程，实现贝叶斯判别对 62 个样本的判别分析得到结果如下：22 个正常组织样本中有 1 个被误判为肿瘤组织；40 个肿瘤组织中有 4 个被误判为正常组织，贝叶斯判别的平均误判率为 0.0797。具体判别情况见图 13.5（数据点在 x 轴以下表示该样本到属于类别距离大于该样本到另外一个类别的距离，即该样本被误判）。

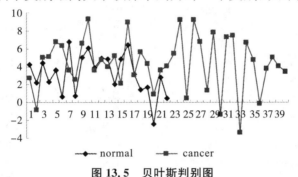

图 13.5 贝叶斯判别图

从图 13.5 可以看出，运用贝叶斯判别法对 62 个样本判别，一共有 5 个样本在 x 轴的下方，即该 5 个样本被误判。其中 normal 2 被误判为肿瘤组织，cancer 2、cancer 30、cancer 33、cancer 36 被误判为正常组织。

4. 模型的灵敏度分析

上述贝叶斯判别的两总体的先验概率是用样本数据的频数代替，即正常组织和肿瘤组织样本的先验概率分别为：$p_1=22/62$，$p_2=40/62$。在实际医学应用过程中，应尽可能避免将肿瘤组织样本判断为正常组织样本，否则会使病人贻误治病的良机。基于此种考虑，我们通过改变先验概率大小的灵敏度分析来改变判别的方法，依次取肿瘤组织样本的先验概率为 $p_2=[0.1, 0.9]$，每隔 0.1 取一个数值，$p_1=1-p_2$，取判别函数 $D=[\ln p_1-0.5d(x,G_1)]-[\ln p_2-0.5d(x,G_2)]$，利用 matlab 软件计算得到 42 个肿瘤组织样本属于肿瘤组织总体与属于正常组织总体的后验概率之差 D 及对应概率的平均误判率如表 13.7 所示。

由表 13.7 可知，$p_2=0.9$ 时，40 个肿瘤组织有 1 个被误判为正常组织，此时的平均误判概率最小，为 0.0416。

当 $p_2=0.9$ 时，$p_1=0.1$，这时的贝叶斯判别图如图 13.6 所示，从图 13.6 中可以看出 40 个肿瘤组织样本有 1 个被误判为正常组织，22 个正常组织样本有 5 个被误判为肿瘤组织。对比根据样本频数确定的先验概率 $p_1=22/62$，$p_2=40/62$ 可知：此时肿瘤组织被误判为正常组织的可能性要小得多，但正常组织被误判为肿瘤组织的概率有所增加，这时被检查出有肿瘤组织的疑似患者可通过再次检查来排除误判的可能，同时保证了最大限度地避免将肿瘤组织患者误判为正常组织的可能性。

表 13.7 后验概率之差与平均误判率

p_2	0.10	0.20	0.30	0.40	0.50	0.60	0.70	0.80	0.90
判别函数 D 的取值	−0.08	0.73	1.27	1.71	2.12	2.52	2.96	3.50	4.31
	−3.66	−2.85	−2.31	−1.87	−1.46	−1.06	−0.62	−0.08	0.73
	2.20	3.01	3.55	3.99	4.39	4.80	5.24	5.78	6.59
	2.32	3.13	3.67	4.11	4.51	4.92	5.36	5.90	6.71
	3.99	4.80	5.34	5.78	6.19	6.59	7.04	7.58	8.39
	3.55	4.36	4.90	5.34	5.75	6.15	6.60	7.13	7.95
	0.82	1.63	2.17	2.61	3.01	3.42	3.86	4.40	5.21
	−0.21	0.60	1.14	1.58	1.99	2.39	2.83	3.37	4.18
	3.78	4.59	5.13	5.57	5.98	6.38	6.83	7.36	8.18
	6.50	7.31	7.85	8.29	8.69	9.10	9.54	10.08	10.89
	0.81	1.62	2.16	2.60	3.00	3.41	3.85	4.39	5.20
	1.93	2.74	3.28	3.72	4.13	4.53	4.98	5.51	6.33
	1.13	1.94	2.48	2.92	3.32	3.73	4.17	4.71	5.52
	2.42	3.23	3.77	4.21	4.61	5.02	5.46	6.00	6.81
	−0.69	0.12	0.66	1.10	1.51	1.91	2.36	2.90	3.71
	6.14	6.95	7.49	7.93	8.34	8.74	9.19	9.73	10.54
	0.25	1.06	1.60	2.04	2.45	2.86	3.30	3.84	4.65
	2.85	3.66	4.20	4.64	5.04	5.45	5.89	6.43	7.24
	1.54	2.35	2.89	3.33	3.73	4.14	4.58	5.12	5.93
	−1.87	−1.06	−0.52	−0.08	0.33	0.73	1.18	1.71	2.53
	0.83	1.64	2.18	2.62	3.03	3.43	3.88	4.42	5.23
	1.26	2.07	2.61	3.05	3.46	3.86	4.30	4.84	5.65
	2.66	3.47	4.01	4.45	4.85	5.26	5.70	6.24	7.05
	6.44	7.26	7.79	8.24	8.64	9.05	9.49	10.03	10.84
	−2.33	−1.51	−0.98	−0.53	−0.13	0.28	0.72	1.26	2.07
	6.42	7.23	7.77	8.21	8.62	9.02	9.47	10.00	10.82
	3.93	4.74	5.28	5.72	6.13	6.54	6.98	7.52	8.33
	−1.47	−0.65	−0.12	0.33	0.73	1.14	1.58	2.12	2.93
	5.06	5.87	6.41	6.85	7.25	7.66	8.10	8.64	9.45
	−4.21	−3.40	−2.86	−2.42	−2.02	−1.61	−1.17	−0.63	0.18
	4.54	5.35	5.89	6.33	6.73	7.14	7.58	8.12	8.93
	4.71	5.52	6.06	6.50	6.91	7.31	7.76	8.29	9.11
	−6.15	−5.33	−4.80	−4.35	−3.95	−3.54	−3.10	−2.56	−1.75
	3.92	4.73	5.27	5.71	6.12	6.52	6.96	7.50	8.31
	1.91	2.73	3.26	3.71	4.11	4.52	4.96	5.50	6.31
	−2.92	−2.11	−1.57	−1.13	−0.72	−0.32	0.13	0.66	1.48
	0.96	1.78	2.31	2.76	3.16	3.57	4.01	4.55	5.36
	2.24	3.05	3.59	4.03	4.43	4.84	5.28	5.82	6.63
	1.28	2.09	2.63	3.07	3.47	3.88	4.32	4.86	5.67
	0.61	1.42	1.96	2.40	2.81	3.21	3.65	4.19	5.00
误判率	0.04	0.06	0.08	0.08	0.08	0.08	0.08	0.06	0.04

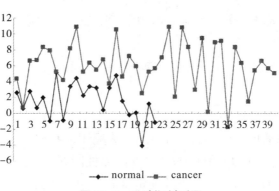

图 13.6　贝叶斯判别图

13.6　模型的误差分析

13.6.1　数据的近似误差

(1)在问题1的求解中,去除重复基因采用了取均值的方法,可能会使基因的表达值产生一定范围内的误差。

(2)对通过基因判别的测度值去除冗余时,对选取的样本比例进行了近似估计,这在之后对数据的进一步提取时必定会造成一定的误差。而因为所用数据为近似数据,所以得到数据的误差是在允许范围内的。

13.6.2　样本缺失的近似误差

相对于基因数目,本文所给样本很少,虽然对基因进行了一系列筛选,但样本相对来说仍然较少,引发机器学习问题,虽然使用支持向量机方法解决了小样本问题,但仍然存在一定的误差。

13.7　模型的评价与推广

13.7.1　模型的优缺点

1. 模型的优点

(1)对于提取基因图谱信息方法的研究,本文针对基因表达谱高维性、高噪声、高冗余的特点,采取适宜的多种方法进行层层降维、去冗余、去噪,提高了分类精确性。

(2)本文利用与模型无关的稳健的基因选择方法,在类内变化小和类间差别大的策略下,方法的鉴别能力通过比较选择敏感的度量函数得以提高,基因标签的选取更具说服力。

(3)对于考察基因表达谱中不可避免地含有的噪声,本文采用阈值去噪法处理数据,该方法实现简单、效果较好,在结果中也得到印证,即去噪后选取的基因标签更具代表性。

(4)本文运用 SPSS、MATLAB、EVIEWS 等多种数学软件进行计算,取长补短,使计算结果更加准确。

2. 模型的缺点

(1)研究样本个数有限,本文选取的建模方法应用上具有一定局限性。

(2)K-means 聚类分析的 k 值依据选取参照冰柱图,受一定主观性影响。

(3)去除冗余基因时,数量和 C 值的确定带有一定主观性。

13.7.2 模型的推广

(1)本文所建模型可以推广到如白血病等其他一系列疾病基因的分类与基因标签的识别上应用。

(2)进一步扩大样本容量,即增加测试集与训练集的样本数目,可提高模型可信度。

参考文献

[1] T. R. Golub, et al. Monitoring and Class Prediction by Gene Expression[J]. *Science*. Vol. 286,1999,531—537.

[2] Guyon I., Weston J., Barnhill S., et al. Gene Selection for Cancer Classification Using Support Vector Machines[J]. *Machine Learning*,2000,46(13): 389—422.

[3] Z. Sun, P. Yang. Gene expression profiling on lung cancer Outcome Prediction: Present Clinical Value and Future Premise[J]. *Cancer Epidemiology Biomarkers & Prevention*,2006,15(11):2063—2068.

[4] 朱建平,殷瑞飞. SPSS 在统计分析中的应用[M]. 北京:清华大学出版社,2007.

[5] 梅长林,范金城. 数据分析方法[M]. 北京:高等教育出版社,2006.

[6] 周伟. MATLAB 小波分析高级技术[M]. 西安:西安电子科技大学出版社,2005.

[7] 李颖新,刘全金,阮晓钢. 急性白血病的基因表达谱分析与亚型分类特征的鉴别[J]. 中国生物医学工程学报,2005(2),240—244.

[8] 申伟科,钟理,葛昆等. 癌症基因表达谱挖掘中的特征基因选择算法 GA/WV[J]. 生物信息学,2010,8(2),98—103.

[9] 黄成玉,阮晓钢,李建更. 基于基因表达谱胃癌特征基因选取研究[J]. 微计算机信息(测控自动化),2009(2—1).

[10] 张娅,饶妮妮,王敏等. 一种基于基因表达谱的结肠癌特征提取方法[J]. 航天

医学与医学工程.2008(4),356—360.

[11] 李泽,包雷,黄英武等.基于基因表达谱的肿瘤分型和特征基因选取[J].生物物理学报,2002(4),413—417.

[12] 段艳华.基于基因表达谱的肿瘤分类特征基因选择研究[D].北京:北京工业大学,1—44.

◈ 建模特色点评

❀论文特色❀

◆标题分析:"结肠癌基因标签的识别模型",标题定位准确、贴切、简洁。

◆方法鉴赏:提取基因图谱信息方法,针对基因表达谱高维性、高噪声、高冗余的特点,采取适宜的多种方法进行层层降维、去冗余、去噪,提高了分类精确性。利用与模型无关的稳健的基因选择方法,在类内变化小和类间差别大的策略下,方法的鉴别能力通过比较选择敏感的度量函数得以提高,基因标签的选取更具说服力。对于考察基因表达谱中不可避免地含有的噪声,采用阈值去噪法处理数据,该方法实现简单、效果较好,在结果中也得到印证,即去噪后选取的基因标签更具代表性。运用 SPSS、MATLAB、EVIEWS 等多种数学软件进行计算,取长补短,使计算结果更加准确。

◆写作评析:论文写作条理性好,层次分明,图文并茂。

❀不足之处❀

不足相对较少,个别图从网上截取,缺少灵敏度分析。

第 14 篇
基因识别问题及其算法实现

◆ **竞赛原题再现**

2012 年 A 题 基因识别问题及其算法实现

一、背景介绍

DNA 是生物遗传信息的载体,其化学名称为脱氧核糖核酸(Deoxyribonucleic acid,缩写为 DNA)。DNA 分子是一种长链聚合物,DNA 序列由腺嘌呤(Adenine,A),鸟嘌呤(Guanine,G),胞嘧啶(Cytosine,C),胸腺嘧啶(Thymine,T)这四种核苷酸(nucleotide)符号按一定的顺序连接而成。其中带有遗传讯息的 DNA 片段称为基因(Gene)。其他的 DNA 序列片段,有些直接以自身构造发挥作用,有些则参与调控遗传讯息的表现。

在真核生物的 DNA 序列中,基因通常被划分为许多间隔的片段,其中编码蛋白质的部分,即编码序列(Coding Sequence)片段,称为外显子(Exon),不编码的部分称为内含子(Intron)。外显子在 DNA 序列剪接(Splicing)后仍然会被保存下来,并可在蛋白质合成过程中被转录(transcription)、复制(replication)而合成为蛋白质。DNA 序列通过遗传编码来储存信息,指导蛋白质的合成,把遗传信息准确无误地传递到蛋白质(protein)上去并实现各种生命功能。

对大量、复杂的基因序列的分析,传统生物学解决问题的方式是基于分子实验的方法,其代价高昂。诺贝尔奖获得者 W. 吉尔伯特(Walter Gilbert,1932—;【美】,第一个制备出混合脱氧核糖核酸的科学家)1991 年曾经指出:"现在,基于全部基因序列都将知晓,并以电子可操作的方式驻留在数据库中,新的生物学研究模式的出发点应是理论的。一个科学家将从理论推测出发,然后再回到实验中去,追踪或验证这些理论假设。"随着世界人类基因组工程计划的顺利完成,通过物理或数学的方法从大量的 DNA 序列中获取丰富的生物信息,对生物学、医学、药学等诸多方面都具有重要的理论意义和实际价值,也是目前生物信息学领域的一个研究热点。

二、数字序列映射与频谱 3-周期性

对给定的 DNA 序列,怎么去识别出其中的编码序列(即外显子),也称为基因预测,是一个尚未完全解决的问题,也是当前生物信息学的一个最基础、最首要的问题。

基因预测问题的一类方法是基于统计学的。很多国际生物数据网站上也有"基因识别"的算法。比如知名的数据网站 http://genes.mit.edu/GENSCAN.html 提供的基因识别软件 GENSCAN(由斯坦福大学研究人员研发的、可免费使用的基因预测软件),主要就是基于隐马尔科夫链(HMM)方法。但是,它预测人的基因组中有 45000 个基因,相当于现在普遍认可数目的两倍。另外,统计预测方法通常需要将编码序列信息已知的 DNA 序列作为训练数据集来确定模型中的参数,从而提高模型的预测水平。但在对基因信息了解不多的情况下,基因识别的准确率会明显下降。

因此在目前基因预测研究中,采用信号处理与分析方法来发现基因编码序列也受到广泛重视。

三、请研究的几个问题

1. 功率谱与信噪比的快速算法

对于很长的 DNA 序列,在计算其功率谱或信噪比时,离散 Fourier 变换(DFT)的总体计算量仍然很大,会影响到所设计的基因识别算法的效率。大家能否对 Voss 映射,探求功率谱与信噪比的某种快速计算方法?

在基因识别研究中,为了通过引入更好的数值映射而获取 DNA 序列更多的信息,除了上面介绍的 Voss 映射外,实际上人们还研究过许多不同的数值映射方法。例如,著名的 Z-curve 映射(参见[5])。试探讨 Z-curve 映射的频谱与信噪比和 Voss 映射下的频谱与信噪比之间的关系;

此外,能否对实数映射,如:A→0,C→1,G→2,T→3,也给出功率谱与信噪比的快速计算公式?

2. 对不同物种类型基因的阈值确定

对特定的基因类型的 DNA 序列,将其信噪比 R 的判别阈值取为 $R_0=2$,带有一定的主观性、经验性。对不同的基因类型,所选取的判别阈值也许应该是不同的。大家还可以从生物数据库下载更多的数据,找你们认为具有代表性的基因序列,并对每类基因研究其阈值确定方法和阈值结果。此外,对按照频谱或信噪比特征将编码与非编码区间分类的有效性,以及分类识别时所产生的分类错误作适当分析。

3. 基因识别算法的实现

我们的目的是要探测、预报尚未被注释的、完整的 DNA 序列的所有基因编码序列(外显子)。目前基因识别方面的多数算法结果还不是很充分。例如前面所列举的某些基因识别算法,由于 DNA 序列随机噪声的影响等原因,还很难"精确地"确定基因外显子区间的两个端点。

对此,你的建模团队有没有更好的解决方法?请对你们所设计的基因识别算法的

准确率做出适当评估,并将算法用于对附件中给出的 6 个未被注释的 DNA 序列(gene 6)的编码区域的预测。

4. 延展性研究

在基因识别研究中,还有很多问题有待深入探讨。比如:

(1)采用频谱或信噪比这样单一的判别特征,也许是影响、限制基因识别正确率的一个重要原因。人们发现,对某些 DNA 序列而言,其部分编码序列(外显子),尤其是短的(长度小于 100 bp)的编码序列,就可能不具有频谱或者信噪比显著性。你们团队能否总结,甚至独自提出一些识别基因编码序列的其他特征指数,并对此做相关的分析?

(2)"基因突变"是生物医学等方面的一个关注热点。基因突变包括 DNA 序列中单个核苷酸的替换、删除或者插入等。那么,能否利用频谱或信噪比方法去发现基因编码序列可能存在的突变呢?

上面提出的基于频谱 3—周期性的基因预测四个方面问题中,"快速算法"与"阈值确定"是为设计基因预测算法做准备的。此外,在最后的延展性研究中,各队也可以对你们自己认为有价值的其他相关问题展开探讨。

原题详见 2012 年全国研究生数学建模竞赛 A 题。

◆ 获奖论文精选

基因识别问题及其算法实现[①]

摘要:本文针对基因识别及其算法实现问题,运用信号处理与分析方法,将碱基序列映射成数值序列,进一步改进原有的 DFT 变换,得到 Voss 映射下功率谱与信噪比的快速计算公式。用快速算法得出已知样本外显子和内含子的信噪比,进一步确定各类基因的阈值。根据确定的阈值和定义的信噪比斜率两个指标,给出了识别基因序列中外显子端点位置的 EPDN 算法,用已知样本来检验,该算法对外显子位置的预测具有较高的精确性,准确度较高。最后用该算法对未知序列编码区的位置进行了预测。

问题一的第一小问:在 Voss 映射下,对原有的 DFT 变换进行改进,得到功率谱与信噪比的快速计算公式,(具体见正文公式(14-10)、(14-11)),这两个公式不再需要对指示序列进行 DFT 变换,而只需要统计四种核苷酸在 3 种不同位置上出现的频数,大大降低了功率谱和信噪比的计算工作量。

问题一的第二小问:根据 Parseval 定理,逐步推导出 Z-curve 映射下,$N/3$ 处频谱 $P_Z[N/3]$ 为 Voss 映射下频谱 $P_V[N/3]$ 的 4 倍,Z-curve 映射下的信噪比 R_Z 是 Voss 映射下信

---

① 本文获 2012 年全国一等奖。队员:王扬眉,方媛媛,陶新新;指导教师:杨桂元等。

噪比 R_V 的 4/3 倍。也即：$P_Z[N/3]=4P_V[N/3]$，$R_Z=4/3R_V$。

问题一的第三小问：针对实数映射 A→0，C→1，G→2，T→3，离散 Fourier 逆变换、Parseval 定理和相关的数学知识，推导出在该实数映射下，N/3 处的总功率谱值为 $P_R[N/3]$（具体见正文公式(14-37)），信噪比 R_R（具体见正文公式(14-42)）

问题二：利用问题一得到的 Voss 映射下信噪比的快速算法，用 Matlab 编程，计算出附件中所给的带有编码外显子信息的 100 个人和鼠类的基因序列的外显子和内含子的信噪比（具体结果见正文表 14.3），根据所给的阈值的确定方法，最后得到人和鼠类基因序列的阈值（具体结果见正文表 14.4）。

建立敏感性指标和专一性指标，对阈值的确定方法进行评判。

问题三：为精确确定外显子的两端点，可以先采用 EPDN 算法实现对外显子区域的识别，然后对识别的区域，往前往后分别扩充 50 个碱基对，逐一寻找，直到找到起始子和终止子，此时就可以精确确定 DNA 序列两端点的位置。用 Java 编程来实现该算法。用已知的样本检验该方法，结果表明该方法可行，最后编程预测附件中给出的 6 个未被注释的 DNA 序列(gene 6)的编码区域。具体的结果见正文中表 14.6。

问题四：由于某些短编码序列可能不具有频谱或者信噪比等显著特征，可以根据终止密码子的相位分布信息与碱基偏性信息相结合，提出两个混合性质特征，根据蛋白质序列信息提取的伪氨基酸组成特征给出 DNA 序列的一组伪碱基特征。混合两种性质的特征提取方式以及伪碱基组成特征能有效地提高编码区的识别精度。

对短编码序列，要提高基因识别的精确度，必须选择精细的模型，径向基函数神经网络是较理想的选择。

关键词：基因识别；信号处理；EPDN 算法；Java

14.1 问题的提出

14.1.1 背景知识

1. 问题概况

DNA 序列由四种不同的碱基组成，分别是：腺嘌呤(A)、胞嘧啶(C)、鸟嘌呤(G)和胸腺嘧啶(T)。绝大多数 DNA 分子都由两条碱基互补的单链构成，两条链上的碱基通过氢键相结合，形成碱基对，其组成遵循碱基互补配对原则：腺嘌呤(A)只能与胸腺嘧啶(T)配对，鸟嘌呤(G)只能与胞嘧啶(C)配对。

从 DNA 链的结构上看，DNA 分为 A-DNA、B-DNA 和 Z-DNA 三类。DNA 通过自身的复制来实现生命的遗传。编码 DNA 序列可通过转录生成 RNA，RNA 通过翻译生成蛋白质，这就是著名的中心法则(见图 14.1)。RNA 也由四种碱基组成：腺嘌呤(A)、胞嘧啶(C)、鸟嘌呤(G)和尿嘧啶(U)，只有第一个碱基与 DNA 不同。在中心法则中，RNA 也可通过反转录生成编码 DNA，还可通过自身的复制生成多条 RNA 链，DNA 也可通过自身复制生成多条DNA链。

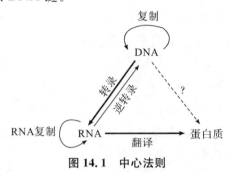

图14.1　中心法则

随着人类基因组测序的顺利完成,生物信息学开始进入蓬勃发展时期,与此同时,基因序列的数量也呈指数级增长。从海量的生物数据中解读、提取和获得有用的生物信息,已成为后基因组时代亟待解决的问题。在遗传学上通常将能编码蛋白质的基因称为结构基因。对真核DNA而言,基因通常由编码区(外显子)和非编码区(内含子)组成,其中外显子携带着主要的遗传信息。因此精确辨识出基因的外显子、内含子的位置是生物信息学中研究的重要方面。对给定的DNA序列,怎么去识别出其中的编码序列(即外显子),也称为基因预测,是一个尚未完全解决的问题,也是当前生物信息学的一个最基础、最首要的问题。

2. 问题原因

在DNA分析方面,预测与识别蛋白编码区或寻找基因是最关键的。由于人类基因组所拥有的DNA序列比编码蛋白质所需的多得多,给定的一段DNA序列可能不为任何蛋白质编码,显然外显子不连续的问题等都是基因预测和识别的一大障碍。

又由于真核生物的基因密度很低,所以辨识任务是很困难的。人类基因组序列中负责编码蛋白质的序列只占三分之一;真核啤酒酵母基因组大约五千比特包含一个基因,也就是基因密度为20%。基因预测算法需要从大量的无用数据中辨识出少量的有用信息,一般会产生很高的假阳性。由于生物研究的快速发展,蛋白质和DNA序列数据库中的数据呈指数增长。序列数量的激增产生了很多的问题。设计高效的算法跟上生物研究的步伐是非常必要的。大量的数据可能占用存储器很大的空间,磁盘访问可能减慢执行速度。在这种情况下,需要设计高效的算法,并不占用大量的工作空间。与生物学数据的海量特征相比,生物学数据的复杂特征则更具有挑战性。从信息科学的角度看,除了少数例外,生物学的实验数据一般是在既无标准词法(semantics)、又无句法(syntax)的条件下生成的。这一情况必然进一步加剧生物学的复杂性。

3. 现状与对策

生物学数据在量(海量)与质(复杂性)方面提出严峻的挑战。如果能够将生物的生命活动过程和基因的结构辨识相结合,对辨识的效果会有很大的提高。但是目前人类对生物生命过程的了解不是很多,所以基因结构辨识主要是应用大量的统计数据提取信号特征,运用机器学习、信号处理等方法进行辨识。由于DNA序列是字符序列,这很大程度上使我们无法使用其他领域的研究方法来对它进行分析。所以如果通过一定的映射规则将它映射为数值序列,这样一条 DNA 序列就可以看作一个离散的时间信号,而 DNA 序列的信息也全部包含其中。在此基础上我们就可以用信号领域的一些分析方法来处理 DNA 序列的信息。在目前基因预测研究中,采用信号处理与分析方法来发现基因编码序列受到广泛重视。

14.1.2 相关试验数据

根据题目给定的基因数据,以及从 NCBI 官方网(http://www.ncbi.nlm.nih.gov/guide/)查询的相关基因数据。由于基因数据较多,这里将不再把原始基因数据附于附录中。

14.1.3 要解决的问题

1. 问题一

对于很长的 DNA 序列,在计算其功率谱或信噪比时,离散 Fourier 变换(DFT)的总体计算量仍然很大,会影响到所设计的基因识别算法的效率。对 Voss 映射,探求功率谱与信噪比的某种快速计算方法;然后进一步探讨 Z-curve 映射的频谱与信噪比和 Voss 映射下的频谱与信噪比之间的关系;最后对实数映射,如:A→0,C→1,G→2,T→3,也给出功率谱与信噪比的快速计算公式。

2. 问题二

依据著名的生物数据网站:http://www.ncbi.nlm.nih.gov/guide/ 的几个基因序列数据,以及另外给出的带有编码外显子信息的 100 个人和鼠类的,以及 200 个哺乳动物类的基因序列的样本数据,给出每类基因研究其阈值的确定方法和阈值结果。此外,对按照频谱或信噪比特征将编码与非编码区间分类的有效性,以及分类识别时所产生的分类错误作适当分析。

3. 问题三

目前基因识别方面的多数算法结果还不是很充分,由于 DNA 序列随机噪声的影响等原因,一些现有的方法,还很难"精确地"确定基因外显子区间的两个端点,对此提出更好的解决方法。请对你们所设计的基因识别算法的准确率做出适当评估,并将算法用于对附件中给出的 6 个未被注释的 DNA 序列(gene 6)的编码区域的预测。

4. 问题四

延展性研究

(1)总结甚至独自提出一些识别基因编码序列的其他特征指数,并对此做相关的分析。(2)利用频谱或信噪比方法去发现基因编码序列可能存在的突变——DNA 序列中单个核苷酸替换、删除或者插入等。

14.2 问题的分析

14.2.1 对问题的具体分析和处理办法

1. 对问题一的分析

对于很长的 DNA 序列,在计算其功率谱或信噪比时,DFT 的总体计算量仍然很大,会影响到所设计的基因识别算法的效率。问题一要求,对 Voss 映射,探求功率谱与信噪比的某种快速计算方法;然后进一步探讨 Z-curve 映射的频谱与信噪比和 Voss 映射下的频谱与信噪比之间的关系;最后对实数映射,如:A→0,C→1,G→2,T→3,也给出功率谱与信噪比的快速计算公式。为了更有效地解决这些问题,我们将问题一分成下面三个小问题,并逐一解决:

(1) 对 Voss 映射,探求功率谱与信噪比的某种快速计算方法;

(2) 试探讨 Z-curve 映射的频谱与信噪比和 Voss 映射下的频谱与信噪比之间的关系;

(3) 最后对实数映射,如:A→0,C→1,G→2,T→3,给出功率谱与信噪比的快速计算公式。

对问题(1),根据 DFT,对其进行改进,从而得到 Voss 映射下,功率谱与信噪比的某种快速计算方法。

对问题(2),利用相关的数学知识,逐步推导出 Z-curve 映射的频谱与信噪比和 Voss 映射下的频谱与信噪比之间的关系。

对问题(3),对实数映射,A→0,C→1,G→2,T→3,根据数学推导,得到该映射下功率谱与信噪比的快速计算公式。

2. 对问题二的分析

问题二要求,对给定的几个基因序列数据,以及另外给出的带有编码外显子信息的 100 个人和鼠类的,以及 200 个哺乳动物类的基因序列的样本数据集合,对每类基因研究其阈值确定方法和阈值结果。此外,对按照频谱或信噪比特征将编码与非编码区间分类的有效性,以及分类识别时所产生的分类错误作适当分析。

对特定的基因类型的 DNA 序列,将其信噪比 R 的判别阈值取为 $R_0=2$,带有一定的主观性、经验性。对不同的基因类型,所选取的判别阈值也许应该是不同的。对给定的外显子序列,按照问题一中问题(1)得到的信噪比的快速计算方法,得出相应的信噪比的值,最后确定阈值。

3. 对问题三的分析

问题三要求设计一个基因识别算法来识别外显子序列,并对所设计的基因识别算法的准确率做出适当评估,最后将算法用于对附件中给出的 6 个未被注释的 DNA 序列(gene 6)的编码区域的预测。

要有效解决这个问题,必须分三步走。第一步:设计一个算法,利用问题二得到的不同类基因的阈值,区分编码序列和非编码序列;第二步:通过建立相应的指标,来对阈值判别结果进行评价,从而对所设计的基因识别算法的准确率做出评估;第三步:利用该算法对附件中给出的6个未被注释的DNA序列(gene 6)的编码区域进行预测。

4. 对问题四的分析

问题四要求:(1)总结甚至独自提出一些识别基因编码序列的其他特征指数,并对此做相关的分析。(2)利用频谱或信噪比方法去发现基因编码序列可能存在的突变——DNA序列中单个核苷酸替换,删除或者插入等。

14.3 模型的假设

(1)所研究物种的基因序列,不存在基因突变;
(2)假设给定的DNA序列没有错误符号;
(3)所有数据均为原始数据,来源真实可靠。

14.4 名词解释与符号说明

14.4.1 名词解释

1. 功率谱

对于具有连续频谱和有限平均功率的信号或噪声,表示其频谱分量的单位带宽功率的频率函数。

2. 信噪比

信噪比(Signal to Noise Ratio,SNR),又称为讯噪比。狭义来讲是指放大器的输出信号的电压与同时输出的噪声电压的比,常常用分贝数表示,设备的信噪比越高表明它产生的杂音越少。一般来说,信噪比越大,说明混在信号里的噪声越小,声音回放的音质越高,否则相反。

3. 信噪比斜率

$$LR = \frac{R[n] - R[n-50]}{50}$$

14.4.2 符号说明

序号	符号	符号说明
1	$S[n]$	$n=0,1,2,\cdots,N-1$,序列长度为N的DNA序列
2	$u_b[n]$	$b \in I=\{A,C,G,T\}$,Voss映射后的0-1序列
3	$U_b[k]$	k位置上长度为N的复数序列

续表

序号	符号	符号说明
4	$P[k]$	k 位置上 DNA 序列的功率谱
5	$\bar{E}$	DNA 序列总功率谱的平均值
6	R	信噪比,即 DNA 序列 $N/3$ 处功率谱与总功率谱均值的比值
7	LR	信噪比斜率

14.5 模型的建立与求解

14.5.1 问题一的分析与求解

1. 对问题的分析

(1)对 Voss 映射,探求功率谱与信噪比的某种快速计算方法;

(2)试探讨 Z-curve 映射的频谱与信噪比和 Voss 映射下的频谱与信噪比之间的关系;

(3)最后对实数映射,如:A→0,C→1,G→2,T→3,给出功率谱与信噪比的快速计算公式。

对问题(1),根据 DFT,对其进行改进,从而得到 Voss 映射下,功率谱与信噪比的某种快速计算方法。

对问题(2),利用相关的数学知识,逐步推导出 Z-curve 映射的频谱与信噪比和 Voss 映射下的频谱与信噪比之间的关系。

对问题(3),对实数映射,A→0,C→1,G→2,T→3,根据数学推导,得到该映射下功率谱与信噪比的快速计算公式。

2. Voss 映射下的频谱与信噪比的快速计算公式探求

(1)需具备的理论知识。

Voss 映射:在 DNA 序列研究中,首先需要把 A、T、G、C 四种核苷酸的符号序列,根据一定的规则映射成相应的数值序列,以便于对其作数字处理。

令 $I=\{A,T,G,C\}$,长度(即核苷酸符号个数,又称碱基对(Base Pair)长度,单位记为 bp)为 N 的任意 DNA 序列,可表达为:

$$S=\{[n]\mid S[n]\in I, n=0,1,2,\cdots,N-1\} \tag{14-1}$$

即 A、T、G、C 的符号序列 $S:S[0],S[1],\cdots,S[N-1]$。现对于任意确定的 $b\in I$,令

$$u_b[n]=\begin{cases}1, & S[n]=b,\\ 0, & S[n]\neq b,\end{cases} \quad n=0,1,2,\cdots,N-1, \tag{14-2}$$

称之为 Voss 映射,于是生成相应的 0−1 序列(即二进制序列)

$$\{u_b[n]\}:u_b[0],u_b[1],\cdots,u_b[N-1] (b\in I) \tag{14-3}$$

频谱 3—周期性:为研究 DNA 编码序列(外显子)的特性,对指示序列分别做离散 Fourier 变换(DFT)

$$U_b[k] = \sum_{n=0}^{N-1} u_b[n] \mathrm{e}^{-\mathrm{j}\frac{2\pi n k}{N}}, k = 0,1,\cdots,N-1 \tag{14-4}$$

以此可得到四个长度均为 N 的复数序列 $\{U_b[k]\}, b \in I$。计算每个复序列 $\{U_b[k]\}$ 的平方功率谱,并相加则得到整个 DNA 序列 S 的功率谱序列 $\{P[k]\}$:

$$P[k] = |U_A[k]|^2 + |U_T[k]|^2 + |U_G[k]|^2 + |U_C[k]|^2, k = 0,1,\cdots,N-1 \tag{14-5}$$

对于同一段 DNA 序列,其外显子与内含子序列片段的功率谱通常表现出不同的特性。

大量实验表明,外显子序列的功率谱曲线在频率 k=N/3 处,具有较大的频谱峰值(Peak Value),而内含子则没有类似的峰值。这种统计现象被称为碱基 3—周期(3-base Periodicity)。

图 14.2 为酿酒酵母一段外显子的频谱图。

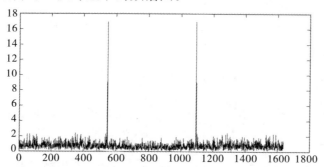

图 14.2 酿酒酵母(AB304259 location=2165..3802)的功率谱曲线

(2)快速公式的探求。

当 DNA 序列较长时,DFT 运算需要耗费大量的计算时间。事实上,DNA 序列在 N/3 处的功率谱值以及信噪比可以由该序列上的 4 种核苷酸的频数分布直接得到。为了推导这一结论,令 x_b, y_b, z_b 表示各核苷酸 b 在 3 个密码子位置上的频数,即核苷酸 b 分别在序列的 $0,3,6,\cdots,N-3$ 和 $1,4,7,\cdots,N-2$ 以及 $2,5,8,\cdots,N-1$ 位置上分别出现的频数,可以得到:

$$\begin{aligned}
\left|U_b\left[\frac{N}{3}\right]\right|^2 &= \left|\sum_{n=0}^{N-1} u_b[n] \cdot \mathrm{e}^{-\mathrm{j}\frac{2\pi n}{N} \cdot \frac{N}{3}}\right|^2 = \left|\sum_{n=0}^{N-1} u_b[n] \cdot \mathrm{e}^{-\mathrm{j}\frac{2\pi}{3}n}\right|^2 = \\
&\left|x_b + y_b \cdot \mathrm{e}^{-\mathrm{j}\frac{2\pi}{3}} + z_b \cdot \mathrm{e}^{-\mathrm{j}\frac{2\pi}{3}}\right|^2 = \\
&\left|x_b + y_b\left(-\frac{1}{2} - \frac{\sqrt{3}}{2}\mathrm{j}\right) + z_b\left(-\frac{1}{2} + \frac{\sqrt{3}}{2}\mathrm{j}\right)\right|^2 = \\
&\left(x_b - \frac{1}{2}(y_b + z_b)\right)^2 + \frac{3}{4}(z_b - y_b)^2 = \\
&(x_b^2 + y_b^2 + z_b^2 - x_b y_b - x_b z_b - y_b z_b) = \\
&(x_b, y_b, z_b)\begin{pmatrix} 1 & -1/2 & -1/2 \\ -1/2 & 1 & -1/2 \\ -1/2 & -1/2 & 1 \end{pmatrix}\begin{pmatrix} x_b \\ y_b \\ z_b \end{pmatrix} = X_b^\mathrm{T} M X_b
\end{aligned} \tag{14-6}$$

其中 $X_b = (x_b, y_b, z_b)^T$, $M = \begin{pmatrix} 1 & -1/2 & -1/2 \\ -1/2 & 1 & -1/2 \\ -1/2 & -1/2 & 1 \end{pmatrix}$。 (14-7)

于是有 DNA 序列的功率谱峰值为：

$$P(N/3) = \sum_{b \in I} X_b^T M X_b \tag{14-8}$$

二次型的系数矩阵 M 为半正定阵，其特征值为 1.5、1.5、0，且当 $x_b = y_b = z_b$ 时，式 (14-8) 值为 0。因此，当碱基在序列的 3 种位置上的频数 $(x_b \ y_b \ z_b)$ 分布偏差越小时，功率谱曲线的峰值 $p[N/3]$ 越接近于 0。功率谱峰值实际上反映了基因密码子出现的某种概率不均衡性。假设 DNA 序列的长度为 N，四个碱基 A、C、G、T 出现的次数分别为 N_A、N_C、N_G、N_T，那么功率谱的序列 $u_{b(n)}$ 为

$$|U_b|^2 = \sum_{k=0}^{N-1} |U_b[k]|^2 = N \cdot N_b \tag{14-9}$$

所以总功率谱的序列为

$$E = \sum_{b \in I} |U_b|^2 = N \cdot N_A + N \cdot N_C + N \cdot N_G + N \cdot N_T = N^2 \tag{14-10}$$

从而总功率的平均值为 $\bar{E} = E/N = N$，因此，简化后的信噪比表达式为

$$R_b = \frac{P(N/3)}{\bar{E}} = \frac{\sum_{b \in I} X_b^T M X_b}{N}, \tag{14-11}$$

也就是说全部 DNA 序列的总信噪比可以被描述成四种碱基序列信噪比的加权求和

$$R = \frac{N_A}{N} R_A + \frac{N_C}{N} R_C + \frac{N_G}{N} R_G + \frac{N_T}{N} R_T \tag{14-12}$$

需要强调说明的是，上式给出的信噪比定义具有以下特点：①在过去基于 DFT 变换的信噪比定义式，序列长度 N 必须是 3 的倍数。而上式定义的信噪比没有这样的限制，因此推广后的定义其适用性更广泛，使用也更方便；②上式定义的信噪比不再需要对指示序列进行 DFT 变换，而只需统计四种核苷酸在 3 种不同位置上出现的频数；其次，这样的计算还有累加功能。信噪比计算工作量锐减。

3. Z-curve 映射和 Voss 映射下的频谱与信噪比之间的关系

Z-curve 映射为

$$\begin{pmatrix} \Delta x[n] \\ \Delta y[n] \\ \Delta z[n] \end{pmatrix} = \begin{pmatrix} 1 & -1 & 1 & -1 \\ 1 & 1 & -1 & -1 \\ 1 & -1 & -1 & 1 \end{pmatrix} \begin{pmatrix} u_A[n] \\ u_C[n] \\ u_G[n] \\ u_T[n] \end{pmatrix} \tag{14-13}$$

令 $F = (\alpha_{ij})_{3 \times 4} = \begin{pmatrix} 1 & -1 & 1 & -1 \\ 1 & 1 & -1 & -1 \\ 1 & -1 & -1 & 1 \end{pmatrix} = \begin{pmatrix} \alpha_1^T \\ \alpha_2^T \\ \alpha_3^T \end{pmatrix} = (\beta_1 \ \ \beta_2 \ \ \beta_3 \ \ \beta_4)$ (14-14)

且 $\tilde{U} = (u_A[n] \quad u_C[n] \quad u_G[n] \quad u_T[n])^T = (u_1[n] \quad u_2[n] \quad u_3[n] \quad u_4[n])^T$

(14-15)

DNA 序列 $\{S[n], n=0,1,2,\cdots,N-1\}$ 中，若 N 为 3 的倍数，将核苷酸符号 $b \in I = \{A,T,G,C\}$ 出现在该序列的 $0,3,6,\cdots,N-3$ 与 $1,4,7,\cdots,N-2$ 以及 $2,5,8,\cdots,N-1$ 等位置上的频数分别记为 x_b, y_b 和 z_b，定义频数矩阵为

$$X = \begin{pmatrix} x_A & x_C & x_G & x_T \\ y_A & y_C & y_G & y_T \\ z_A & z_C & z_G & z_T \end{pmatrix} = \begin{pmatrix} x_1 & x_2 & x_3 & x_4 \\ y_1 & y_2 & y_3 & y_4 \\ z_1 & z_2 & z_3 & z_4 \end{pmatrix} = (X_1 \quad X_2 \quad X_3 \quad X_4) \quad (14\text{-}16)$$

由于 $\Delta x[n] = \alpha_1^T \tilde{U} = \alpha_1^T \cdot (u_1[n] \quad u_2[n] \quad u_3[n] \quad u_4[n])^T$，则序列 $\{\Delta x[n]\}$ 在 $N/3$ 处的功率谱值为

$$\left|\Delta X\left(\frac{N}{3}\right)\right|^2 = \left|\sum_{n=0}^{N-1} \Delta x[n] \cdot e^{-j\frac{2\pi n \cdot \frac{N}{3}}{N}}\right|^2 = \left|\sum_{n=0}^{N-1} \Delta x[n] \cdot e^{-j\frac{2\pi}{3}n}\right|^2 =$$
$$\left|\sum_{j=1}^{4} \alpha_{1j}\left(\sum_{n=0}^{N-1} u_j[n] \cdot e^{-j\frac{2\pi}{3}n}\right)\right|^2 =$$
$$\left|\sum_{j=1}^{4} \alpha_{1j} \cdot (x_j + y_j \cdot e^{-j\frac{2\pi}{3}} + z_j \cdot e^{j\frac{2\pi}{3}})\right|^2 =$$
$$\left|\sum_{j=1}^{4} \alpha_{1j} x_j + \left(\sum_{j=1}^{4} \alpha_{1j} y_j\right) \cdot e^{-j\frac{2\pi}{3}} + \left(\sum_{j=1}^{4} \alpha_{1j} z_j\right) \cdot e^{j\frac{2\pi}{3}}\right|^2 =$$
$$\alpha_1^T X^T M \alpha_1, \quad (14\text{-}17)$$

其中 $M = \begin{pmatrix} 1 & -1/2 & -1/2 \\ -1/2 & 1 & -1/2 \\ -1/2 & -1/2 & 1 \end{pmatrix}$。

同理有

$$|\Delta Y(N/3)|^2 = \alpha_2^T X^T M X \alpha_2, \quad |\Delta Z(N/3)|^2 = \alpha_3^T X^T M X \alpha_3, \quad (14\text{-}18)$$

则 Z-curve 映射在 $\frac{N}{3}$ 处的总功率谱值为

$$P_Z\left[\frac{N}{3}\right] = \left|\Delta X\left(\frac{N}{3}\right)\right|^2 + |\Delta Y(N/3)|^2 + |\Delta Z(N/3)|^2 = \sum_{i=1}^{3} \alpha_i^T X^T M X \alpha_i$$

(14-19)

由于

$$X^T M X = \begin{pmatrix} X_1^T \\ X_2^T \\ X_3^T \\ X_4^T \end{pmatrix} M(X_1 \quad X_2 \quad X_3 \quad X_4) = \begin{pmatrix} X_1^T M X_1 & \cdots & X_1^T M X_4 \\ \vdots & \ddots & \vdots \\ X_4^T M X_1 & \cdots & X_4^T M X_4 \end{pmatrix} \quad (14\text{-}20)$$

则

$$P_Z\left[\frac{N}{3}\right] = \sum_{i=1}^{3} \alpha_i^T X^T M X \alpha_i = 3\left(\sum_{i=1}^{4} X_i^T M X_i\right) - 2\sum_{i<j}^{4} X_i^T M X_j (i \neq j, \text{且 } i,j = 1,2,3,4)$$

(14-21)

变换得

$$P_Z\left[\frac{N}{3}\right] = 4\left(\sum_{i=1}^{4} X_i^T M X_i\right) - \left[\sum_{i=1}^{4} X_i^T M X_i + 2\sum_{i<j}^{4} X_i^T M X_j\right], \quad (14\text{-}22)$$

由于

$$X_1 + X_2 + X_3 + X_4 = \begin{pmatrix} x_A + x_C + x_G + x_T \\ y_A + y_C + y_G + y_T \\ Z_A + Z_C + Z_G + Z_T \end{pmatrix} = \left(\frac{N}{3} \quad \frac{N}{3} \quad \frac{N}{3} \quad \frac{N}{3}\right)^T, \quad (14\text{-}23)$$

且由于

$$MX(1 \quad 1 \quad 1 \quad 1)^T = M(X_1 \quad X_2 \quad X_3 \quad X_4)(1 \quad 1 \quad 1 \quad 1)^T = M(X_1 + X_2 + X_3 + X_4) =$$

$$\begin{pmatrix} 1 & -\frac{1}{2} & -\frac{1}{2} \\ -\frac{1}{2} & 1 & -\frac{1}{2} \\ -\frac{1}{2} & -\frac{1}{2} & 1 \end{pmatrix} \begin{pmatrix} \frac{N}{3} \\ \frac{N}{3} \\ \frac{N}{3} \end{pmatrix} = 0 \text{。} \quad (14\text{-}24)$$

则

$$\sum_{i=1}^{4} X_i^T M X_i + 2\sum_{i<j}^{4} X_i^T M X_j = (1 \quad 1 \quad 1 \quad 1)\begin{pmatrix} X_1^T \\ X_2^T \\ X_3^T \\ X_4^T \end{pmatrix} M(X_1 \quad X_2 \quad X_3 \quad X_4)$$

$(1 \quad 1 \quad 1 \quad 1)^T = 0$ 由于 Voss 映射在 $\frac{N}{3}$ 处的总功率谱值

$$P_V\left[\frac{N}{3}\right] = \sum_{i=1}^{4}(x_i^2 + y_i^2 + z_i^2 - x_i y_i - x_i z_i - y_i z_i) = \sum_{i=1}^{4} X_i^T M X_i \text{。} \quad (14\text{-}25)$$

故而

$$P_Z[N/3] = 4\left(\sum_{i=1}^{4} X_i^T M X_i\right) = 4 P_V[N/3] \text{。} \quad (14\text{-}26)$$

DNA 序列在 Z-curve 映射下的总功率谱

$$E_Z = \sum_{k=0}^{N-1} P_Z[k] = \sum_{k=0}^{N-1}(|\Delta X[k]|^2 + |\Delta Y[k]|^2 + |\Delta Z[k]|^2), \quad (14\text{-}27)$$

根据 Parseval 定理,得到

$$\sum_{k=0}^{N-1}|\Delta X[k]|^2 = \frac{1}{N}\sum_{k=0}^{N-1}|\Delta X[k]|^2 \text{。} \quad (14\text{-}28)$$

推导出

$$\sum_{k=0}^{N-1}|\Delta X[k]|^2 = N \cdot \sum_{k=0}^{N-1}|\Delta X[k]|^2 =$$

$$N \cdot \sum_{k=0}^{N-1}|\alpha_1^T \cdot (u_A[k] \quad u_C[k] \quad u_G[k] \quad u_T[k])^T|^2, \sum_{k=0}^{N-1}|\Delta Y[k]|^2 = N^2 \text{。} \quad (14\text{-}29)$$

同理有

$$\sum_{k=0}^{N-1}|\Delta Y[k]|^2=N^2, E_Z=3N^2, \sum_{k=0}^{N-1}|\Delta Z[k]|^2=N^2 \tag{14-30}$$

于是 DNA 序列在 Z-curve 映射下的平均总功率谱

$$\overline{E_Z}=E_Z/N=3N, \tag{14-31}$$

最终 DNA 序列在 Z-curve 映射下的信噪比

$$R_Z=\frac{P_Z[N/3]}{\overline{E_Z}}=\frac{P_z[N/3]}{3N}, \tag{14-32}$$

而由第一小问中推导出的 Voss 映射下的信噪比快速算法知此时的信噪比为

$$R_V=\frac{P_V[N/3]}{N}, \tag{14-33}$$

可以得出:

$$\frac{R_Z}{R_V}=\frac{P_Z[N/3]}{3N}\cdot\frac{N}{P_V[N/3]}=\frac{P_Z[N/3]}{3P_V[N/3]}=\frac{4}{3}\text{。} \tag{14-34}$$

4. 实数映射的功率谱与信噪比的快速计算公式

针对实数映射 A→0,C→1,G→2,T→3,我们定义以下序列

$$r[n]=0\cdot u_A[n]+1\cdot u_C[n]+2\cdot u_G[n]+3\cdot u_T[n], n=0,1,\cdots,N-1$$

令 $\alpha=(0\quad 1\quad 2\quad 3)^T$,则

$$r[n]=\alpha^T\widetilde{U}=\alpha^T(u_A[n]\quad u_C[n]\quad u_G[n]\quad u_T[n])^T \tag{14-35}$$

将核苷酸符号 $b\in I=\{A,T,G,C\}$ 出现在该序列的 $0,3,6,\cdots,N-3$ 与 $1,4,7,\cdots,$ $N-2$ 以及 $2,5,8,\cdots,N-1$ 等位置上的频数分别记为 x_b,y_b 和 z_b,频数矩阵依然采取 (14-16) 的定义。则序列 $\{r[n]\}$ 在 $N/3$ 处的功率谱值为:

$$\left|R\left(\frac{N}{3}\right)\right|^2=\left|\sum_{n=0}^{N-1}r[n]\cdot e^{-j\frac{2\pi n}{N}}\right|^2=\left|\sum_{n=0}^{N-1}r[n]\cdot e^{-j\frac{2\pi}{3}n}\right|^2=$$

$$\left|\sum_{j=1}^{4}\alpha_j\left(\sum_{n=0}^{N-1}u_j[n]\cdot e^{-j\frac{2\pi}{3}n}\right)\right|^2=$$

$$\left|\sum_{j=1}^{4}\alpha_j(x_j+y_j\cdot e^{-j\frac{2\pi}{3}}+z_j\cdot e^{j\frac{2\pi}{3}})\right|^2=$$

$$\left|\sum_{j=1}^{4}\alpha_jx_j+\left(\sum_{j=1}^{4}\alpha_jy_j\right)\cdot e^{-j\frac{2\pi}{3}}+\left(\sum_{j=1}^{4}\alpha_jz_j\right)\cdot e^{j\frac{2\pi}{3}}\right|^2=$$

$$\alpha^T X^T M X \alpha=$$

$$(0\quad 1\quad 2\quad 3)(X_1^T\quad X_2^T\quad X_3^T\quad X_4^T)M(X_1\quad X_2\quad X_3\quad X_4)(0\quad 1\quad 2\quad 3)^T=$$

$$(X_2^T+2X_3^T+3X_2^T)M(X_2+2X_3+3X_4),$$

$$\tag{14-36}$$

其中 $M=\begin{bmatrix}1 & -1/2 & -1/2 \\ -1/2 & 1 & -1/2 \\ -1/2 & -1/2 & 1\end{bmatrix}$,

即此实数映射下在 $N/3$ 处的总功率谱值：

$$P_R[N/3] = \alpha^T X^T M X \alpha 。 \quad (14\text{-}37)$$

另外 DNA 序列在实数映射下的总功率谱：

$$E_R = \sum_{n=0}^{N-1} P_R[k] = \sum (|R[n]|^2) 。 \quad (14\text{-}38)$$

由离散 Fourier 逆变换和 Parseval 定理，得到

$$\sum_{n=0}^{N-1} |r[n]|^2 = \frac{1}{N} \sum_{n=0}^{N-1} |R[n^2]|, \quad (14\text{-}39)$$

从而推导出

$$\sum_{n=0}^{N-1} |R[n]|^2 = N \cdot \sum_{N=0}^{N-1} |r[n]|^2 =$$

$$N \cdot \sum_{n=0}^{N-1} |\alpha^T \cdot (u_A[n] \quad u_C[n] \quad u_G[n] \quad u_T[n])^T|^2 = \quad (14\text{-}40)$$

$$N(0 \cdot N_A + 1 \cdot N_C + 4 \cdot N_G + 9 \cdot N_T) = N(N_C + 4N_G + 9N_T),$$

其中 $N_A、N_C、N_G、N_T$ 分别为 DNA 序列中四种核苷酸出现的总次数。

于是 DNA 序列在实数映射下的平均总功率谱：

$$\overline{E_R} = E_R/N = N_C + 4N_G + 9N_T, \quad (14\text{-}41)$$

则实数映射下的信噪比为：

$$R_R = \frac{P_R[N/3]}{\overline{E_R}} = \frac{\alpha^T X^T M X \alpha}{N_C + 4N_G + 9N_T} = \frac{(X_2^T + 2X_3^T + 3X_4^T)M(X_2 + 2X_3 + 3X_4)}{N_C + 4N_G + 9N_T} 。$$

$$(14\text{-}42)$$

表 14.1 DNA 三种映射方法对比表

映射方法	DNA 表示	$S(n) = [CGAT]$	维数
VOSS	$X_n = 1 \quad S(n) = X$ $X_n = 0 \quad S(n) \neq X$	$C_n = (1 \ 0 \ 0 \ 0)$ $G_N = (0 \ 1 \ 0 \ 0)$ $A_n = (0 \ 0 \ 1 \ 0)$ $T_n = (0 \ 0 \ 0 \ 1)$	4
Z-curve	$\begin{pmatrix}x_n\\y_n\\z_n\end{pmatrix} = 2 * \begin{pmatrix}1&0&1&0\\1&1&0&0\\1&0&0&1\end{pmatrix} * \begin{pmatrix}x_A[n]\\x_C[n]\\x_G[n]\\x_T[n]\end{pmatrix} - \begin{pmatrix}1\\1\\1\end{pmatrix}$	$\begin{pmatrix}x_n\\y_n\\z_n\end{pmatrix} = \begin{pmatrix}-1&1&1&-1\\1&-1&1&-1\\-1&-1&-1&-1\end{pmatrix}$	3
Real Number	$A \to 0, C \to 1, G \to 2, T \to 3$	$(1 \ 2 \ 0 \ 3)$	1

14.5.2 问题二的分析与求解

1. 对问题的分析

问题二要求，对附中给出的几个基因序列数据，以及另外给出的带有编码外显子信息的 100 个人和鼠类的，以及 200 个哺乳动物类的基因序列的样本数据集合，对每类基因研究其阈值确定方法和阈值结果。此外，对按照频谱或信噪比特征将编码与非编码

区间分类的有效性,以及分类识别时所产生的分类错误作适当分析。

对特定的基因类型的 DNA 序列,将其信噪比 R 的判别阈值取为 $R_0=2$,带有一定的主观性、经验性。对不同的基因类型,所选取的判别阈值也许应该是不同的。对给定的外显子序列,按照问题一中问题(1)得到的信噪比的快速计算方法,得出相应的信噪比的值,最后确定阈值。

2. 基因序列信噪比阈值的研究

现有的对外显子信噪比特性的一些文献中均指出,区分和识别外显子和内含子的一种普遍方法是,外显子的信噪比 $R \geqslant 2$。然而对特定的基因类型的 DNA 序列,将其信噪比 R 的判别阈值取为 $R_0=2$,带有一定的主观性、经验性,不同的基因类型,所选取的判别阈值应该是不同的。对附件中所给的酵母、线虫粘粒、拟南芥、人体线粒子四种生物的 DNA 序列的外显子和内含子进行信噪比统计、比较与分析见表 14.2。

表 14.2 四类基因外显子与内含子信噪比的均值与标准差统计

基因种类	外显子			内含子		
	数量	R 均值	R 标准差	数量	R 均值	R 标准差
酵母	4914	4.905	0	30491	1.5315	0.9169
线虫粘粒	17947	1.4596	0.6389	13971	0.9513	0.6635
拟南芥	13651	1.4845	0.9059	7341	0.4684	0.2265
人线粒体	11297	7.1683	4.6574	5273	0.8330	0.6074

注:因题中所给的酵母的外显子三段基因序列完全一样,以至于其外显子 R 均值为 0,表中采用均值平均法判断的总正确率为 100%。

表 14.3 附件中 gene 100 里人和鼠类外显子和内含子信噪比统计

基因种类	外显子			内含子		
	数量	R 均值	R 标准差	数量	R 均值	R 标准差
人	67	3.0209	3.1269	45	0.8213	0.5694
褐家鼠	38	3.3119	5.7203	54	0.9703	2.0874
小家鼠	322	2.5122	2.6208	439	0.6844	0.3973

由表 14.2 和表 14.3 可知:不同种类基因的信噪比均值有较大的差异,且内含子的标准差一般大于外显子,且四类基因中外显子信噪比均值大于 2 的只有酵母和人体线粒。因此,对基于阈值的外显子判别方法而言,怎样去选择分类阈值指数,阈值的选择显得尤其重要。就这四种基因而言,每种基因的外显子和内含子 R 均值差异很大。

3. 信噪比阈值的确定方法

以下提出区分基因外显子和内含子的信噪比阈值的两种计算方法。

(1)均值平均法。

对已有的均值基因数据,设所有外显子的信噪比均值为 $\bar{x}_1$,所有内含子的信噪比均值为 $\bar{x}_2$,我们可以采用一种简单的确定阈值的方法是求二者的算术平均值,以此来作为外显子和内含子的阈值。即令:

$$R_0 = (\bar{x}_1 + \bar{x}_2)/2 \tag{14-43}$$

(2) 带标准差的加权平均法。

设所有外显子的信噪比标准差为 σ_1，所有内含子的信噪比标准差为 σ_2。确定阈值的带标准差的加权平均法即为

$$R = \frac{\bar{x}_1 \sigma_1 + \bar{x}_2 \sigma_2}{\sigma_1 + \sigma_2}。 \tag{14-44}$$

4. 对阈值确定方法的评价与分析

现假设所选定的信噪比分类阈值为 R_0，即 $R \geqslant R_0$ 作为外显子的判别，$R < R_0$ 则作为内含子的判别。通过阈值判别外显子与内含子的效果可用式指标表示。

敏感性

$$S_n = T_P/(T_P + F_N), \tag{14-45}$$

专一性

$$S_P = T_N/(T_N + F_P), \tag{14-46}$$

式(14-45)、(14-46)中：T_P 表示被正确判为外显子的个数；T_N 表示被正确判为内含子的个数；F_N 表示被错误地判为内含子的个数，F_P 表示被错误的判为外显子的个数。最后，阈值判别的总正确率(A_C)定义为：

$$A_C = (S_n + S_P)/2。 \tag{14-47}$$

用上述两种方法，分别计算人、褐家鼠及小家鼠的信噪比，结果如下：

表 14.4　人及两种鼠的信噪比阈值

基因种类	均值平均法	带标准差的加权平均法
人	1.9211	2.6821
褐家鼠	2.1411	2.6859
小家鼠	1.5983	2.2716

由表 14.4 看出，由均值平均法算出的人、褐家鼠及小家鼠的信噪比阈值有些差别，但用第二种带标准差的加权平均法算出这三种物种基因的信噪比差距不是很大，表明这三种物种的基因有某种相似的特征。

为评价两种阈值确定方法的性能，将其与固定阈值为 2 的方法比较，基于四种基因数据的判别正确统计率如表 14.5 所示。

表 14.5　三种阈值确定方法的判别正确率统计

基因种类	均值平均法		带标准差的加权平均法		以 2 为信噪比阈值	
	$R_0^{(1)}$	A_c	$R_0^{(2)}$	A_c	$R_0^{(3)}$	A_c
酵母	3.2183	1.0000	1.5315	0.8750	2	0.8750
线虫粘粒	1.2055	0.7083	1.2007	0.6333	2	0.0904
拟南芥	0.9765	0.8889	1.2813	0.8500	2	0.5417
人线粒体	4.0007	0.8929	6.4374	0.8438	2	0.8611

由表 14.4 数据分析可知，从总体上而言，采用均值平均法和带标准差的加权平均法

确定阈值的判别正确率高于以 2 为信噪比阈值的判别正确率。尤其是线虫粘粒基因,若采用 2 为信噪比阈值的判别正确率仅为 9.04%,误差较大。而拟南芥基因采用均值平均法和带标准差的加权平均法来确定阈值的判别正确率均在 80% 以上,若采用以 2 为信噪比阈值,其判别的正确率为 54.17%,效果没有前两者好。

14.5.3 问题三的分析与求解

1. 对问题的分析

本题需要探求原题附件中所给的 6 个未被注释的 DNA 序列(gene 6)的编码区域的预测,精准地确定基因外显子区间的两个端点。

一条完整的 DNA 序列,基因通常被划分为许多间隔的片段,其中编码蛋白质的部分,即编码序列片段,称为外显子,不编码的部分称为内含子。外显子是真核生物基因的一部分,它在剪接后仍会被保存下来,并可在蛋白质生物合成过程中被表达为蛋白质。在 DNA 序列中,内含子和外显子的交替排列构成了割裂基因。我们利用通过核苷酸分布外显子预测(EPND)法来查找一个完整的 DNA 序列上外显子区域。

2. 基因识别方法

基于 DNA 序列上"移动序列"信噪比曲线的基因识别方法:

设已知 DNA 序列 S 和它的指示序列 $\{u_b[n]\}, b\in I, n=0,1,\cdots,N-1$。对任意 $n(0<n\leqslant 1)$,通常 n 取 3 的倍数并逐步增大。在 n 的左边一个长度为 n 的序列片段 $[0, n-1]$ 上,相应的子序列 $S_{0\sim n-1}$ 称为 DNA 序列 S 的"移动子序列",作该移动子序列对应的四个指示序列的离散 Fourier 变换(DFT)

$$U_b[k] = \sum_{i=0}^{i=n-1} u_b[i] e^{-j\frac{2\pi k}{n}}, k=0,1,\cdots,n-1,$$

并求出移动子序列 $S_{0\sim n-1}, n=0,1,\cdots,N-1$ 上的信噪比 $R[n]$

$$R[n] = \frac{P\left[\frac{n}{3}\right]}{\overline{E}[n]} = \frac{\left|U_A\left[\frac{n}{3}\right]\right|^2 + \left|U_T\left[\frac{n}{3}\right]\right|^2 + \left|U_G\left[\frac{n}{3}\right]\right|^2 + \left|U_C\left[\frac{n}{3}\right]\right|^2}{\overline{E}[n]}, 0<n\leqslant N-1$$

(14-48)

其中 $\overline{E}[n]$ 为移动子序列 $S_{0\sim n-1}$ 的功率谱的平均值。

$$\overline{E}[n] = \frac{\sum_{k=0}^{n-1} p[k]}{n}。$$

(14-49)

在问题一的求解中,我们对 Voss 映射下,探求过功率谱与信噪比的某种快速计算方法。在这里我们利用问题一得到的快速算法,求出该移动子序列的信噪比值。

由于在一个 DNA 序列中,绝大多数的外显子序列和内含子序列都超过 50 个碱基对,因此这里定义在位点 n 上信噪比的斜率为 LR,LR 的表达式为:

$$LR = \frac{R[n] - R[n-50]}{50} \quad n=50,51,\cdots,N-1。$$

(14-50)

对于给定的阈值 α，当 $R[n] \geq \alpha, LR > 0$ 时，认为该点核苷酸为外显子，否则为内含子。

3. 算法的实现

EPND算法：EPND(Algorithm for exon prediction by nucleotide distribution)是通过核苷酸分布来预测外显子的位置，算法的具体步骤如下：

① n 取值为 0。

② 利用问题一得到的快速算法，n 逐步递增 1，计算 DNA 移动子序列 $S_{0 \sim n-1}$ 的信噪比值 $R[n]$，$n=0,1,\cdots,N-1$。

③ n 从 50 开始，逐步递增 1，计算 $LR = \dfrac{R[n] - R[n-50]}{50}$ 的值（$n=50,51,\cdots,N-1$）。

④ 当 $n=i$ 时，如果 $LR > 0, R[i] \geq R_0^i$，其中，R_0^i 为第 i 条 DNA 序列的最佳信噪比。我们就认为该核苷酸为外显子，也即第 $i+1$ 个位置的核苷酸为外显子。否则为内含子。

⑤ 消噪：当 DNA 区域少于 50 个碱基对时，如果在第④步被判断为内含子，且该区域夹在两个外显子之间，那么这个判断就是假负的，这时就认为这个区域也是外显子。同样，如果在第④步被判断为外显子，且该区域夹在两个内含子之间，那么这个判断就是假正的，这时就重置这个区域为内含子。

具体实际操作的流程图如图 14.3 所示。

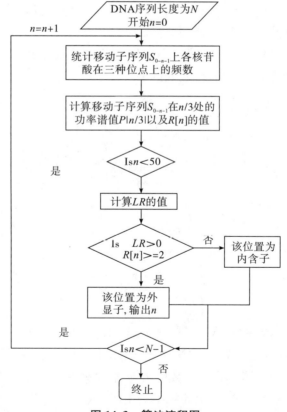

图 14.3 算法流程图

4. 结果分析

对所需要判断的 6 个基因,经过编程运算,我们能具体判别每一个基因的外显子的起始点位置和终止点位置,具体结果分析如下:

对于基因 1,根据最佳信噪比公式确定其信噪比 $R_0^1=1$,共找到其有 10 段外显子区间,其信噪比移动曲线如图 14.4 所示,对于基因 2,根据最佳信噪比公式确定其信噪比 $R_0^2=0.5$,其信噪比移动曲线如图 14.5 所示。

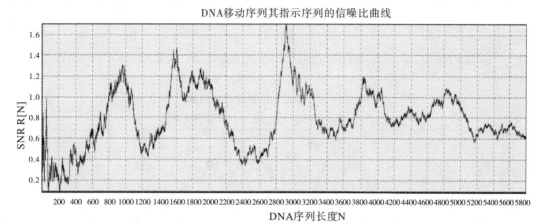

图 14.4　基因 1 信噪比移动曲线

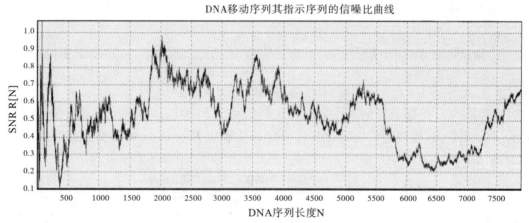

图 14.5　基因 2 信噪比移动曲线

对于基因 3,根据最佳信噪比公式确定其信噪比 $R_0^3=1$,其信噪比移动曲线如图 14.6 所示,对于基因 4,根据最佳信噪比公式确定其信噪比 $R_0^4=1$,其信噪比移动曲线如图 14.7 所示。

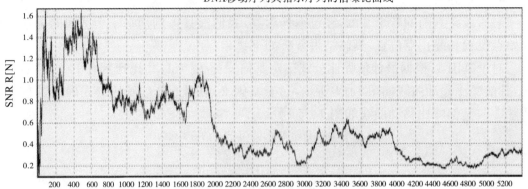

图 14.6　基因 3 信噪比移动曲线

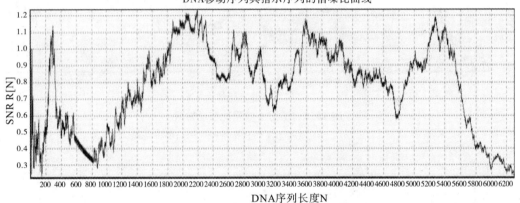

图 14.7　基因 4 信噪比移动曲线

对于基因 5，根据最佳信噪比公式确定其信噪比 $R_0^5=1$，其信噪比移动曲线如图 14.8 所示，对于基因 6，根据最佳信噪比公式确定其信噪比 $R_0^6=0.4$，其信噪比移动曲线如图 14.9 所示。

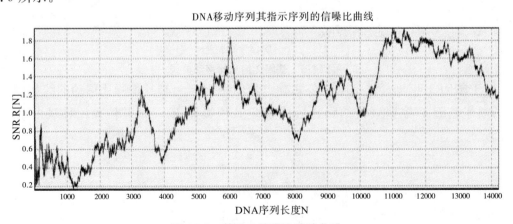

图 14.8　基因 5 信噪比移动曲线

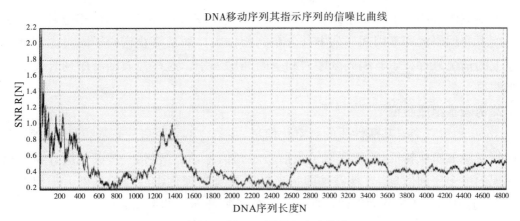

图 14.9 基因 6 信噪比移动曲线

最终确定的 gene 6 中各基因序列中外显子的识别端点位置如表 14.6 所示。

表 14.6 gene6 中各基因序列外显子识别的端点位置

基因序号	外显子位置
基因 1	Location(804…996；1518…1614；724…1818；1869…2026；2821…2946；；3139…3224；3805…3863；3915…3945；4000…4050；4797…4864)
基因 2	Location(160…223；616…670；1003…1097；1582…1667；1803…1909；1987…2081；2557…2626；2682…2734；3118…3242；3389…3579；3810…3935；4272…4323；4429…4487；4947…5006；5048…5124；5173…5230；7429…7498)
基因 2	Location(241…501；562…668；1782…1861)
基因 4	Location1844…2090；2159…2381；2763…2827；3544…3625；4021…4072；5178…5304；5361…5424)
基因 5	Location(3207…3346；4755…5074；5130…5193；5730…5780；5855…6098；6274…6350；6477…6528；6582…6782；7141…7294；7511…7570；7671…7747；8548…8803；9016…9125；9215…9442；9526…9633；11414…11463；11771…11904；12349…12597；12653…12705；12802…12966；13067…13354；13636…13699)
基因 6	Location(145…199；297…347；1179…1300；2629…2774；2926…2982；3288…3353)

14.5.4 问题四的拓展

常用的识别基因编码区的方法包括两种，分别为：内在的以统计预测为基础的算法和外在的基于同源性搜索的算法。内在算法又称为从头预测算法，包含信息类型以及预测模型两个方面。这些算法用到的信息类型多为统计特征，如密码子使用偏性，密码子频率，碱基偏性等。除了频谱或者信噪比等特性外，目前基因的新特征还有如下几个。

1. 包含碱基偏性信息和终止密码子相位分布信息的新特征

一个开放阅读框或者 DNA 的一段序列中，按照密码子中三个碱基的位置，可以将序列分成三个相位(phase)，比如相位 $p=1$，即为：1,4,7,…；$p=2$，即为：2,5,8,…。

在人类编码区序列中，四个核苷酸 A，T，C 和 G 在三个相位上有不同分布频率，在大量的统计数据下，这种分布呈现一定的规律性，即碱基偏性，而在非编码区中及基因间序列碱基分布是随机的，可以看成在三个相位上四个核苷酸的出现频率是相同的。

研究表明编码区序列上终止密码子 STOP={TGA,TAA,TAG} 分布的相位数一般少于同长度下非编码区和基因间序列中其分布的相位数。而且终止密码子同时出现在三个相位上的情况只存在于非编码区序列和基因间序列中。这些性质随着序列长度的增长更加明显。

2. 伪碱基组成特征

Chou 利用三个氨基酸性质的数字化信息，并结合不同距离下氨基酸间的作用信息，提出伪氨基酸组成特征，不仅提取了蛋白质序列信息，而且包括一些其他统计特征所无法体现的更多序列信息。[13] 伪碱基即是在碱基自身物理化学性质的基础上提出的一组特征，用于识别人类短基因编码区序列。伪碱基特征向量的形式表达如下：

$$x = (x_1, \cdots, x_4, x_{4+1}, \cdots, x_{4+\lambda}),$$

其中，λ 为人工设置的偏移最大值。向量的各个元素具体表达如下：

$$x_u = \begin{cases} \dfrac{f_u}{\sum_{i=1}^{4} f_i + w \sum_{j=1}^{\lambda} \theta_j}, & (1 \leqslant u \leqslant 4), \\ \dfrac{w\theta_{u-4}}{\sum_{i=1}^{4} f_i + \sum_{j=1}^{\lambda} \theta_j}, & (4+1 \leqslant u \leqslant 4+\lambda), \end{cases}$$

其中，f_i 表示各种碱基在序列中出现的频率，w 为权值，$\theta_1, \theta_2, \cdots, \theta_\lambda$ 表示在相应偏移 i 下获得的偏序数值。

选择碱基分子量和 π 电子共核能值两种碱基物理化学性质。$H_k(R_i)$ 表示碱基 R_i 在性质 k 下的数值。$H_k(R_i)$ 的公式如下：

$$\begin{cases} H_1(i) = \dfrac{H_1^0(i) - \sum_{i=1}^{4} \dfrac{H_1^0(i)}{4}}{\sqrt{\sum_{i=1}^{4} \left[H_1^0(i) - \sum_{i=1}^{4} \dfrac{H_1^0(i)}{4} \right]^2}}, \\ H_2(i) = \dfrac{H_2^0(i) - \sum_{i=1}^{4} \dfrac{H_2^0(i)}{4}}{\sqrt{\sum_{i=1}^{4} \left[H_2^0(i) - \sum_{i=1}^{4} \dfrac{H_2^0(i)}{4} \right]^2}}, \end{cases}$$

其中 $H_1^0(i), H_2^0(i)$ 表示碱基分子量和 π 电子共振能性质下碱基 i 的具体数值。取 $\lambda=5$，得到一个 9 维的伪碱基组成向量，即为 $F_7 \cdots F_{15}$。

结合终止密码子的相位分布特性和碱基偏性，可以提出两个新的用于识别编码区的特征：

$$F_1 = \text{Max}(T_1(p)) - \text{Min}(T_1(p)), p \in \{1,2,3\},$$
$$F_2 = \text{Max}(T_2(p)) - \text{Min}(T_2(p)), p \in \{1,2,3\},$$

其中 $T_1(p) = f_{G(p)} f_{A(p+1)} / ((f_{STOP(p)} + w_1) \times (1 + 0.9 \times S))$,

$$T_2(p) = f_{A(p)} f_{G(p)} / (f_{C(p)} f_{G(p+1)} f_{A(p+2)} + f_{STOP(p)} + w_2) \times (1 + 0.1 \times S)$$

其中 p、$p+1$、$p+2$ 分别表示样本序列的三个相位。S 表示终止密码子 STOP 分布的相位数,$S \in \{0,1,2,3\}$。$f_b(p)$ 表示在序列的相位为 p 时碱基 b 或者密码子 b 的频率。两个阈值 w_1,w_2 取值分别为:0.006,0.009。

3. 改进的终止密码子变量

对终止密码子变量的系数进行调整。即把 $F_3' = (1+S^2) \times N_{STOP}$,$S \in \{0,1,2,3\}$,改进为:$F_3 = (1+3^S) \times f_{STOP}$,$S \in \{0,1,2,3\}$,

其中,f_{STOP} 表示序列中终止密码子出现的频率。研究结论表明,改进特征在各个长度的识别结果都优于终止密码子变量 F_3'。

可以结合终止密码子三相位分布信息与碱基偏性信息,得到两个新的混合性质特征。同时将它们与改进的终止密码子变量一起组成新的三维特征向量用于识别人类短基因编码序列。由于碱基间的相互作用可能包含重要的结构信息。Chou 等提出伪氨基酸组成方法提取蛋白质序列上不同位置氨基酸的相互作用信息,可以将伪氨基酸组成方法用于 DNA 序列结构信息提取,得到一组新的特征——伪碱基组成特征。该组特征包含了碱基自身的三个物理化学性质信息。最终得到一组具有较强生物意义的 15 维特征向量,并采取支持向量机为分类器进行编码区的识别。支持向量机(SVM)是统计学习理论的一个分支。它采用结构风险最小化原理,具有很强的泛化能力。当训练集数量有限时,支持向量机仍可得到一个分类性能良好的分类模型。支持向量机的提出是基于二分类线性可分的情况,通过求最优分类线(面)进行分类。当输入为非线性可分时,它通过核函数把数据映射到高维空间,再寻找最优分类超平面进行样本分类。对于非线性分类来说,高斯径向基核函数的性能比较优秀。

4. 径向基函数神经网络

对短编码序列,要提高基因识别的精确度,必须选择精细的模型,径向基函数神经网络是较理想的选择。径向基函数神经网络拥有与传统的神经网络相同的基本结构。其区别处在于径向基函数网络的隐层神经元采用径向基函数作为传输函数。隐层通过径向基函数,对输入向量与另一个参考向量进行内积操作,实现输入向量非线性映射操作。其中参考向量被称为径向基函数的中心向量。隐层中心向量的优化对神经网络的分类性能至关重要。其优化的目标是在保证神经网络良好的泛化性能下,尽可能减少隐层中心的数目。最常用的径向基函数为高斯核函数,公式如下:

$$\varphi_i(x) = \exp\left(-\frac{\|u_i - c_i\|}{2\sigma_i^2}\right),$$

其中 c_i 表示第 i 个节点的中心,u_i 表示输入向量,σ_i 表示函数的宽度参数,控制了函数的径向作用范围。通过隐层径向基函数的非线性映射,输出变成线性的。输出层的输出值为隐层输出的线性组合,具体表示如下:$y = Wq + b$,其中 y 表示输出层的输出值,W 表示隐层和输出层间的权值矩阵,q 表示隐层的输出,b 为阈值向量。

14.6 模型的评价与推广

14.6.1 模型的优缺点

1. 优点

(1)利用 Java 软件编出相应的计算软件,处理某些问题简便、快捷。

(2)本文建立的模型及一些处理方法,对具有物种信息的基因片断均可以使用,从而方便了今后的研究。

(3)运用多种数学软件进行计算,取长补短,使计算结果更加准确。

2. 缺点

(1)由于基因数据下载对软件要求较高,从而使得数据收集和处理方面不是很完善。

(2)由于时间匆忙,未能探求多种物种的基因序列特征。

14.6.2 模型的推广

本文针对 DFT 变换在计算功率谱和信噪比的算法中,提出的快速算法,减少了今后类似问题的计算量,并且在基因识别的问题上,建立的 EPND 算法,为今后识别基因编码区外显子和内含子提供了一些指导意义。

参考文献

[1] Burge C., Karlin S., Prediction of Complete Gene Structures in Human Genomic DNA[J]. *J. Mol. Biol.* 268,1997,78—94.

[2] Anastassiou D., Frequency-domain Analysis of Biomolecular Sequences[J]. *Bioinformatics* 16,2000,1073—1081.

[3] Kotlar D., Lavner Y., Gene Prediction by Spectral Rotation Measure: a New Method for Identifying Protein-coding Regions[J]. *Genome Res.* 13,2003,1930—1937.

[4] Berryman M. J., Allison A., Review of Signal Processing in Genetics[J]. *Fluctuation and Noise Letters.* 5(4),2005,13—35.

[5] Sharma S. D., Shakya K., Sharma S. N., 2011. Evaluation of DNA mapping schemes for exon detection[J]. *International Conference on Computer, Communication and Electrical Technology-ICCCET*,2011, 18th & 19th.

[6] Yin C., Yau S. S. T. Prediction of protein coding regions by the 3-base periodicity analysis of a DNA sequence[J]. *Journal of Theoretical Biology*,2007,247, 687—694.

[7] Yin C., Yau S. Prediction of Protein Coding Regions by the 3-base Periodicity Ananlysis of a DNA Sequence[J]. *J Theor Biol*,2007,247:687—694.

[8] Yin C., Yau S. A Fourier Characteristic of Coding Sequences: Origins and a Non-Fourier Approximation [J]. *J Comput Biol*, 2005, 9: 1153—1165.

[9] Jianfeng Shao, Xiaohua Yan, Shuo Shao. SNR of DNA Sequences Mapped by General Affine Transformations of the Indicator Sequences [J]. *J Math Biol*, 2012.

[10] 吴宪明, 吴松峰, 任大明等. 密码子偏性的分析方法及相关研究进展 [J]. 遗传, 2007(04): 420—426.

[11] 齐立省, 王永红, 展永等. 蛋白质编码区碱基分布与终止密码子的关系 [J]. 山东理工大学学报(自然科学版), 2004(02): 95—98.

[12] Wang Y., Zhang C., Dong P. Recognizing shorter coding regions of human genes based on the statistics of stop codons [J]. *Biopolymers*, 2002, 63(3): 207—216.

[13] Chou K. C. Prediction of protein cellular attributes using pseudo amino acid composition [J]. *Proteins*, 2001, 43(3): 246—255.

[14] 何海峰. 基于新型特征的基因识别方法研究 [D]. 长沙: 湖南大学, 2011.

[15] 邵建峰, 严晓华, 邵伟等. DNA 序列信号 3—周期特性 [J]. 南京工业大学学报(自然科学版), 2012, 34(04): 133—137.

[16] 周立前. 基因识别算法研究与基因组进化分析 [D]. 湘潭: 湘潭大学, 2008.

[17] 赵亚宁, 赵彦晖. 基于统计分析法的肿瘤特征基因提取和分类研究 [J]. 襄樊学院学报, 2011, 32(08): 13—16.

[18] 林岩. 原核生物基因识别新算法研究及 DNA 序列分析 [D]. 天津: 天津大学, 2006.

[19] 沈志军. 原核生物基因识别 [D]. 天津: 河北工业大学, 2007.

[20] 饶妮妮, 俞心博. DNA 序列分析的现代信号处理方法初探 [J]. 中国科协第 81 次青年科学家论坛, 2003.

建模特色点评

❋论文特色❋

◆标题分析: 标题为"基因识别问题及其算法实现", 原始标题没动, 涉及算法实现。

◆方法鉴赏: 利用 Java 软件编出相应的计算软件, 处理某些问题简便、快捷; 建模及一些处理方法, 对具有物种信息的基因片断均可以使用, 从而方便了今后的研究。多种数学软件进行计算, 取长补短, 使计算结果更加准确。

◆写作评析: 论文写作条理性好, 层次分明, 图文并茂。

❋不足之处❋

缺少误差分析, 没有对模型进行改进, 缺少灵敏度分析。

第15篇 中等收入定位与人口度量模型研究

◆ 竞赛原题再现

2013年E题 中等收入定位与人口度量模型研究

居民收入分配关系到广大民众的生活水平,分配公平程度是广泛关注的话题。其中中等收入人口比重是反映收入分配格局的重要指标,这一人口比重越大,意味着收入分配结构越合理,称之为"橄榄型"收入分配格局。在这种收入分配格局下,收入差距不大,社会消费旺盛,人民生活水平高,社会稳定。一般经济发达国家都具有这种分配格局。我国处于经济转型期,收入分配格局处于重要的调整期,"橄榄型"收入分配格局正处于形成阶段。因此,监控收入分配格局的变化是经济社会发展的重要课题,例如需要回答,与前年比较,去年的收入分配格局改善了吗?改善了多少?可见实际上需要回答三个问题:什么是"橄榄型"收入分配格局?收入分配格局怎样的变化可以称之为改善?改善了多少?直观上,中间部分人口增加,则收入分配格局向好的方向转化。于是基本问题回答什么是中间部分。

中等收入人口的多少与两极分化(polarization)的程度有关,所谓两极分化,用密度函数表示时,例如严重右偏且厚尾,也即中间部分空洞化。两极分化与收入不平等(inequality)是不同的概念,文献对这两个概念进行了准确阐述。建立了一种指数,这种指数说明两极分化的大小或严重程度,该指数扩大意味着两极分化严重了,这时表示中等收入人口缩小了。反之若该指数缩小了,则意味着中等收入人口扩大了。但该文献并没有给出测算中等收入人口比例大小的方法。

为此,需要研究中等收入定位与人口度量问题,请你根据表15.1中给出的分组数据,用数学模型研究给出的问题。

表 15.1 收入分配分组数据

x_j	x_{j+1}	f_j	p_j	L_j
0.00	999.00	0.0780	0.0780	0.0059
1000.00	1499.00	0.0560	0.1340	0.0165
1500.00	1999.00	0.0420	0.1760	0.0276
2000.00	2499.00	0.0470	0.2230	0.0436
2500.00	2999.00	0.0420	0.2650	0.0611
3000.00	3499.00	0.0440	0.3090	0.0828
3500.00	3999.00	0.0410	0.3500	0.1061
4000.00	4999.00	0.0860	0.4360	0.1647
5000.00	5999.00	0.0920	0.5280	0.2413
6000.00	6999.00	0.0880	0.6160	0.3279
7000.00	7999.00	0.0800	0.6960	0.4188
8000.00	8999.00	0.0650	0.7610	0.5024
9000.00	9999.00	0.0520	0.8130	0.5772
10000.00	11999.00	0.0780	0.8910	0.7071
12000.00	14999.00	0.0560	0.9470	0.8216
15000.00	24999.00	0.0430	0.9900	0.9453
	25000.00	0.0100	1.0000	1.0000

请研究如下问题：

一、构造满足式(15-9)的新模型 $L(p,\tau)$，使其能很好地拟合上述分组数据、反映经济规律。

注意，本题中最好能构造新模型，而不是通过简单处理（例如加权）文献中的已有模型而得到的模型。

二、研究可否改进上述提到的收入空间法，这时需要研究确定中等收入的范围、中等收入人口的范围的科学方法，以克服中等收入区间取法的任意性；研究可否改进上述提到的人口空间法，例如研究在各年中 p_1 与 p_2 取不同的值时，纵向比较各年中等收入人口与收入变动的方法。

提示：目前经济理论界将中等收入人口定义为中位收入附近的人口，于是若中间部分比前一年隆起得更高，则认为中等收入人口扩大了；若两边人口扩大了，则中等收入人口减少了。所提出的原理与模型应与这一直观相符。其他有关价值取向方面的示例性提示见问题四。

三、利用最后表二～表五所附 A，B 两个地区前后两个不同年份的收入分配分组数据，请研究：(1)对各地区、各年份的中等收入的数量（或范围）、中等收入人口的数量或范围进行定量描述，说明中等收入人口的变化趋势；(2)比较两个地区的中等收入人口、收入等变化情况。

四、除二题中所述方法外,提出中等收入人口的定义、原理及经济学意义,并提出与之相应的中等收入人口的测算方法、模型或指数,说明其经济学意义。

原题详见 2013 年全国研究生数学建模竞赛 E 题。

◆ 获奖论文精选

中等收入定位与人口度量模型[①]

摘要: 本文首先构造了新的洛伦兹函数,运用最小二乘原理结合所给数据中的和值,建立数据模拟模型,然后从洛伦兹曲线的经济含义及模型本身入手,分别从改进的"收入空间法"及"人口空间法"建立中等收入定位与人口度量模型。从总体上就题目中给出的四组数据,求解出不同地区不同年份的中等收入范围及中等收入人口比例,并对不同地区不同年份的收入分配格局作出了优劣评判。最后提出了一种新的中等收入人口测算方法,并对模型作出分析、评价和改进。

对于问题一,我们建立了模型Ⅰ——基于非线性最小二乘的洛伦兹曲线拟合模型。先从题中所给参考文献中找出 10 种洛伦兹曲线模型,通过分析这 10 种模型的函数特性,构造出一种新的洛伦兹函数,利用最小非线性二乘法原理,根据题中已给出的收入分配分组数据,对新模型进行拟合,并从模型优劣的三个判别准则判断出新构造的模型比其他 10 种模型要好。

对于问题二,主要从新构造的洛伦兹曲线模型入手,建立了模型Ⅱ——基于"收入空间法"的中等收入定位模型,计算出确定的中等收入区间范围,这样就规避了一般"收入空间法"的缺陷;从影响收入分配格局变化的因素——两极分化指数着手,运用改进的模糊综合评价方法建立了模型Ⅲ——基于"人口空间法"的中等收入定位模型,纵向比较某个地区各年收入分配格局的变化或横向比较某年各个地区的收入分配格局,规避了一般"人口空间法"的片面性。

对于问题三,只需要在问题二研究的基础上,结合题目中所给的四组收入分配数据分别对模型Ⅱ、模型Ⅲ进行实证检验。基于模型Ⅱ可以计算出各地区、各年份中等收入人口的比例依次为 55%,基于模型Ⅲ可以计算出各地区、各年份中等收入范围依次为 56.5% 和 54.99%。A 区的后一年较前一年的中等收入人口比例下降不明显,B 区的后一年中等收入人口所占比例为 55%,较前一年相比有所下降。B 区后一年中等收入人口所拥有的收入占总收入的比例较前一年相比有所上升,且 B 区收入分配格局明显要比 A 区好。

[①] 本文获 2013 年全国二等奖。队员:吴青青,李丽,石琼强;指导教师:杨桂元等。

对于问题四,综合考虑各种普遍的价值取向和一般的经济规律,我们建立模型Ⅳ,构建中等收入人口比例指标,衡量区间范围,确定区间长度,利用该指标建立一个收入分配合理性指数,衡量收入分配格局。

关键词:中等收入;洛伦兹曲线;人口空间法;收入空间法;Matlab

15.1 问题的重述

15.1.1 背景知识

党的十七大报告提出,要形成合理有序的收入分配格局,"中等收入者占多数"。2010年4月1日温家宝在《关于发展社会事业和改善民生的几个问题》中,提出"橄榄型"社会是稳定的社会,我国要形成"橄榄型"分配格局。"橄榄型"收入分配格局指的是低收入者和高收入者相对较少,中等收入者占多数的收入分配结构。在这种收入分配格局下,收入差距不大,社会消费旺盛,人民生活水平高,社会稳定。一般经济发达国家都具有这种分配格局。我国处于经济转型期,收入分配格局处于重要的调整期,"橄榄型"收入分配格局正处于形成阶段。因此,中等收入的定位与人口定量模型的研究受到当下经济学者们的青睐。

15.1.2 要解决的问题

(1)构造满足一定条件的新模型 $L(p,\tau)$,使得能很好地拟合上述分组数据、反映经济规律。并在现有参考文献中(参考文献[1]-[3]有关的文献)找出至少10种模型,与提出的新模型进行比较。通过比较,说明新构造的模型不差。

(2)研究可否改进上述提到的收入空间法,这时需要研究确定中等收入的范围、中等收入人口的范围的科学方法,以克服中等收入区间取法的任意性;研究可否改进上述提到的人口空间法,例如研究在各年中 p_1 与 p_2 取不同的值时,纵向比较各年中等收入人口与收入变动的方法。

(3)利用最后附录中所附 A,B 两个地区前后两个不同年份的收入分配分组数据,请研究:(1)对各地区、各年份的中等收入的数量(或范围)、中等收入人口的数量或范围进行定量描述,说明中等收入人口的变化趋势;(2)比较两个地区的中等收入人口、收入等变化情况。

(4)除第二题中所述方法外,提出中等收入人口的定义、原理及经济学意义,并提出与之相应的中等收入人口的测算方法、模型或指数,说明其经济学意义。

15.2 问题的分析

居民收入分配关系到广大民众的生活水平,分配公平程度是广泛关注的话题。其中中等收入人口比重是反映收入分配格局的重要指标,这一人口比重越大,意味着收入分配结构越合理,确定这一指标要分为两个步骤进行:第一,对所给的分组数据所表示的洛伦兹曲线进行拟合,寻找累计人口比重与累计收入比重之间的具体关系;第二,确定既可以衡量中等收入水平,又可以根据收入与人口关系反映收入分配合理程度的指标。

要合理准确处理好本问题,关键必须弄清问题的相关知识并对问题作出深入的分析。

15.2.1 相关知识的介绍

中等收入的变化受到众多方面因素的影响,因此对中等收入的定位与定量研究也就十分复杂,很难在一个模型中综合考虑到各个因素的影响。为了更好地解决此问题,我们分析了题目以及参考文献中所给的相关模型,经过理论推理证明,确定洛伦兹曲线为解决每一问题的主线,因此有必要对洛伦兹曲线进行相关介绍。

洛伦兹曲线的弯曲程度有重要意义,一般来讲,它反映了收入分配的不平等程度。弯曲程度越大,收入分配越不平等,反之亦然。特别是,如果所有收入都集中在一人手中,而其余人口均一无所获时,收入分配达到完全不平等。图 15.1 中曲线是某收入分配的洛伦兹曲线。其中横轴表示人口比例,纵轴表示总收入比例。显然,图中曲线位置越高,所代表的收入分配越平等。其中 45°线可以理解为平等收入线,这时,任何低收入端人口比例为 p 的人口拥有的总收入比例也是 p,从而必定是完全平等的收入分配。因此定义 45°线与 $L(p)$ 之间面积的 2 倍为基尼系数。于是基尼系数定义为:

$$G = 1 - 2\int_0^1 L(p)\mathrm{d}p \tag{15-1}$$

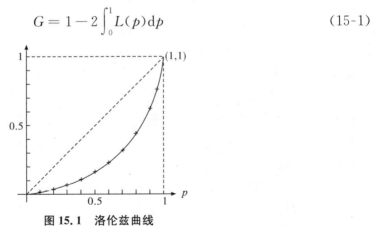

图 15.1　洛伦兹曲线

15.2.2 对问题的具体分析

1. 对问题一的分析

由于要构造新的函数即洛伦兹曲线模型(Lorenz curve)来拟合表 15.1 中的分组数据,并与文献[1]中已有的至少 10 种模型进行比较,说明新构造的模型并不差。这就需要我们利用洛伦兹函数的性质来构造一种拟合精度高的洛伦兹曲线模型,要具有洛伦兹曲线的函数性质,并且经过非线性最小二乘拟合的均方误差、平均绝对误差、最大绝对误差与已有模型相比要绝对的小。由此我们建立了模型Ⅰ：基于非线性最小二乘的洛伦兹曲线模型。

2. 对问题二的分析

对"收入空间法"进行分析，找出该方法的缺陷是区间取法的任意性，通过制定科学的计算方法，保证中等收入区间取法的一致有效性。对"人口空间法"进行分析，发现此种方法存在的缺陷是该种方法描述收入分配格局变化不够全面，必须与其他收入分配格局的影响因素配合使用，制定一个统一的评价标准以确保结论的有效性。

3. 对问题三的分析

基于对问题二的研究基础上，对A，B两个地区前后两个不同年份的收入分配分组数据，运用问题二中建立的模型Ⅱ、模型Ⅲ进行实证分析求解，最后定量分析各地区、各年份的中等收入的数量（或范围）、中等收入人口的数量（或范围）进行定量描述，说明中等收入人口的变化趋势；并且比较两个地区的中等收入人口、收入等变化情况。

4. 对问题四的分析

基于前面三个问题的求解与结果分析，对中等收入人口重新定义，该定义必须满足一般的价值取向与经济规律，同时建立一个指标可以全面衡量分配收入格局变化。

15.3 模型的假设

（1）模糊综合评价过程中，设某地区的中等收入人口拥有总收入占总收入的比重的理想值为0.6，两极分化指数理想值为0。

（2）在模型Ⅱ中，由于将平均收入附近一定范围人口数定义为中等收入人口数，我们认为中等收入者所拥有总收入占总人口收入的60%。

（3）在较短的时间内，平均收入变化较小，可以认为不变。

（4）所有数据均为原始数据，来源真实可靠。

15.4 符号说明

表 15.2 符号说明

序号	符号	符号说明
1	x	表示居民收入实际值
2	$F(x)$	表示居民收入为x时的累计概率
3	p	表示累计人口比重
4	$L(p)$	表示累计人口比重为p时的收入累计比重
5	$L_i(i=0,1,\cdots,10)$	依次表示为新构造的模型和参考文献中的10种模型
6	p_l	表示中等收入区间的下界对应的累计人口比重

续表

序号	符号	符号说明
7	p_h	表示中等收入区间的上界对应的累计人口比重
8	L_l	表示中等收入区间的下界对应人口拥有总收入的累计比重
9	L_h	表示中等收入区间的上界对应人口拥有总收入的累计比重
10	A_i, B_i	分别表示 A、B 两个地区第 i 年的状态
11	$Q(x)$	表示总体人口收入的两极分化指数
12	$G(x)$	表示总体的模糊综合评价指数
13	$H(x)$	表示某地区的收入分配的合理性指标

15.5 模型的建立与求解

从所要解决的问题和对问题所做的假设出发，我们对问题一建立了模型Ⅰ，对问题二建立了模型Ⅱ和模型Ⅲ，对问题四建立了模型Ⅳ。

模型Ⅰ 基于非线性最小二乘的洛伦兹曲线模型

本模型是首先构造新的洛伦兹曲线函数形式，根据所给数据进行非线性拟合，计算参数值，据此可以给出洛伦兹曲线的方程。同时根据曲线拟合评价指标，将拟合方程与从参考文献中找出的各类洛伦兹曲线方程进行对比。

模型Ⅱ 基于收入空间法的中等收入定位与定量模型

由于原来的收入空间法对不同地区中等收入区间取法具有任意性，因此本模型主要是克服原有的缺陷，确定统一的标准来制定中等收入区间，同时确定中等收入区间人口比例是否合理。

模型Ⅲ 基于人口空间法的中等收入定位与定量模型

本模型针对人口空间法本身局限，在确定在收入中位数附近范围人口拥有总收入占所有人口总收入 60% 时的人口比重，然后结合两极分化指标建立模糊综合评价指数，来衡量收入分配格局是否发生改善或恶化。

模型Ⅳ 变区间中等收入定位模型

对于不同地区而言，由于不同的收入分配格局，我们要从总体上进行考虑，为此，本模型是从收入分配格局上，对每个地区确定模型Ⅲ中模糊综合评价指数时，设定与该指数有关的中等收入人口上下限，最终根据理论推导出中等收入人口定义，并确定新的指标衡量该地区收入分配的合理性。

15.5.1 问题一的分析与求解

1. 对问题的分析

尽管可根据收入分配的统计数据加以描绘，但至今却未能找到一种有效的方法，准确地拟合洛伦兹曲线方程并由此求出精确的基尼系数。目前常被使用的方法主要有几何计算法、间接拟合法、曲线拟合法。

一般地，利用第一种方法不能得到洛伦兹曲线的表达式，只能用来计算基尼系数，但由于在计算分块面积时用直线近似地代替曲线，所估计的基尼系数要小于实际值，尤其在数据点较少时，误差较大。第二种方法由于计算收入分配的概率密度的复杂性，很难提出合适的密度函数。至于第三种方法，即直接用曲线方程去拟合洛伦兹曲线，应该不失为一种较好的方法，但目前主要的问题在于现有常用的曲线并不适用，曲线含义不明确，或拟合误差较大。

为了构造一个能很好拟合的收入分配分组数据的新模型并评价其优劣，需要对构造的新模型与文献中找到的 10 种模型[1]进行比较。根据已给出的收入分配分组数据分别进行拟合，依据模型拟合的结果，分别测算各模型的均方误差、平均绝对误差和最大绝对误差并以此作为判别模型优劣的三个标准。若从这三个判别标准来看，新模型优于已选取的 10 种模型，则构造的新模型是可行的。

2. 模型 I 基于非线性最小二乘的洛伦兹曲线拟合模型

(1)模型的理论准备。

经济学界常采用洛伦兹曲线模型 $L(p,\tau)$ 拟合收入分配分组数据，其中 τ 是一组参数，使用非线性最小二乘法求解

$$\min \sum_{i=1}^{n} (L(p_i,\tau) - L_i)^2 \tag{15-2}$$

确定其中参数向量 τ 的估计值 $\hat{\tau}$，然后用 $L(p,\hat{\tau}) = \hat{L}(p)$ 作为近似的洛伦兹曲线来进行收入分配分析。同时要求 $L(p,\tau)$ 是定义在区间 $[0,1]$ 上且取值于 $[0,1]$ 区间的函数，满足

$$L(0,\tau) = 0, L(1,\tau) = 1, L'(p,\tau) \geqslant 0, L''(p,\tau) \geqslant 0 \tag{15-3}$$

即 $L(p,\tau)$ 在 $[0,1]$ 上是凸增函数。

因此我们可以建立一个满足上述式(15-3)的模型，并根据非线性最小二乘法对洛伦兹曲线模型 $L(p,\tau)$ 进行拟合。同时选取 10 种具有代表性的模型，同样采用最小二乘法对洛伦兹曲线进行拟合，并对模型结果进行比较判别优劣。

本文选取了 10 种具有代表性的模型，分别为：

$$L_1(p;b) = \frac{(1+b)p}{1+bp} \tag{15-4}$$

$$L_2(p;\theta) = \frac{(1-2\theta)(1+\theta p)p}{(1+\theta)(1-2\theta p)}, \theta \in [0,12) \tag{15-5}$$

$$L_3(p;\lambda) = \frac{(\lambda-1)p}{\lambda-p}, \lambda \in (-1,0] \tag{15-6}$$

$$L_4(p;b) = (b+(1-b)p)p, b \in (0,1] \tag{15-7}$$

$$L_5(p;a,b) = \frac{(1-a)(1+bp)p}{(1+b)(1-ap)} \tag{15-8}$$

$$L_6(p;w) = \frac{(1+wp)p}{1+w} \tag{15-9}$$

$$L_7(p;a) = 1-(1-p)^a, a \in (0,1] \tag{15-10}$$

$$L_8(p;\alpha,\lambda) = \frac{1-\lambda\cos\alpha}{1+\lambda\sin\alpha} \cdot \frac{1+p\lambda\sin\alpha}{1-p\lambda\cos\alpha} \tag{15-11}$$

$$L_9(p;a) = \frac{e^{ap}-1}{e^a-1}, a > 0 \tag{15-12}$$

$$L_{10}(p;a,b,c) = ap^2+bp+c, b \in (-\infty,1], a \in [0,+\infty) \tag{15-13}$$

3. 模型的建立与求解

(1)新模型的构造。

根据洛伦兹曲线的性质,我们构造函数:

$$L(p;a,b) = 1-\frac{(1-p)^a}{(1+p)^b} (a \leqslant 0, b \leqslant -a) \tag{15-14}$$

满足:

① $L(0;a,b)=0$;

② $L(1;a,b)=1$;

③ $L'(p;a,b)=(1-p)^{a-1}[-a(1+p)-b(1+p)]/(1+p)^{b+1} \geqslant 0$;

④ $L''(p;a,b)=(1+p)^{(1-p)a-2}[(1-a)(-a(1+p))-b(1-p)+(1-p)(b-a)]$
$+(1+b)^{(1-p)a-1}[a(1+p)+b(1-p)] \geqslant 0$;

$$\tag{15-15}$$

即函数满足[0,1]区间上为凸增函数的要求,记函数为 $L_0(p;a,b)$,且参数的取值范围为 $(a \leqslant 0, b \leqslant -a)$,证明过程如下:

$$L' = \frac{a(1-p)^{a-1}(1+p)^b + b(1-p)^a(1+p)^{b-1}}{(1+p)^{b+1}} = \frac{a(1-p)^{a-1}}{1+p} + \frac{b(1-p)^a}{(1+p)^2} \geqslant 0$$

即: $\frac{(1-p)^{a-1}}{(1+p)^2}[a(1+p)+b(1-p)] \geqslant 0$,

$\because 0 \leqslant p \leqslant 1, \therefore 1-p \geqslant 0, \therefore (1-p)^{a-1}/(1+p)^2 \geqslant 0, \therefore$ 上式只需满足: $[a(1+p)+b(1-p)] \geqslant 0$,

即:

$$b \geqslant a(1+p)/(p-1), (p \neq 1), \tag{15-16}$$

$$L'' = \frac{-a(a-1)(1-p)^{a-2}-a(1-p)^{a-1}}{(1+p)^2}$$

$$-\frac{ab(1-p)^{a-1}(1+p)^2-2b(1+p)(1-p)^a}{(1+p)^3} \geqslant 0,$$

即: $a(a-1)(1-p)^{a-2}+a(1-p)^{a-1}+ab(1-p)^{a-1}(1+p)+2b(1-p)^a \geqslant 0$,

得: $(1-p)^{a-1}\left[\frac{a^2-a}{1-p}+ab(1+p)+2b(1-p)\right] \geqslant 0$,

$\because 0 \leqslant p \leqslant 1, \therefore 1-p \geqslant 0, \therefore (1-p)^{a-1}/(1+p)^2 \geqslant 0$,

所以只需满足:$\frac{a^2-a}{1-p}+ab(1+p)+2b(1-p)=\frac{a^2-a}{1-p}+b[a(1+p)+2(1-p)]\geqslant 0$

即:
$$b \geqslant \frac{a^2-a}{(p-1)[a(1+p)+2(1-p)]} \quad (p \neq 1) \tag{15-17}$$

由式(15-16)和式(15-17),且当$0 \leqslant p < 1$时,可得:$a \leqslant 0$, $b \leqslant -a$ 或者 $b \leqslant \frac{a^2-a}{a+2}$,即得证。

(2)对洛伦兹曲线模型的非线性最小二乘拟合。

对模型$L_0(p;a,b)$的拟合。根据统计调查获取的经济人口相关统计收入分配分组数据,得到数组(p_j, L_j)共17组数据。结合非线性最小二乘估计原理,利用Matlab软件计算出关于参数(a,b)的最优估计,得到a和b的估计值分别为$\hat{a}=0.765$, $\hat{b}=-0.7148$,即模型估计结果为:

$$L_0(p) = 1 - \frac{(1-p)^{0.765}}{(1+p)^{-0.7148}} \tag{15-18}$$

进一步可以得到:

均方误差:$\frac{1}{n}\sum_{i=1}^{n}(L_0(p)-L_i) = 1.78 \times 10^{-5}$;

平均绝对误差:$\frac{1}{n}\sum_{i=1}^{n}|L_0(p)-L_i| = 0.0037$;

最大绝对误差:$\max_{1 \leqslant i \leqslant n}|L_0(p)-L_i| = 0.0082$;

根据均方误差的数量级达到10^{-5},且平均绝对误差与最大绝对误差也较小,可初步认为模型拟合效果较理想。模型拟合曲线如图15.2所示,拟合值与真实值达到几乎完全重合,进一步说明了模型拟合精度较高。

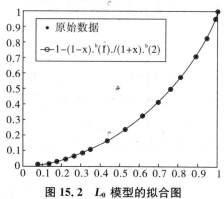

图15.2 L_0模型的拟合图

(3)对文献中10种模型的非线性最小二乘拟合。

同样对统计调查获取的经济人口相关统计数据,结合非线性最小二乘原理,利用Matlab软件计算出关于各个参数的最优估计量,最终确定各个模型中的参数如表15.3所示。

由表15.3可知,从MSE指标来看,模型L_0明显小于模型L_1—L_{10},从MAE指标和

MAS指标来看,模型L_0明显小于L_1-L_9且略大于模型L_{10},但模型L_0较模型L_{10}函数形式更加简单。整体来看,模型L_0要优于其他10种模型。

表15.3 10种模型的参数估计结果

模型	L_1	L_2	L_3	L_4	L_5
参数值	(−0.7124)	(−0.3345)	(−1.404)	(−0.183)	(0.552, 2.003)
MSE	0.0005	0.0003557	0.0005	0.001	0.000223
MAE	0.02	0.017	0.02	0.029	0.012
MAS	0.0329	0.0263	0.033	0.069	0.036
模型	L_6	L_7	L_8	L_9	L_{10}
参数值	(−6.4651)	(0.8031, 0.6794)	(1.571, −544.76)	(−2.689)	(1.162, −0.249, 0.004)
MSE	0.0012	0.000009383	0.00043	0.00027	0.00038
MAE	0.0288	0.0027	0.017	0.014	0.002
MAS	0.069	0.0055	0.048	0.036	0.005

15.5.2 问题二的分析与求解

1. 对问题的分析

(1)对收入空间法的分析。

由于题目中提到的"收入空间法",存在收入区间取法的任意性,为了准确确定中等收入区间,需要用定量的方法来计算中等收入区间。在这里,需要解决的问题是确定中等收入占总收入60%比重的人口比例,即确定中等收入区间内的人口比重。显然,洛伦兹曲线$L(p)$上任一点(p_j, L_j)的横坐标表示收入在$[0, x_{j+1}]$范围内的人口比例,纵坐标表示收入在$[0, x_{j+1}]$中的人口所拥有的收入占总收入的比例,$L_h - L_l$表示$[x_l, x_h]$收入区间内人口拥有的总收入占总人口收入比例,那么$p_h - p_l$表示收入在$[x_l, x_h]$区间内的人口占总人口的比例。我们需要从洛伦兹曲线模型着手,通过数学推理来计算出确定的中等收入区间范围。

(2)对人口空间法的分析。

由于一般"人口收入法"是在已知中等收入人口比例的情况下来确定中等收入范围,但是这种方法不能很好地判断出模型收入分配格局的变化,往往不能反映出人们收入分配的真实情况。因此需要改进原有的"人口收入法",可以从影响收入分配格局变化的因素着手,再结合中等收入人口比例来判断收入分配格局是变好还是变差。这就需要寻找一种能测度某个收入分配格局的两极分化程度[4]的指标,构造出一种能度量两极分化的指数。最后运用改进的模糊综合评价方法来综合考虑收入分配格局的影响因素,纵向比较某个地区各年收入分配格局的变化或某年各个地区的收入分配格局。

2. 模型Ⅱ 基于收入空间法的中等收入定位模型

(1)模型准备。

经济理论界一般考虑取收入落在中位收入m的一个范围内的人口为中等收入人

口,可以视这种方法为"收入空间法"。显然,这种方法中 x_l 与 x_h 的取法具有任意性,由于经济进步,通货膨胀等因素的影响,收入的区间是变化的,更多的情形是所有人口的收入都提高了,即全社会的收入区间右移,可见 x_l 与 x_h 的任意性使纵向比较各年的中等收入人口时出现困难。

(2)改进的收入空间法。

为了规避一般"收入空间法"中 x_l 与 x_h 的任意性,需要通过严密的数学推导来定量地确定中等收入范围。根据问题一的解答,我们已经构造了新的洛伦兹曲线模型:

$$L_0(p) = 1 - \frac{(1-p)^a}{(1+p)^b}。$$

假定中等收入人口表示为在已知洛伦兹曲线情况下,求 L 在(0.2,0.8)之间的人口比例,当 $L_l = 0.2$ 时,根据洛伦兹曲线解得 p_l。

已知,

$$L'_0(p) = \frac{b(1-p)^a + a(1+p)(1-p)^{a-1}}{(1+p)^{b+1}} \tag{15-19}$$

$$L'(p) = x/\mu \tag{15-20}$$

由式(15-19)可以解得点(p_l, L_l)的一阶导数 $L'_0(p_l)$ 的值。由式(15-20)可求得此时的收入 x_l。

同理,当 $L_h = 0.8$ 时,计算得出 p_h,$l'(p_h)$,x_h。

根据问题要求确定的 $L_l = 0.2$,$L_h = 0.8$,那么 $L_h - L_l$ 为中等收入区间(拥有的总收入占总收入比例为60%),则 $p_h - p_l$ 表示收入在中等收入区间人口占总人口的比例。由该模型计算的结果(x_l, x_h)即为中等收入的范围,规避了 x_l、x_h 的任意性。

通过比较 $L_1(p)$ 与 $L_2(p)$ 两个不同收入分配分布的洛伦兹曲线,若对(x_l, x_h)区间上 $\Delta p = p_h - p_l$ 都有 $\Delta p_1 \geqslant \Delta p_2$,则 $L_1(p)$对应的收入分配情况显然更优,因为在 $L_1(p)$中,收入处于中等收入区间的人口比例更大。

3. 模型Ⅲ 基于人口空间法的中等收入定位模型

(1)模型准备。

与"收入空间法"相对应的另一种方法可以视为"人口空间法",即选择 $F(m) = 1/2$ 邻近的一个范围为中等收入人口,例如,取范围 $p_1 = 20\%$ 到 $p_2 = 80\%$,按定义中等收入人口比例已经取定为60%。再用此60%的人口所拥有的收入占总收入的比例来描述中等收入人口的状态,此时中等收入人口的收入范围$[x_l, x_h]$很容易算得。例如当范围选取为20%到80%时,中等收入人口的状态即定义为:

$$S = L(0.8) - L(0.2) = \frac{1}{\mu}\int_{0.2}^{0.8} F^{-1}(p)\mathrm{d}p \tag{15-21}$$

这种方法计算过程简单严谨,似乎有道理,但是也存在一定的缺陷性。设收入分配是[10000,30000]上的均匀分布,这时中位收入是 $m = 20000$。此时,中间60%人口拥有总收入的60%,收入范围为14000到26000。考虑收入分配发生了变化,变成了

[0,40000]上的均匀分布,这时收入范围拉大了,低端人口收入下降了,高端收入人口收入增加了,直观上两极分化扩大了,也即中等收入人口应该是下降了,但按"人口收入法",中间60%的人口拥有的总收入比例仍是60%,没有反应出两极分化程度,这与经济直观不符,需要进行完善。

(2)改进的人口空间法。

一般"人口收入法"最大的缺陷就是不能很好地反映出收入分配格局的变化,两极分化是影响收入分配格局的重要因素[4]。在这里,我们把收入分配格局的主要影响因素考虑进来,构造一种测度两极分化的指数,结合一般"人口收入法"的中等收入定位模型,运用模糊综合评价方法评价各样本数据收入分配格局的优劣,既可以纵向比较各年中等收入人口与收入变动,也可以横向比较各地区之间的收入分配格局变化。同理还可以研究在各年中 p_1 与 p_2 取不同的值时,纵向比较某地区各年中等收入人口与收入范围的变动。

(3)两极分化指数。

现有的两极分化测度方法,基本可以分为两大类。一类方法由 Wolfson(1994)提出,由 Wang 和 Tusi(2000)、Chakravart 和 Majumde(2001)以及 Rodriguez 和 Salas(2003)等进行了拓展,我们将这一类型的测度指数统称为 W 型指数。该类指数是以中位数为界限将所有成员分为高收入和低收入两组,分别测算两组中各成员的收入对中位数收入的偏差,最后将所有偏差加总。另一类方法是由 Esteban 和 Ray(1994)提出,后经 Esteban、Gradn 和 Ray(1999)及 Duclos、Esteban 和 Ray(2003)等改进,我们将这种类型的指数统称为 ER 型指数。该类方法首先按照一定标准对所有成员进行分组,然后测定组与组之间的差异程度以及各个组内成员的相似程度,最后采用一定的形式构造测算指数[4]。

两极分化有两个基本因素,即"增加的延展性"和"增加的两极分化",分别用"第一类极化曲线"和"第二类极化曲线"来描述[5]。在两类极化曲线的基础上综合成两极分化指数,通常把中位数作为分组的基本临界点,将收入水平位于中位数(即满足 $F(m)=1/2$ 的收入数量 m)附近的人口理解为中等收入群体。具体计算公式如式(15-22)。

$$S_F(q) = \frac{|F^{-1}(q) - F^{-1}(0.5)|}{m}, F(m) = \frac{1}{2} \tag{15-22}$$

第一类极化曲线如式(15-19)所示,描述的是第 q 分位数位置上某人收入与中位数(即二分位数)收入间的距离,也即"增加的延展性"。S 值越大,说明中位数收入附近的人群越少,换言之,S 值越小越好。

第二类极化曲线为:

$$B_F = |\int_q^{0.5} S_F(p) dp| \tag{15-23}$$

描述的是第一极化曲线下从中位数到第 q 分位点间的面积,也即"增加的两极分化",是各个分位点收入与中位数收入间距离的平均数,描述的是收入的两极分化。在两列极

化曲线的基础上,可以得到两极分化指数:

$$Q(q) = \int_0^1 2B_F(q)\mathrm{d}q \qquad (15\text{-}24)$$

这样就可以用 $Q(q)$ 来描述每个分位数所对应的两极分化指数,且中等收入人口比例为 $\Delta p = q_h - q_l = Q^{-1}(q_h) - Q^{-1}(q_l)$。化简两极分化指数得:

$$Q(q) = \int_0^1 2B_F(q)dq = 2\frac{\mu}{m}\left[\frac{1-g}{2} - L_F(0.5)\right] \quad (0 \leqslant q \leqslant 1) \qquad (15\text{-}25)$$

由式(15-25)可看出,对同一收入分配的形成的分布函数或密度函数,两极分化指数 Q 与分位数 q 的取值无关,只与基尼系数、$L_F(0.5)$ 有关。

(4) 改进的模糊综合评价法。

综合评价法[6]是根据各个评价指标建立一种评价体系来判断被评价方案的优劣程度。一般有对偏差模糊矩阵评价法、主成分评价法及熵值评价法。相对偏差模糊矩阵评价方法的优点在于不需要先对原始数据进行预处理,所建立的相对偏差矩阵在消除量纲的同时得到了一个成本型模糊矩阵;缺陷在于理想方案并不一定是实际中的某个方案,在这里为避免其缺陷,我们先根据经验数据添加一个虚拟的理想方案 $u = (u_1^0, u_2^0, \cdots, u_n^0)$,再建立相对偏差模糊矩阵,具体步骤如下:

①建立理想方案:$u = (u_1^0, u_2^0, \cdots, u_n^0)$,其中 $u_i^0 = \begin{cases} \max(a_{ij}) & \text{当 } a_{ij} \text{ 为效益型指标,} \\ \min(a_{ij}) & \text{当 } a_{ij} \text{ 为成本型指标。} \end{cases}$

②建立相对偏差模糊矩阵 $\underset{\sim}{R}$:

$$\underset{\sim}{R} = \begin{bmatrix} r_{11} & r_{12} & \cdots & r_{1n} \\ r_{21} & r_{22} & \cdots & r_{2n} \\ \cdots & \cdots & \cdots & \cdots \\ r_{m1} & r_{m2} & \cdots & r_{mn} \end{bmatrix}, 其中 r_{ij} = \frac{|a_{ij} - u_i^0|}{\max\{a_{ij}\} - \min\{a_{ij}\}}。$$

③建立各评价指标的权数 $\bar{w}_i (i=1,2,\cdots,m)$:$\bar{w}_i = v_i / \sum_{i=1}^m v_i$,其中 $v_i = s_i / |\bar{x}_i|$,$\bar{x}_i = \frac{1}{n}\sum_{j=1}^n a_{ij}$,$s_i^2 = \frac{1}{n-1}\sum_{j=1}^n (a_{ij} - \bar{x}_i)^2$。

④建立综合评价模型:$G_j = \sum_{i=1}^m \bar{w}_i r_{ij}$。

计算出各个方案的评价得分,若 $G_t < G_s$,则第 t 个方案排在第 s 个方案前。

(5) 人口空间法的改进。

一般的"人口空间法"推理计算严谨,只是不能反映出其分配收入格局的变化及改善程度。在这里,基于一般"人口空间法"的计算原理,定义中等收入人口所占比例为60%,即 $p_l = 0.2, p_h = 0.8$,由式(15-19)可以解得点 (p_l, L_l) 的一阶导数 $L'(p_l)$ 的值,根据式(15-20),求得收入 x_l。同理,当 $p_h = 0.8$ 时,计算得出 $L'(p_h), x_h$,则 $L(p_h) - L(p_l)$ 为中等收入人群拥有的总收入占总收入比例,记为 ΔL。结合已经构造的两极分化指数 Q,这样可以组成一个二维指标的评价体系,其中 Q 为成本型指标,ΔL 为效益性指标。

若人口收入呈均匀分配时,则60%的人口所拥有的收入占总收入的比例也为60%,即 $\Delta L=60\%$,两极分化指数 $Q=0$,故可以添加虚拟的理想地区 $U=(0,0.6)$,应用模糊综合评价法计算出评价得分 G_j,则可以判断出某个地区 j 年的收入分配格局变化情况及改善程度,或 j 个地区之间收入分配格局的优劣。

该模型不仅能研究在各年中 p_1 与 p_2 取不同的值时,纵向比较各年中等收入人口与收入的变动,由于其基本计算原理还是依据于一般的人口空间法,中等收入人口定义为中位收入附近的人口,于是若中间部分比前一年隆起得更高,则认为中等收入人口扩大了;若两边人口扩大了,则中等收入人口下降了。所提出的原理与模型应与这一直观相符。

15.5.3 问题三的分析与求解

1. 对问题的分析

要对 A,B 两个地区前后两个不同年份的分组数据进行定量描述,确定各地区、各年份的中等收入的数量(或范围)、中等收入人口的数量(或范围),只需要在问题二研究的基础上,将统计调查获取的经济人口相关统计数据分别对基于收入空间法和人口空间法的中等收入定位模型进行实证检验,并判断出中等收入人口的变化趋势,再对两个地区的中等收入人口、收入等变化情况进行比较。

2. 对问题的求解

(1)基于收入空间法的中等收入定位模型的求解

根据统计调查获取的经济人口相关统计数据,选取 A、B 两个地区两个不同年份的共四个样本数据 (p_j,L_j)。要验证问题二中提出的基于收入空间法的中等收入定位模型,首先运用 Matlab 软件来估计模型 $L_0(P;a,b)$ 中的待估参数 $\hat{a}$、$\hat{b}$,估计结果如式(15-26~15-29)所示:

$$L_0A_1(P;a,b) = 1 - \frac{(1-p)^{0.7447}}{(1+p)^{-0.5159}} \tag{15-26}$$

$$L_0A_2(P;a,b) = 1 - \frac{(1-p)^{0.7492}}{(1+p)^{-0.5478}} \tag{15-27}$$

$$L_0B_1(P;a,b) = 1 - \frac{(1-p)^{0.7280}}{(1+p)^{-0.3070}} \tag{15-28}$$

$$L_0B_2(P;a,b) = 1 - \frac{(1-p)^{0.8054}}{(1+p)^{-0.4230}} \tag{15-29}$$

由表15.4可以看出,每个模型的均方误差都达到 10^{-6} 数量级,模型 $L_0(P;a,b)$ 对四个样本数据的拟合精度都很高,我们可以运用已构造的新模型 $L_0(P;a,b)$ 进行中等收入的定量计算。

表15.4　A、B两个小区两年的模型拟合评判结果

地区	MSE	MAE	MAS
A_1	0.00000306	0.0014	0.0039
A_2	0.000000303	0.0015	0.0035
B_1	0.000000349E	0.0016	0.0036
B_2	0.000000924	0.00074	0.0025

其次,根据问题二中基于"收入空间法"的中等收入定位模型来计算(x_l, x_h)及(p_l, p_h),即中等收入的范围及对应的中等收入人口和比例,计算结果如表9所示,此时同样假定中间收入人口表示为在已知洛伦兹曲线情况下,求L在$(0.2, 0.8)$之间的人口比例。若知道每个小区每年的总人口数N,则可以根据$n = N * \Delta p$计算出中等收入范围内的人口数。

表15.5　基于收入空间法计算中等收入范围及对应的中等收入人口的比例

样本	A_1	A_2	B_1	B_2
(p_l, p_h)	(0.418, 0.927)	(0.428, 0.928)	(0.352, 0.917)	(0.353, 0.903)
Δp	0.5088	0.5	0.565	0.5499
(x_l, x_h)	(4604.2, 12458.4)	(6583.2, 17942.2)	(12143.2, 29081.6)	(15966.8, 36425.9)
Δx	7854.188	11359.02	16938.46	20459.14

由表15.5计算结果可以看出,在这里,规避了"收入空间法"中x_l、x_h的任意性,对每个小区每年的中等收入都有指定中等收入范围(x_l, x_k)。A区后一年中等收入人口所占比例约为50%,较前一年相比有所下降,可能是高端收入人口比例增大,也有可能是低端收入人群增多。A区后一年较前一年的中等收入人口比例下降不明显,并且中等收入区间明显右移且区间范围扩大;B区后一年中等收入人口所占比例为55%,较前一年相比也有所下降,中等收入区间明显右移且区间范围扩大。A、B两个小区的中等收入区间范围的明显右移及扩大,其本质原因可能是由于通货膨胀引起的人民名义收入增加,人民实际生活水平并没有提高。

若统计调查获取的经济人口相关统计数据是A、B两小区同一年份的数据,由中等收入区间范围可以知道B区两年的中等收入人口比例分别为56.5%和54.99%,明显大于A区两年的中等收入人口比例,且其平均收入水平明显大于A区,说明B区收入分配较A区来说要均匀的多,人民生活水平要高于同期A区的人民生活水平。

(2)基于人口空间法的中等收入定位模型的求解。

"人口空间法"指的是拥有中等收入的人口比例为一定时确定其中等收入范围,我国一般定义中等收入人口比例为60%,即$p_l = 0.2, p_h = 0.8$。依据"人口空间法"求解出来的洛伦兹曲线模型,如式(15-26~15-29),基于模型Ⅲ,计算出A、B两区中等收入水平的范围及中等收入人口的比例,如表15.6所示。

表 15.6　A、B 两小区中等收入范围及中等收入人口的收入比例

样本	A_1	A_2	B_1	B_2
$(L(0.2),L(0.8))$	$(0.70,0.592)$	$(0.065,0.587)$	$(0.101,0.629)$	$(0.098,0.649)$
ΔL	0.5219	0.5217	0.5279	0.5517
(x_l,x_h)	$(2927.7,8818.4)$	$(3989.9,12642.8)$	$(9961.5,218010)$	$(13126,29566.2)$
Δx	5890.641	8652.841	11848.45	16440.22

表 15.6 可以看出，A 区前后两年中等收入人口所拥有的收入占总收入的比例几乎没有变化，但中等收入区间明显右移且扩大。B 区后一年中等收入人口所拥有的收入占总收入的比例为 55.17%，较前一年相比上升约 3%，说明 B 区人民收入分配水平后年较前一年来说要均匀些。

基于模型Ⅲ，根据式(15-25)计算出 A、B 两区的两极分化指数 Q，已知收入分配呈均匀分布时，则 60% 的人口拥有的收入占总收入的比例也为 60%，因此添加虚拟的理想地区 Au，使得该地区的两个评价指标 Q、L 分别为 0、0.6，两个指标在评价体系中所占的权重 $w=(0.0934,0.9006)$，再用模糊综合评价方法计算出各个地区不同年份的能反映出其收入分配格局的得分 G，其结果如表 15.7 所示。

表 15.7　基于模型Ⅲ的评价得分

样本	A_1	A_2	B_1	B_2	A_u
Q	0.1566	0.162	0.1153	0.1158	0
ΔL	0.5219	0.5217	0.5279	0.5517	0.6
G	0.9695	1	0.7313	0.7057	0

由表 15.7 可直观判断出，$G_{B2}<G_{B1}<G_{A1}<G_{A2}$，总体上来说，B 区后一年的收入分配格局是最好的，A 区后一年的收入分配格局最差。$G_{A1}>G_{A2}$，则 A 区后一年相对于前一年来说，收入分配格局变差；$Q_{A1}<Q_{A2}$，$\Delta L_{A1}\approx\Delta L_{A2}$ 说明导致其收入分配格局变差的主要原因是两极分化指数变大，低端收入人群增加或其低端收入人口比例增大。$G_{B1}>G_{B2}$，B 区后一年相对于前一年来说，收入分配格局变好；由 $Q_{B1}<Q_{B2}$，$\Delta L_{B1}<\Delta L_{B2}$，可以说明导致其收入分配格局变好的主要原因是 ΔL 变大，即中等收入人群所拥有的收入占总收入的比例变大。虽然两极分化指数变大，阻滞收入分配格局变好的程度，但其阻滞分配收入格局变好的幅度，不如由于中等收入比例增大而改善分配收入格局的幅度，最终表现为 B 区后一年的收入分配格局相对于前一年来说有所改善。

基于收入空间法可以计算出各地区、各年份中等收入人口的范围，A 区的后一年较前一年的中等收入人口比例下降不明显，B 区的后一年中等收入人口所占比例为 55%，较前一年相比有所下降。基于人口空间法可以计算出各地区、各年份中等收入的范围及中等收入的比例，A 区前后两年中等收入人口所拥有的收入占总收入的比例几乎没有变化；B 区后一年中等收入人口所拥有的收入占总收入的比例较前一年相比有所上升，且 B 区收入分配格局明显要比 A 区好。

15.5.4 问题四的分析与求解

1. 对问题的分析

如果我们根据统计调查结果的分组数据可以拟合洛伦兹曲线,利用模型Ⅲ评价我们获得样本地区收入分配状态,但这样评价只是根据我们对模型Ⅲ的假设:而对于某地区评价时,总希望该地区60%的人口所拥有的收入占总收入的比例也为60%。然而这只是我们的假定,事实总是背道而驰,所以我们利用模型Ⅲ对样本地区收入分配状态的评价是一个大致估计,并且得到模糊综合评价指数来说明该地区收入分配状态,尽管这是存在误差的,但我们可以利用这一指数比较准确的确定中等收入人口范围,定义中等收入人口。

中等收入一般在平均收入附近一定范围,平均收入 μ 是根据 $p=F(u)=1/2$ 计算得到的。根据目标收入分配状态是中等收入人口比例为60%且在平均收入人口比例在中位数左右各占30%。由已知的普遍价值规律[7],在收入分配状态较好的国家,其中等收入人口比例范围较窄,而收入分配状态较差的国家其中等收入人口比例范围一般较宽。但一般不会出现如下情况:即在平均收入比例左右各占比例小于30%;此时说明该地区平均收入大部分集中于较小收入区间内,这种良好的收入分配状态一般不会出现。因此我们只要构造的一个中等收入人口比例函数必须满足以上条件。

2. 模型Ⅱ 变区间中等收入定位模型

(1)模型的准备。

在提出中等收入人口的定义下,必须符合普遍的价值判断和一般的经济规律,一般的价值取向简单归纳为以下几种[8]:

①中等收入人口比例接近于60%。

②中等收入的概念是动态的,这里说的动态不是指每年都要变化,而是指不同的发展阶段会有不同的中等收入标准。

③中等收入人口是一个社会的组成部分,但中等收入并不意味着是社会的平均收入,不能简单地把达到社会平均收入水平的人群都划分为中等收入的范畴。

④收入分配格局较为合理的地区,其中等收入人口拥有的总收入占全部人口总收入比重越接近于中等收入人口比重。

鉴于对问题的分析,我们构造的中间收入人口比例指标应该与模型Ⅲ模糊综合评价指数 G 有关,应满足 G 越大中间人口收入比例指标越大,两者具有正向关系;并且根据经济学意义中等收入人口比例指标应该在[0,1]区间上,并且该指标是比例指标不能简单地加减计算,应与指数函数或者对数函数有关。

(2)变区间中等收入定位模型。

模型的设定:一般人口空间法所得结论 ΔL 和收入两极分化指数 Q,模糊综合评价指数 G,p_l 为中等收入人口比重的下界,p_u 为中等收入人口比重的上界,中等收入范围

的上下界分别为 x_h 和 x_l,某地总人口为 N_0。

根据模型分析与模型准备,我们可以构造简单地中等收入人口比例指标

$$\Delta p(G) = 0.6e^{-G}。$$

根据对问题的分析可以知道,中等收入人口比例应该均匀分布在平均收入人口比例左右两侧,则中等收入上下界为 x_h 和 x_l 所对应的人口比重分别为:$p_l = 0.5 - 0.3e^{-G}$,$p_h = 0.5 + 0.3e^{-G}$。

根据中等收入人口经济学含义:中等收入人口比例与某地区总体人口数的乘积。则定义:中等收入人口是指在收入区间 (x_l, x_h) 上,所有居民总数,即 $N(G) = 0.6e^{-G} N_0$。

在该定义性质下:

①显然符合普遍价值取向第一条;

②每一个样本地区可以对应唯一一个 G 指数,每一个 G 指数又对应唯一一个中等收入人口,则样本地区与中等收入人口定义呈一一对应关系,这点符合普遍价值取向第二条;

③中等收入是在平均收入附近一定范围,与平均收入有明显的区别;

④定义中等收入人口比例指数函数 p 与自变量 G 两者之间存在正向的关系。

根据定义,利用模型Ⅲ计算中等收入人口的总收入占总收入的比例 $\Delta L' = L(p_h) - L(p_l)$,具体应该怎样衡量该地区居民收入分配格局是否合理,这与 ΔL 接近 Δp 程度有关。可以建立指标 $H(p) = \Delta L'/\Delta p$,定义该指标为收入分配合理指数。

该指数越接近于 1 说明该地区收入分配越合理。这样就可以利用 $H(p)$ 指数衡量不同地区不同时间的收入分配格局问题。可以根据该指标验证该模型建立的指标是否符合普遍价值取向第四条。

15.6 模型的误差分析、检验和进一步讨论

15.6.1 误差分析

1. 考虑非线性拟合过程

在模型Ⅰ中,由于统计调查获取的经济人口相关统计数据在拟合预测过程中一定会产生拟合误差,模型Ⅰ已经给出评价拟合函数的精度。

在模型Ⅱ、Ⅲ中,由于统计调查获取的经济人口相关统计数据中给出的是 A, B 两个地区前后两个不同年份的收入分配分组数据,利用新构造洛伦兹曲线函数拟合预测过程中一定会产生拟合误差,如图 15.3 所示。

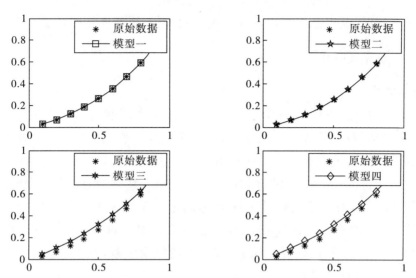

图 15.3　关于不同地区不同年份下的拟合曲线

2. 考虑数据收集过程

统计调查获取的经济人口相关统计数据一般都是统计调查得出的结果,在调查过程中,居民的隐性收入或者被调查者隐瞒的收入,都会导致模型结果与实际情况的偏离。

15.6.2　模型的检验

对于模型中中等收入人口的划分,标准是我们自定的,根据实际观察,选择两个收入分配格局明显不同的地区,把这两个地区的统计调查数据进行带入计算结果,将其结果与具体事实进行对比,可以对原模型进行检验。

15.6.3　模型的进一步讨论

对于模型Ⅱ,可以进一步联系两极分化指数,建立模糊综合评价指数,进行综合分析。对于模型Ⅲ,还可以考虑另外的一些指标,例如恩格尔系数、基尼系数等进行评价。

15.7　模型的评价与推广

15.7.1　模型的优缺点

1. 优点

(1)利用 Matlab 软件对数据进行处理并作出各种平面图,简便、直观、快捷;

(2)本文建立的模型与实际紧密联系,充分考虑现实情况的不同阶段,从而使模型更贴近实际,通用性强;

(3)运用多种指标对收入分配格局进行分析,取长补短,使结果更加准确。

2. 缺点

(1)对于模型Ⅰ,由于时间限制,选择与构造函数进行比较的函数数量仅限于10种;

(2) 对于模型Ⅳ,仅仅是根据理论知识提出的,没有进行实证验证。

15.7.2 模型的推广

1. 模型Ⅰ的推广

对于模型Ⅰ为了追求模型的高精度,可以选择几种精度较高的洛伦兹曲线函数形式综合考虑,结合模型自身特点,扬长避短以达到提高模型精度的目的。根据 n 个不同模型的函数,进行目标函数的设定,利用 lingo9.0 软件进行最优化问题的求解,计算结果赋予某些函数的权重 $w_i=1$,某些函数的权重 $w_j=0$,其中 $i\neq j$。但是并不是普通的给有些函数赋予权重1,另一些函数赋予权重0;而是某个模型的函数有些权重是1,有些是0,某些模型的函数的权重有些是1,有些是0,给1和0的标准是取决于其精度,当期谁拟合效果好就取谁。根据计算的权重向量,构造分段函数,其拟合精度肯定比 n 个模型中任何一个的精度都要高。

2. 模型Ⅱ的推广

从所建立的模型Ⅲ可以得到启示,一般人口空间法的计算虽然正确合理,但是其不能反映出人口收入分配格局的变化,故改进的收入空间法也不能全面反映出该变化,可以通过引进两极分化指数 Q 综合反映该变化。

参考文献

[1] Wang Z. X., Y-K Ng, R. Smyth, 2011. A General Method for Creating Lorenz Curves[J]. *The Review of Income and Wealth* 57, 561—582.

[2] Ibrhim M. Abdalla, Mohamed Y. Hassan, Maximum Likelihood Estimation of Lorenz Curves Using Alternative Parametric Model[Z]. 2004.

[3] Wang, Z. X. and R. Smyth, A Hybrid Method for Creating Lorenz Curves with an Application to Measuring World Income Inequality[Z]. 2013.

[4] 龙莹. 2012,中国中等收入群体规模动态变迁与收入两极分化:统计描述与测算[J]. 财贸经济,2012.2.

[5] Foster J. E., M. C. Wolfson, 2009. Polarization and the Decline of the Middle Class[J]: *Canada and the U. S. Journal of Economic Inequality* 8, 247—273.

[6] 杨桂元,黄己立. 数学建模[M]. 合肥:中国科学技术大学出版社,2008.

[7] 徐建华. 对中等收入的界定研究[J]. 上海统计,2003,3.

[8] 胡荣华等. 南京城市具名中等收入界定及分析[J]. 地方经济社会发展研究,2006,1.

[9] 胡守信,李柏年. 基于 MATLAB 的数学实验[M]. 北京:科学出版社,2004.

◆ 建模特色点评

❋论文特色❋

◆标题分析:"基因识别问题及其算法实现",是原始标题,涉及算法实现。

◆方法鉴赏:利用 Matlab 软件对数据进行处理并作出各种平面图,简便、直观、快捷;所建的模型与实际紧密联系,充分考虑现实情况的不同阶段,从而使模型更贴近实际,通用性强;运用多种指标对收入分配格局进行分析,取长补短,使结果更加准确。

◆写作评析:论文写作条理性好,层次分明,图文并茂。

❋不足之处❋

由于时间限制,选择与构造函数进行比较的函数数量仅限于 10 种;个别模型根据理论知识提出,没有进行实证验证。